14th International Conference on Fluid Control, Measurements and Visualization (FLUCOME 2017)

IOP Conference Series: Materials Science and Engineering
Volume 249

Notre Dame, Indiana, USA
8 – 12 October 2017

ISBN: 978-1-5108-4996-9
ISSN: 1757-8981

Printed from e-media with permission by:

Curran Associates, Inc.
57 Morehouse Lane
Red Hook, NY 12571

Some format issues inherent in the e-media version may also appear in this print version.

Printed by Curran Associates, Inc. (2017)

For permission requests, please contact the Institute of Physics
at the address below.

Institute of Physics
Dirac House, Temple Back
Bristol BS1 6BE UK

Phone: 44 1 17 929 7481
Fax: 44 1 17 920 0979

techtracking@iop.org

Additional copies of this publication are available from:

Curran Associates, Inc.
57 Morehouse Lane
Red Hook, NY 12571 USA
Phone: 845-758-0400
Fax: 845-758-2633
Email: curran@proceedings.com
Web: www.proceedings.com

Table of contents

Volume 249

14th International Conference on Fluid Control, Measurements and Visualization (FLUCOME 2017)

8–12 October 2017, University of Notre Dame, USA

Accepted papers received: 28 September 2017
Published online: 16 October 2017

Preface

Flow/Fluid Control

Measurements/Visualization

Multiphase Flows

Measurements/Environment

Combustion

IOP Conf. Series: Materials Science and Engineering **249** (2017) 011001 doi:10.1088/1757-899X/249/1/011001

Preface

The 14th International Conference on Fluid Control, Measurements and Visualization (FLUCOME 2017) was hosted by the Department of Aerospace and Mechanical Engineering on the campus of the University of Notre Dame, USA, in October 8-12, 2017. This well established Conference was collocated with the Notre Dame Workshop on Plasma-Based Flow Control.

The FLUCOME conference was contemplated as a continuation of the previous thirteen meetings. The 1st FLUCOME symposium was organized by Professor Yasuki Nakayama (1916-2016) in Tokyo, Japan (1985). The 2nd symposium was held in Sheffield, UK (1988), the 3rd in San Fransisco, CA, USA (1991), the 4th in Toulouse, France (1994), the 5th in Hayama, Japan (1997), the 6th in Sherbrooke, Canada (2000), the 7th in Sorrento (2003), the 8th in Chengdu, China (2005), the 9th in Tallahassee, FL, USA (2007), the 10th in Moscow, Russia (2009); the 11th in Keelung/Kenting, Taiwan (2011); 12th in Nara, Japan (2013); and the 13th in Doha, Qatar (2015).

The scientific program covered a wide range of topics, including Flow/Fluid Control, Flow Measurements, Non-Intrusive Methods, Visualization, Multiphase Flows, etc. The Workshop on Plasma-Based Flow Control (Plasma Aerodynamics) was organized as an entirely invited one-day meeting. Three keynote, fourteen invited and over fifty oral papers were presented during four Conference days. A special session was dedicated to a memory of Professor Nakayama, the FLUCOME founder and an outstanding leader in the international visualization community, who passed away in 2016.

This volume of IOP Journal Conference Series: Material Science and Engineering includes 21 selected papers presented in sections "Flow/Fluid Control", "Measurements", "Visualization", and "Multiphase Flow". We believe these manuscripts well represent the Conference topics and will raise readers' interest to the area considered.

We would like to thank the members of both the International Steering Committee and the Local Organizing Committee, who greatly helped in a challenging process of the meeting organization. We are also pleased to thank the Academic Conferences Office in the University of Notre Dame and the Engineering Communications and Marketing Group in the College of Engineering for their everyday assistance in the Conference preparation. Finally, we are grateful to the College of Engineering and the Institute of Flow Physics and Control for the financial support of FLUCOME 2017.

Thomas Corke and Sergey Leonov, FLUCOME 2017 Chairs

IOP Conf. Series: Materials Science and Engineering **249** (2017) 011002 doi:10.1088/1757-899X/249/1/011002

FLUCOME 2017

14th International Conference on Fluid Control, Measurements, and Visualization
October 8-12, 2017, University of Notre Dame, Notre Dame, IN 46556, USA

The Department of Aerospace and Mechanical Engineering, Institute of Flow Physics and Control of the University of Notre Dame, USA, hosts the 14th International Conference on Fluid Control, Measurements, and Visualization (FLUCOME) in the famous campus of Notre Dame, Indiana, USA.

This well-established Conference will be collocated with the Notre Dame Workshop on Plasma-Based Flow Control.

The FLUCOME conference is contemplated as a continuation of the previous thirteen meetings. The 1st FLUCOME symposium was held in Tokyo, Japan (1985), the 2nd in Sheffield, UK (1988), the 3rd in San Fransisco, CA, USA (1991), the 4th in Toulouse, France (1994), the 5th in Hayama, Japan (1997), the 6th in Sherbrooke, Canada (2000), the 7th in Sorrento (2003), the 8th in Chengdu, China (2005), the 9th in Tallahassee, FL, USA (2007), the 10th in Moscow, Russia (2009); the 11th in Keelung/Kenting, Taiwan (2011); 12th in Nara, Japan (2013); and the 13th in Doha, Qatar (2015).

The 14th FLUCOME will provide a forum for researches and scholars to exchange the latest scientific and technical achievements in multiple disciplines, including engineering, physics, chemical sciences, astrophysics, and meteorology. Conference topics include fluid processes across a wide range of scales: micro-scale biomechanical engineering flows, small-scale machine flows, large-scale civil and environmental fluid flows, and global-scale geophysical and meteorological fluid flows.

The FLUCOME-2017 program will include three themes A, B, C and the Notre Dame Workshop on Plasma-Based Flow Control dedicated to a special topic

Theme A: Flow/Flight Control

- ❑ Active flow control technique
- ❑ MHD and EHD
- ❑ Environmental pollution control
- ❑ Aeroacoustics
- ❑ Microfluidics
- ❑ Combustion and detonation
- ❑ Wind engineering

Theme B: Flow/Fluid Measurements and Visualization

- ❑ Shock-dominated flows
- ❑ Multiphase flows
- ❑ PIV and imaging
- ❑ PSP, TSP, and liquid crystals
- ❑ Aero-optics
- ❑ Geophysical and environmental flows
- ❑ Remote and nonintrusive sensing

Theme C: Visualization as a scientific art (new)

- ❑ Computer fiction
- ❑ Lightning
- ❑ Novel metamaterials
- ❑ Fractals
- ❑ Astrophysics

Notre Dame Workshop on Plasma-Based Flow Control

IOP Conf. Series: Materials Science and Engineering **249** (2017) 011002 doi:10.1088/1757-899X/249/1/011002

FLUCOME 2017

14th International Conference on Fluid Control, Measurements, and Visualization
October 8-12, 2017, University of Notre Dame, Notre Dame, IN 46556, USA

Committees

FLUCOME Honorary Board

Yasuki Nakayama – founder	Future Technology Research Institute	Japan
Giovanni Maria Carlomagno	University of Naples Federico II	Italy
Qing-Ding Wei	Peking University	China
Ching-Jen Chen	Florida State University	USA
Sergey B Leonov	University of Notre Dame	USA
Jiahn-Horng Chen	National Taiwan Ocean University	Taiwan
Toshiharu Kagawa	Tokyo Institute of Technology	Japan
Yousef Haik	Hamad Bin Khalifa University	Qatar

International Steering Committee

Igor Adamovich	Nicolas Benard	Mark Capelli
Gennaro Cardone	Campbell Carter	Antonio Delgado
Ivan Egorov	Deog-Hee DOH	Lihao Feng
Song Fu	Nobuyuki Fujisawa	Wenrui Huang
Shinichiro Ito	Toshio Kobayashi	Christian Kaehler
Doyle Knight	Victor Kozlov	Richard Miles
Gerard Pimeau	Ethirajan Rathakrishnan	Mo Samimy
Tatsuo Sawada	Jan Stasiek	Sonny H. Winoto
Yu Zhou		

Local Organizing Committee

Sergey Leonov – Chair	Thomas Corke – Chair	Harindra Joe Fernando
Stanislav Gordeyev	Thomas Juliano	Eric Jumper
Seong-Kyun Im	Ahsan Kareem	Scott Morris
Joseph Power	Hirotaka Sakaue	Flint Thomas
Gretar Tryggvason	Meng Wang	Melissa Neill – Secretary

Publication

After reviewing, selected papers will be published in the IOP Conference Series: "Materials Science and Engineering" http://iopscience.iop.org/journal/1757-899X

Location

The location of the **FLUCOME 14** will be the Notre Dame Conference Center: http://morrisinn.nd.edu/conference-center/notre-dame-conference-center/

Important Dates

Deadline for abstract submission:	April 21, 2017
Notification of abstract acceptance:	May 19, 2017
Deadline for full paper submission:	July 14, 2017

Contacts

Web Site: **https://engineering.nd.edu/flucome2017**
E-mail: flucome2017_info@nd.edu

IOP Conf. Series: Materials Science and Engineering **249** (2017) 011003 doi:10.1088/1757-899X/249/1/011003

Peter Garnavich

Professor, Department Chair

Astrophysics/Cosmology Physics

University of Notre Dame

Title:

The Accelerating Universe and the Discovery of Dark Energy

Summary. One of the most dramatic scientific revolutions of the last fifty years is the discovery that the rate of expansion of the universe is accelerating. The source of acceleration is thought to be a mysterious "dark energy" that makes up 3/4 of the content of the universe, but more exotic causes are possible. This is a story that begins with Einstein and Hubble but shifts to two research teams racing to find the ultimate fate of the universe using exploding stars and new detector technology.

Bio. Peter Garnavich is a Professor and Chair of Physics at the University of Notre Dame. Previously he was a Research Fellow at the Harvard-Smithsonian Center for Astrophysics and received a M.Sc. from the Massachusetts Institute of Technology and a Ph.D. from the University of Washington. His research interests include cataclysmic variable stars, supernovae and cosmology. He is co-recipient of the 2007 Gruber Prize in Cosmology and the 2015 Breakthrough Prize in Fundamental Physics. Garnavich was a key member of the "High-Z" team that discovered the universal expansion rate is accelerating. The discovery of the accelerating universe was awarded the 2011 Nobel Prize in Physics.

IOP Conf. Series: Materials Science and Engineering **249** (2017) 011003 doi:10.1088/1757-899X/249/1/011003

William Saric
Distinguished Professor of Aerospace Engineering
George Eppright '26 Chair in Engineering

Texas A&M University
Aerospace Engineering

Title:

Flight experiments on discrete roughness element technology for laminar flow control

William S. Saric has been Professor of Aerospace Engineering at Texas A&M University since January 2005. He received his PhD in Mechanics from the Illinois Institute of Technology in 1968 and has held appointments at Sandia Laboratories (Atomic and Fluid Physics, 1963-1966, 1968-1975), Virginia Polytechnic Institute and State University (Engineering Science and Mechanics, 1975-1984), Tohoku University, Japan (Aeronautics and Astronautics, 1991-1992), and Arizona State University (Mechanical and Aerospace Engineering, 1984-2005).
He is a registered Professional Engineer in Virginia and was an Aerospace Engineering Evaluator for ABET.
He is a Fellow of AIAA, APS, and ASME. He received the AIAA Fluid Dynamics Award in 2003, the G.I. Taylor Medal from SES in 1993, the Scientific Achievement Award from AGARD (NATO) in 1996, and the Alumni Research Award from V.P.I.&S.U in 1984. In 2006, he was elected to the National Academy of Engineering for "contributions to the fundamental understanding and control of shear-flow and boundary-layer transition".
He has established two major wind tunnel research facilities and a flight research center at Texas A&M University and was the Director of the AFOSR/NASA National Center for Hypersonic Laminar-Turbulent Transition Research.
Most recently, he has conducted experimental, and flight research on stability, transition, and control of two-dimensional and three-dimensional boundary layers for subsonic aircraft, supersonic aircraft and reentry vehicle applications.

IOP Conf. Series: Materials Science and Engineering **249** (2017) 011003 doi:10.1088/1757-899X/249/1/011003

Gennaro Cardone

Professor, Dr.,
Chair of Aerospace Engineering Study Course

Università degli Studi di Napoli "Federico II", Italy
Department of Industrial Engineering –
Aerospace division

Title:

Thermo-fluid-dynamic analysis of innovative synthetic jet devices

Summary: Synthetic jets are largely used, especially in the field of electronic cooling; indeed their heat transfer performances have been widely investigated and some of the work performed at University of Naples Federico II is herein described. Heat transfer coefficients have been enhanced through the design of innovative synthetic jet devices; in particular, twin synthetic jets, multi orifice nozzles, chevron exit nozzles are considered. Obviously, the heat transfer performances of both the classic and innovative devices are strictly related to their impinging flow field on the surface to be cooled. In this work the behavior of innovative impinging synthetic jets is experimentally investigated by using Particle Image Velocimetry (PIV) and IR thermography leading to both time average and phase average flow velocity and heat transfer measurements. Three-dimensional coherent vortex structures, time-averaged and phase-averaged means, as well as turbulent statistics of flow fields and wall heat transfer data are presented and discussed.

Bio. Born in Naples (Italy) on 1963. Graduated in Mechanical Engineering with summa cum laude (1988). PhD in Aerospace Engineering (1991). Assistant professor (1992). Associate professor (1998). Actual position: Full Professor of Fluid Mechanics (2010) and Chair of the Aerospace engineering study courses, at the University of Naples Federico II, Department of industrial engineering, aerospace division. Scientific interests: two-beam and reference beam interferometry; velocimetry techniques such as PIV, Stereo-PIV and Tomo-PIV and their developments; heat transfer measurements with the infrared thermography technique, particularly focused on heat flux sensors and thermal image restoration; steady and unsteady natural convection; boundary layer diagnostics; transition detection and flow field visualization from subsonic to hypersonic flow regimes; free and impinging synthetic jets; swirling jets; aerodynamic heating and its thermal visualization; thermal measurements in investment casting.

IOP Conf. Series: Materials Science and Engineering **249** (2017) 011004 doi:10.1088/1757-899X/249/1/011004

FLUCOME 2017 Invited Lectures

Prof. **Mo Samimy**, The Ohio State University, USA
Title: Active Flow Control in High-Speed Flows.
Session: Flow/Fluid Control

Dr. **Li-Hao Feng**, Beihang University, Beijing, China
Title: Active control of bluff body wake using synthetic jets.
Session: Flow/Fluid Control. Special Recognition: Young Researcher

Dr. **Campbell Carter**, Air Force Research Laboratory, USA
Title: Advances in kHz Imaging Diagnostics for Practical Problems.
Session: Measurements

Prof. **Stanislav Gordeyev**, University of Notre Dame, USA
Title: Aero-Optical Effects to Study Flow Dynamics at Hypersonic Speeds.
Session: Measurements

Prof. **Nobuyuki Fujisawa**, Niigata University, Japan
Title: Visualization of Liquid Droplet Impingement and Erosion Mechanism on Rough Surface.
Session: Visualization

Prof. **Shinichiro Ito**, Kogakuin University, Japan
Title: Memories of Prof. Nakayama and Dedicated presentation to Prof. Nakayama, "PIV analysis and aerodynamic vibration on the soccer ball wake".
Session: Visualization

Prof. **Jan Stasiek**, Gdansk University of Technology, Poland
Title: Liquid Crystals, PIV and IR-Photography in Selected Technical and Biomedical Applications.
Session: Visualization

Dr. **Christopher Limbach**, Texas A&M University, USA
Title: Visualization and Applications of Femtosecond, Nanosecond and Dual-Pulsed Laser-Induced Plasmas.
Session: Visualization. Special Recognition: Young Researcher

IOP Conf. Series: Materials Science and Engineering **249** (2017) 011004 doi:10.1088/1757-899X/249/1/011004

FLUCOME 2017 Invited Lectures
(Notre Dame Workshop on Plasma Aerodynamics)

Dr. **Nicolas Benard**, University of Poitiers, France
Title: Linear surface plasma actuator, a useful device for manipulation of coherent vortices in turbulent flows.
Session: ND Workshop

Prof. **Thomas Corke**, University of Notre Dame, USA
Title: Cross-Flow Transition Control at Mach 6 Using Patterned Discrete Roughness.
Session: ND Workshop

Dr. **Pavel Georgievsky**, Institute of Mechanics MSU, Moscow, Russia
Title: Thermal Spike Conception for Supersonic Flow Control and Wave Drag Reduction of Bodies.
Session: ND Workshop

Dr. **Marios Kotsonis**, Delft University of Technology, Netherland
Title: Towards control of swept wing transition using plasma actuators.
Session: ND Workshop

Prof. **Jonathan Poggie**, Purdue University, USA
Title: Controlling Compressible, Turbulent Flow with Plasmas: A Review.
Session: ND Workshop

Dr. **Andrey Starikovsky**, Princeton University, USA
Title: Dynamic Stall Control at High Angle of Attack by NS Pulsed Actuator.
Session: ND Workshop

IOP Conf. Series: Materials Science and Engineering **249** (2017) 011005 doi:10.1088/1757-899X/249/1/011005

Peer review statement

All papers published in this volume of *IOP Conference Series: Materials Science and Engineering* have been peer reviewed through processes administered by the proceedings Editors. Reviews were conducted by expert referees to the professional and scientific standards expected of a proceedings journal published by IOP Publishing.

Thermo-fluid-dynamic analysis of innovative synthetic jet devices

Carlo Salvatore Greco, Gerardo Paolillo, Gennaro Cardone*
University of Naples "Federico II", Naples, Italy

*corresponding author: gennaro.cardone@unina.it

Abstract. Synthetic jets are largely used, especially in the field of electronic cooling; indeed their heat transfer performances have been widely investigated and some of the work performed at University of Naples Federico II is herein described. Heat transfer coefficients have been enhanced through the design of innovative synthetic jet devices; in particular, twin synthetic jets and multi orifice nozzles are considered. Obviously, the heat transfer performances of both the classic and innovative devices are strictly related to their impinging flow field on the surface to be cooled. In this work, the behaviour of innovative impinging synthetic jets is experimentally investigated by using Particle Image Velocimetry (PIV) and IR thermography leading to both time average and phase average flow velocity and heat transfer measurements. Three-dimensional coherent vortex structures, time-averaged and phase-averaged means, as well as turbulent statistics of flow fields and wall heat transfer data are presented and discussed.

1. Introduction

The continuous technology improvement is causing a proliferation of electronic systems in our society. Nowadays human society is completely dependent on such electronic devices; hence, the reliability of these systems is mandatory. The failure of these systems can be caused by many factors such as high temperature, electrical overstress and material fatigue.

The system failure is mostly due to the thermal overstress caused by a continuous request of intense work. Although most of the electronic devices have a high efficiency, a continuous use can cause an increase in temperature which can affect the device performance until the failure. Electronic components are designed to operate within a certain temperature range thus need for a cooling system. Such a cooling system has to be designed to dissipate all the thermal power to maintain a suitable temperature and guarantee electronic component safety without affecting working performance.

According to Moore's law [1], the number of transistors on an integrated circuit doubles every one and a half year and the transistor density is expected to be very high. Moreover, the size of these transistors decreases exponentially as silicon microfabrication processes are improved. These two components lead to an increase in the heat power which must be carried away by cooling devices in order to let the electronic systems work properly within a suitable temperature range.

Conventional cooling techniques such as heat sink, fans and liquid cooling need to be continuously optimized to face this continuous request of higher heat power dissipation and miniaturization. Such needs have led researchers to focus their efforts on the design of new cooling devices.

In the last decades, researchers discovered a new technology which has promising heat transfer performances. Such a new technology is the synthetic jet actuator, as named for the first time by Coe et al. [2] and then Allen and Glezer [3]. The synthetic jet actuator, depicted in Figure 1, is a device that releases a jet in the external environment without the need of an external piping. In particular, the synthetic jet is a zero-net-mass-flux jet which transfers linear momentum to the flow system without net mass injection across the flow boundary. It is usually produced by a sinusoidal oscillating membrane (e.g. piezoelectric or loudspeaker) or piston which alternatively forces the fluid through an orifice or a slot into the external flow field and entrain fluid back. In this periodic behavior, one can identify two stages that totally describes the phenomenon: the blowing phase, when fluid is ejected from the orifice, and the suction phase, when fluid is entrained back. During the blowing phase, the ejected fluid separates at the sharp edges of the orifice and rolls up to form a vortex ring or vortex pair, respectively in the 2D or 3D case. During the suction phase of the cycle, the vortex pair is quite far from the orifice and keeps on propagating away due to its self-induced velocity. However, the vortex pair is not entrained into the cavity only if its convection velocity is larger enough than the averaged velocity during the suction cycle. In this case [4] the vortex structure will propagate downstream and dissipate at some distance from the exit, synthesizing a jet with momentum transfer to the embedding flow. Thus, a train of vortex structures is created by the actuator and, in the mean, the velocity profile appears like a steady jet. Over a single period of oscillation of the membrane, there is, therefore, zero net mass-flux into or out of the cavity, yet there is a non-zero mean jet velocity. The parameters that govern the synthetic jet behaviour are the Reynolds number and the Strouhal number defined as:

$$Re = U_0 D/\upsilon \tag{1}$$

$$St = D/L_0 = D/fU_0 \tag{2}$$

where D is the characteristic length, υ is the fluid kinematic viscosity, f is the actuation frequency, U_0 is the characteristic velocity [5] and L_0 is the stroke length (equal to fU_0).

Since these actuators are easy to be miniaturized and are low-cost devices, they have been widely used in many fields of application such as: flow control [6], jet vectoring [7], mixing enhancement [8], thrust generation [9] and electronic cooling.

In the last decades, the application of this technology in the electronic cooling field has been widely investigated. The first literature work on synthetic jets used as cooling devices was presented in 1982 by Gutmark et al. [10]. They presented a work where synthetic jet were used to enhance both natural and forced convection. They obtained an increase of 400% of the overall heat transfer coefficient with this acoustically excited airflow. Later Mahalingam and Glezer [11] designed a heat sink with synthetic jet impingement for high power dissipation in electronics. They studied the thermal performances of this device obtaining a temperature decrease from 71.5 to 36°C with synthetic jets operation and a power dissipation of 20-40% higher with respect to the same heat sink with a fan in the flow rate range of 3-5 cubic feet per minute. Chaudhari et al. [12] studied the thermal performance of synthetic jets, issued by a circular orifice, used to cool a flat plate. Such experiments were undertaken under the following operating conditions: Reynolds number in the range 1,500-4,200 and nozzle to plate distance in the range 0-25*D*. The results show that the Nusselt number is comparable with that of continuous axisymmetric jets at low Reynolds number (up to 4,000), expecting it to be higher at greater values of Reynolds number. Valiorgue et al. [13] studied the influence of the stroke length on the thermal behaviour of impinging synthetic jets. They found a critical stoke length which is a boundary between two different flow regimes. This critical stoke length is L_0/H equal to 2.5. The heat transfer rate (that obviously increases with Reynolds number increasing) is found to be linearly proportional with L_0 up to $L_0/H = 2.5$ than constant for increasing L_0 values.

As for steady jets, heat transfer correlations have been developed also for synthetic jets [14,15]. Arik and Icoz [14] found an empirical correlation to determine the heat transfer coefficient of impinging synthetic jets. Such a correlation depends on several parameters: Reynolds number, axial distance, orifice size and jet driving frequency. In particular, the range of validity of this correlation are $Re <$

2,900, $5 < H/D < 20$ and actuation frequency between 0.16 times the resonance frequency and the resonance frequency. Persoons et al. [15] accounts, in their general correlation, for several scaling parameters such as Reynolds number ($500 < Re < 1{,}500$), jet to surface spacing ($2 < H/D < 16$) and stroke length ($2 < L_0/D < 40$). Based on this study, they defined four heat transfer regimes, each one identified by a different range of values acquired by the ratio L_0/H, whose flow field characteristics were later described by McGuinn et al. [16]. They [16] undertook high speed PIV and single point hot wire anemometry experiments to understand the influence of the dimensionless stroke length on the impinging synthetic jet flow field for a wide range of nozzle-to-surface spacing ($2 < H/D < 16$) and a single Reynolds number (1,500).

Early studies on synthetic jets, in the heat transfer field, were focused on the assessment of the effects of all the characteristics parameters on their heat transfer behavior. Differently, the most recent research is aimed at the design of innovative configurations in order to further enhance the cooling rate. Rylatt et al. [17] proposed a confined synthetic jet device stating that, with such a configuration, cold air is drawn from a remote location into the jet flow during the suction phase. Such a configuration was experimentally found to achieve a heat transfer enhancement of up to 36% in the stagnation region. Chaudhari et al. [18] designed a new synthetic jet characterized by a centre orifice surrounded by multiple satellite orifices. They tested this device varying both the Reynolds number ($1{,}000 < Re < 2{,}600$) and the normalized axial distance ($1 < H/D < 30$). This innovative configuration achieves a heat transfer of approximately 30% higher than that of the conventional single orifice jet. Luo et al. [19] proposed a new generation of synthetic jet actuators consisting in two cavities sharing the same oscillating piezoelectric diaphragm with two exit slots located at an appropriate distance. The numerical simulation showed that this innovative device doubles the function of the existing single synthetic jet and also solves the problems of pressure loading and energy inefficiency. Luo et al. [20] also carried out PIV measurements of this new device at Reynolds number and Strouhal number equal to 2500 and 0.17. The results reveal, in the near field, a more complex flow field characterized by a "self-support" phenomenon between the two synthetic jets, while, in the far field, the two jets merge a single and more stable synthetic jet. Persoons et al. [21] studied two adjacent synthetic jets, with slot orifice, which allows to direct the flow by changing the phase between the jets. They studied the flow field and the heat transfer performances of this device through PIV and IR thermography experiments. The device was characterized by a Reynolds number of 600 and a dimensionless stroke length equal to 29. Several values of phase and jet-to-surface spacing (6, 12 and 24) were investigated. The results showed a 90% enhancement of the maximum and overall cooling rate, compared to a single jet, for a phase equal to 120° and a jet-to-surface spacing H/D equal to 12. Also Lasance et al. [22] investigated a device belonging to the new synthetic jet generation proposed by Luo et al. [19]. Lasance et al. [22] device has a double circular exit. Such a double configuration is found to be advantageous because it improves the heat transfer performances [24] and, above all, it reduces the noise emission [23].

One of the main drawbacks of synthetic jet actuators is the generation of noise due to the membrane oscillation and to the aerodynamic noise generated by the synthetization of the jet. Among the several solutions proposed in literature are modification of the orifice shape and use of multiple interacting jets. Chaudhari et al. [25] investigated, for the first time, the influence of the orifice shape on the heat transfer performances of synthetic jets. Three configurations were analysed (square, circular, and rectangular exits) and the synthetic jet was characterized by a Reynolds number between 950 and 4500 and normalized impingement distance between 1 and 25. They found that the square orifice is more effective than the other shapes at $H/D > 5$ at the same exit flow conditions, while rectangular orifices with aspect ratio between 3 and 5 gives the best performance at smaller distances. Bhapkar et al. [26] experimentally investigated the acoustic aspects and average heat transfer characteristics of the elliptic shape and found that the latter performs better than the square, circular, and rectangular only at low H/D. Jeyalingam and Jabbal [27] studied the acoustic and flow behaviour of synthetic jets with 4-lobed and 6-lobed orifices and showed that the 6-lobed orifice offers a good broadband noise reduction over mid to high frequency range with respect to the baseline circular orifice.

IOP Conf. Series: Materials Science and Engineering **249** (2017) 012001 doi:10.1088/1757-899X/249/1/012001

Multiple jet configurations are by far the most efficient technique towards aerodynamic noise abatement: this goal may be achieved by phasing neighboring jets in such a way to obtain acoustic destructive interference. For this reason, at the University of Naples Federico II the design of innovative synthetic jet devices involving multiple jet configuration has been developed. In particular, two innovative devices have been investigated through Particle Image Velocimetry and InfraRed Thermography: twin synthetic jets [27-29] and multi orifice nozzles [30-31]. In the present work, the flow field the heat transfer performances and enhancement provided by these two devices is analysed and discussed.

2. Experimental setup and technique

In this section, the two innovative devices, the setup for the flow field and heat transfer measurements are described.

2.1. Twin synthetic jets device

Figure 1 shows the schematic of the twin circular synthetic jets (TSJ) device. The oscillating membrane is a loudspeaker (CIARE HS250), with a diameter of 270 mm. This loudspeaker splits the cavity in two sub-cavities with the same volume (V=2x10^{-3}m^3). Two pipes are attached to these sub-cavities. The pipe's length L is equal to 210 mm and their thickness and their inner diameter D are 1 mm and 21 mm, respectively. The parameter l is the jet axes distance and its corresponding non-dimensional parameter is Σ, defined as l/D. Such jet axes distance is varied between 1.1 and 5 nozzle diameters. The configuration Σ=1.1 corresponds to the condition for which the two pipes are adjacent. The two sub-cavities have the same resonance frequency, because the pipe length and diameter are equal. The same volume of the two sub-cavities is achieved by filling the upper and the bottom sub-cavity using particular geometrical items. The single synthetic jet is obtained by using a bended tube at the pipe exit in order to deflect one of the two synthetic. The device is operated at a frequency of 4Hz by supplying the loudspeaker with a sinusoidal input signal, which is generated by a wave generator (Digilent Analog DiscoveryTM), coupled with a power amplifier. Under these operating conditions, the synthetized jet has a Reynolds number of 5100 and a Strouhal number of 0.024.

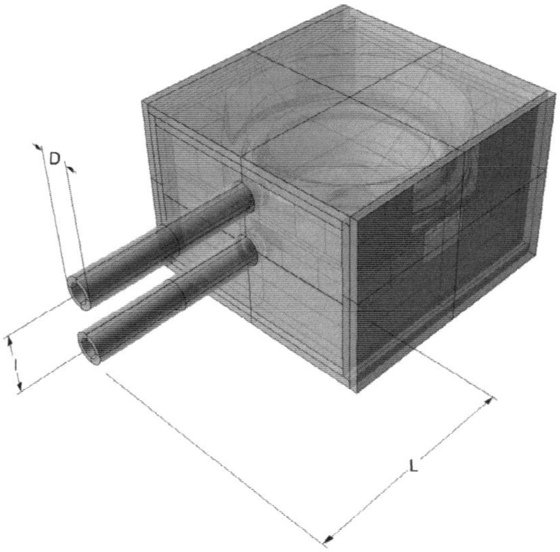

Figure 1. Twin synthetic jets device.

IOP Conf. Series: Materials Science and Engineering **249** (2017) 012001 doi:10.1088/1757-899X/249/1/012001

2.2. *Quadruple synthetic jet device*

A schematic of the quadruple synthetic jet (QSJ) device is shown in Figure 2a. The device consists of an aluminum cubic box of 115 mm side with four recesses in its lateral faces. In these recesses, four loudspeakers (CIARE PA065) are located and, therefore, four cavities are formed. These four cavities are connected to the upper face of the aluminum box by four identical L-shaped interior nozzles. The speakers, screwed on the bottom of the box recesses, cause the formation of synthetic jets at the nozzle outlets thorugh their oscillation. The nozzles are placed at an axis-to-axis distance l of $1.1D$ (D=10mm) in order to cause a strong near-field interaction between neighboring synthetic jets [27]. The four loudspeakers are supplied with 120 Hz sinusoidal input signals by using the same equipment employed for the other device. In the following, the four cavities and the corresponding nozzles and synthetic jets are named after the cardinal points. Moreover, each jet is associated with a phase angle, which is indeed the phase angle of the corresponding cavity pressure response.

Figure 2b shows the four investigated quadruple synthetic jet configurations: monopole-like (MP), dipole-like (DP), quadrupole-like (QP), and 90-degrees-circularly-shifted-jets (CSP) in analogy with acoustics (more details are reported in Paolillo et al. [30]). Each of the four single synthetic jets have the same Reynolds (4000) and Strouhal numbers (0.2). In order to obtain identical exit conditions for the single synthetic jets, a calibration of both the amplitudes and the phases of the speaker driving signals is performed, because of the slightly different mechanical features of the employed commercial speakers. All the details about the calibration procedure can be found in Paolillo et al. [30].

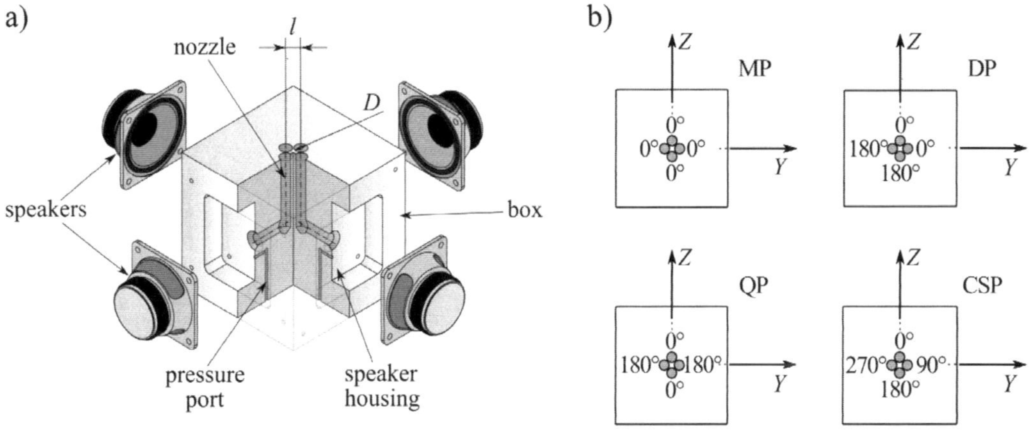

Figure 2. Quadruple synthetic jet actuator: a) schematic of the device, b) investigated configurations: the jet phase angles are relative to that of the jet in the Z > 0 hemispace (north jet).

2.3. *Particle Image Velocimetry setup and data reduction*

The impinging TSJ flow field has been investigated by using the experimental apparatus sketched in Figure 3. The operating conditions are: Reynolds number equal to 5100, Strouhal number equal to 0.024 and the jet-axes-distance of 1.1, and 3 nozzle diameters for five different values of dimensionless nozzle-to-plate distance H/D (2, 4, 6, 8 and 10). Illumination is provided by a Quantel Evergreen laser (Nd-YAG, 200 mJ/pulse) while particle images are acquired with a Zyla 5.5 sCMOS camera lens with 50 mm focal length and with a relative aperture *f/#* equal to 11 (more details are reported in Greco et al. [29]). In order to perform phase average measurements, the imaging system is synchronized with the synthetic jet. The synthetic jet is sampled at each 12° and the average of each phase is performed on 400 instantaneous flow fields. All the 12000 double frame images taken for each test are used to obtain reliable turbulence statistics. The vector field has been obtained processing images with a multiple pass algorithm with window deformation by using the Blackman weighting windows according to Astarita and Cardone [32] and Astarita [33-35]. The detailed employed process is clearly described by Greco et al. [29].

Differently from the TSJ device, the QSJ impinging flow field has not investigated yet. Anyway, important information, to deeply understand the heat transfer behaviour of this innovative device, can be inferred by its free flow field. This flow field is experimentally investigated by using the stereoscopic particle image velocimetry (SPIV). A sketch of the SPIV apparatus is illustrated in Figure 4. The employed equipment (cameras and laser) is the same used in the previous described experiment. In this experiment, the cameras work with two different relative apertures (f-stop): 5.6 for camera 1 and 8 for camera 2. Furthermore, the cameras are arranged in an angular displacement configuration with an angle, between the two optical axes, of about 50°. The Scheimpflug condition is satisfied with the installation of tilt-axis adapters. The two-dimensional (2D) two-components (2C) displacement fields in each image plane of the two cameras are computed through a multiple-pass algorithm with window deformation. The employed PIV algorithm is the same previously used. The PIV process details are described by Paolillo et al. [30].

All the velocimetry measurements are then analysed by applying the triple decomposition technique [36]. For further details, the reader is referred to Greco et al. [29, 37].

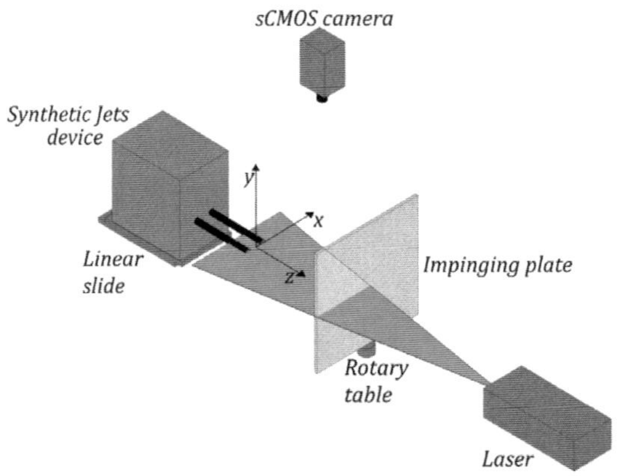

Figure 3. PIV setup for the investigation of the TSJ impinging flow field.

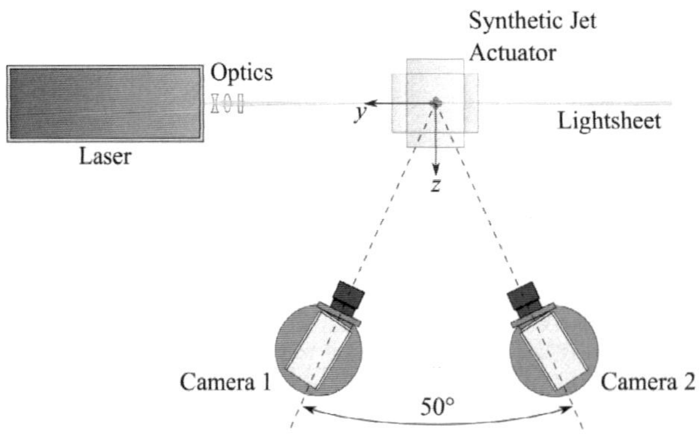

Figure 4. StereoPIV setup for the investigation of the QSJ free flow field.

2.4. Heat transfer measurements setup and data reduction

Figure 5 shows a schematic of the classic thermal test apparatus. The heat flux sensor can be a constantan foil of 50 µm thickness, 200 mm width and 450 mm length (employed in QSJ's experiments) or a stainless steel foil 243 mm wide, 715 mm long and 40 µm thick (used in TSJ's experiments). The foil is clamped with two copper conducting bars pairs, mounted on the shortest side of the foil and provided with a constant voltage using a 0 V to 250 V DC stabilized power supply. The stabilized current control ensures uniform and constant heating by Joule effect. The very large equivalent cross section of the bus bars, with respect to that of the heated foil, ensures that the voltage drop along them is very small.

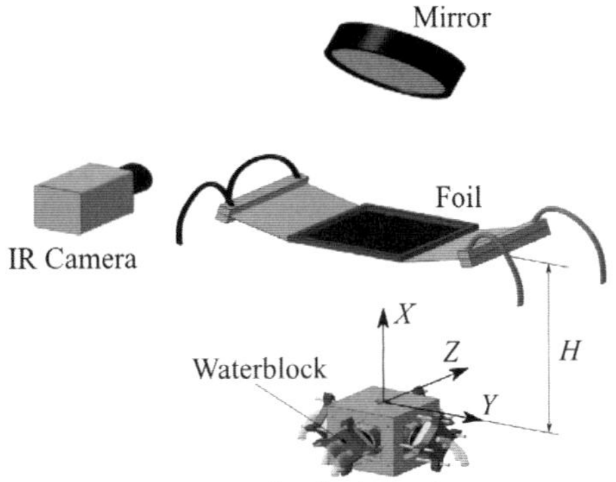

Figure 5. Schematic of thermal test apparatus.

The foil is positioned horizontally with the synthetic impinging on its lower (QSJ) or upper (TSJ) face. The distance H between the nozzle exit plane and the target foil (impingement distance) is adjusted by means of a linear slide and varied between 2.5D and 20D (QSJ case) with a spacing of 2.5D or between 2D and 10D (TSJ case) with a spacing of 2D.

In the QSJ device, four waterblocks are employed to cool the speakers. Such waterblocks are connected with a thermostatic bath to avoid the jet temperature increase due to the Joule heating caused by the electric currents flowing through the speaker coils. This effect is not negligible for this device because the cavity (volume delimited by the speaker cone and the bottom of the recess in the box face) is very small compared to the overall box dimensions and thus the aluminum body acts to insulate the heat dissipated in the speaker coil and transferred to the cavity.

An infrared camera (FLIR SC6000 for QSJ case while CEPID JADE III for TSJ) measures the temperature of the foil face, imaged by means of a mirror, that is opposite to that where the synthetic jet impinges. Temperature across the foil is considered constant as the Biot number ($Bi = hs/\lambda_f$ where h is the time-averaged convective heat transfer coefficient and s and λ_f are the foil thickness and thermal conductivity, respectively) and the inverse of modified Fourier number ($Fo = \alpha/\pi f s^2$ where α is the thermal diffusivity) are small with respect to unity. The imaged foil face is coated by high-emissivity black paint (hemispherical emissivity ε = 0.95) in order to increase the accuracy of temperature measurements. The IR camera is used in conjunction with the unsteady [38] and steady [39] heated thin foil heat transfer sensor, whose details, not reported here for the sake of brevity, can be found in Greco et al. [28].

The experimental data are reduced in dimensionless form in terms of time average Nusselt number $\overline{Nu} = \bar{h}D/k$ (k is the thermal conductivity of air at film temperature), phase average Nusselt number $Nu(\varphi) = h(\varphi)D/k$, standard deviation of Nusselt number phase averaged fields Nu' and percentage area-weighted standard deviation of the Nusselt number $\sigma_{\overline{Nu}}$ (non-uniformity parameter).

3. Results

3.1. Twin Synthetic Jets device: impinging flow field and heat transfer

In this section, for the sake of brevity, only the impinging flow field of SSJ and TSJ at $\sum=1.1$ are analysed and discussed. This configuration (i.e. $\sum=1.1$) is that which shows the strongest interaction of the two synthetic jets giving a complete different behaviour with respect the classic single synthetic jet. Indeed, the configurations at $\sum$ equal to 3 and 5 show the behaviour of two superimposed SSJ whose exit centres are shifted of $3D$ or $5D$, respectively. Anyway, for all the other configurations, the reader is referred to Greco et al. [29].

In Figures 6 all the velocity components are depicted for H/D equal to 2. For the sake of brevity, only eight phase-average flow fields (representing the ejection phase) are reported. The phases range from $\varphi = 0°$ up to $\varphi = 168°$. For each phase the velocity vector arrows are placed every 0.08 D in x and z direction.

In Figure 6a) the phase-average axial and radial velocity maps are shown. At $\varphi = 0°$ the synthetic jet starts being issued by the nozzle and at $\varphi = 24°$ the vortex ring is formed. Then the synthetic jet impinges and spreads over the plate as observable at $\varphi = 48°$. For this phase, it is possible to detect the formation of a counter rotating vortex ring near the impinging plate and also two regions of high radial velocity at x/D equal to about 0.75 (where the complete rotation of the impinging synthetic jet occurs) and at the position of the convective primary vortex ring. From this phase to $\varphi = 120°$ the effect of the impinging plate on the impinging synthetic jet is clearly observable. Indeed, the morphology of the axial velocity is characterized by a hemispherical axial velocity decreasing, near the plate, caused by the strong adverse pressure gradient. For $48° \leq \varphi \leq 96°$, the vortex ring continues sweeping the impinging plate and the position of high radial velocity is always located at x/D equal to 0.75 (while the second region shifts towards higher radial position with the vortex ring and its value decreases). For this range of phases the axial velocity attains the maximum value. For $\varphi > 96°$ the vortex ring is out of the measurement zone and the values of axial and radial velocity start decreasing.

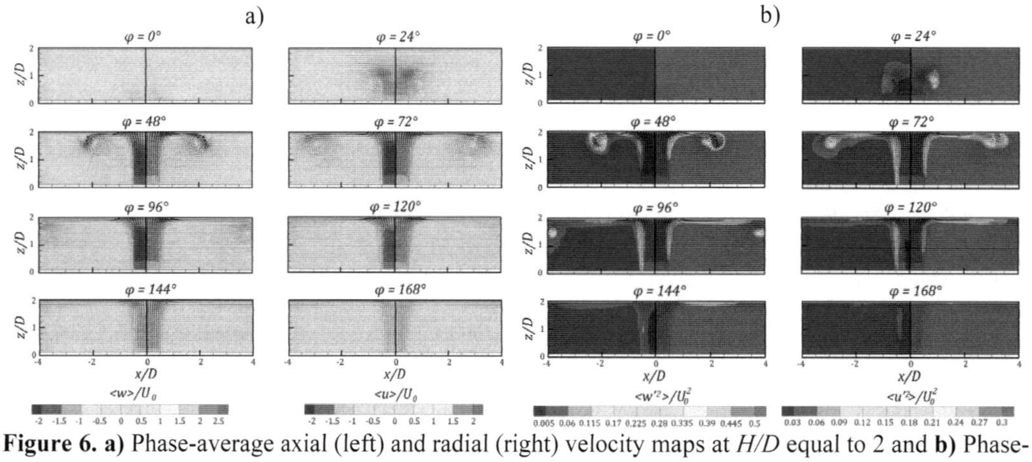

Figure 6. a) Phase-average axial (left) and radial (right) velocity maps at H/D equal to 2 and **b)** Phase-average mean–squared axial (left) and radial (right) turbulent velocity maps at H/D equal to 2.

In Figure 6b) the phase-average mean-squared axial and radial turbulent velocity maps are depicted. At $\varphi = 0°$ the axial values are zero because, obviously, the synthetic jet is coming out. At $\varphi = 24°$ the axial turbulent velocity acquires non-zero values on the front boundary of the vortex ring while the radial one attains its maximum near the vortex ring core. At $\varphi = 48°$ the maximum value for both components are located in the middle between the vortex ring and the counter rotating one. Moreover, for the axial turbulent velocity a value different from zero is also detected at x/D equal to -0.5 and along the shear layer (only for $z/D > 1$). The radial turbulent velocity shows high values also along the last zone ($z/D > 1$) of the shear layer. At $\varphi = 72°$ the axial component has maxima located in the vortex ring core, near the impinging wall at x/D between 1.5 and 2 and along all the shear layer up to the nozzle exit. Differently, the radial component shows a zero turbulent value near the nozzle ($0 < z/D < 0.4$) and maxima along the wall ($1 < x/D < 2.25$) and ahead of the vortex ring (near the wall) at a radial position equal to approximately 3.3 D. At $\varphi = 96°$ the vortex ring is completely detached from the impinging plate (as visible from the peak of axial and radial turbulence) but near the plate still high value of axial ($1.5 < x/D < 2.5$) and radial ($1 < x/D < 2.5$) are observable. For $\varphi > 96°$ the two maxima, located in vortex ring core, are out of the measurement zone and the other high values start decreasing (still visible is the maximum of axial turbulence near the impinging plate at x/D equal to approximately 2). It is important to highlight that for this nozzle-to-plate configuration, from φ equal to 48° until φ equal to 120° the zone between $-0.5 < x/D < 0.5$ is characterized by no turbulence and a constant value of the velocity, resembling the potential core behaviour of a continuous jet (defined as a potential core-like region).

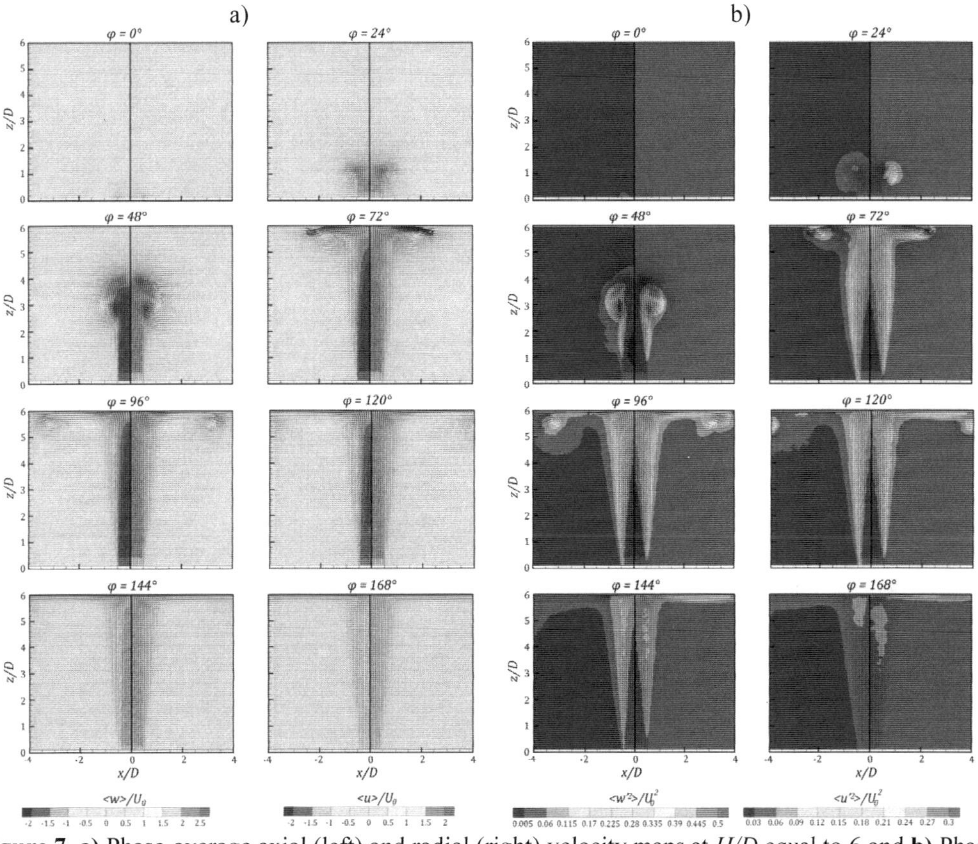

Figure 7. a) Phase-average axial (left) and radial (right) velocity maps at H/D equal to 6 and **b)** Phase-average mean–squared axial (left) and radial (right) turbulent velocity maps at H/D equal to 6.

In Figure 7 all the velocity components are depicted for H/D equal to 6. In Figure 7a) the phase-average axial and radial velocity maps are reported. It is possible to see how the vortex ring is generated (by the roll-up of the shear layer sheet) and convects downstream (first three phases). Then (at $\varphi = 72°$) it impinges on the plate and still generates a smaller counter rotating vortex ring (than that generated in the shortest nozzle-to-distance configuration, as also visible in Figure 6). For this latter phase, it is possible to highlight that a maximum in the radial velocity, near the impinging plate, is detectable at a radial position of about 1.3 *D* from the jet axis. The axial velocity, approaching the impinging plate, shows a bell-shape distribution that is completely different from the behaviour detected for the shortest configuration. At $\varphi = 96°$ the vortex is sweeping the plate and a local radial maximum is located between the vortex ring and the impinging plate. Differently from the shortest case, the second radial velocity peak has approximately the same value of the first one. At this phase, the axial component shows a region of high value that has the greatest extension (toward the impinging plate). As φ increases, the radial and axial values decrease.

In Figure 7b) the phase-average mean-squared axial and radial turbulent velocity maps are shown. The behavior, in the first two phases, is similar to what shown for the case at *H/D* equal to 2. Then, at $\varphi = 48°$, high value of the axial turbulence is located mainly along the shear layer and near the front boundary of the vortex ring. The radial turbulence shows its maximum along the shear layer, as well, and near the vortex ring core. At $\varphi = 72°$ it is possible to point out that the mean-squared axial turbulent velocity has high value all along the shear layer (until the nozzle exit) and in the counter rotating vortex ring. The radial turbulence shows, basically, the same characteristics of the axial one. Differently from the shortest nozzle-to-plate configuration, the axial and radial turbulence start merging on the jet axis at *z/D* equal to about 3.8. That zone, featured by a zero-turbulence value, is the so-called potential core-like region. At $\varphi = 96°$ the axial turbulence has its maximum along the shear layer and in the vortex ring core but, differently from *H/D* equal to 2, the zone of high turbulence near the wall at *x/D* equal to about 2 shows lower values. The radial turbulence still shows (as for *H/D* equal to 2) a high value region near the plate (which covers the zone between the shear layer and the vortex ring) apart from the maxima present along the shear layer and inside the vortex ring core. At $\varphi \geq 120°$ the turbulence values start decreasing and the vortex ring is moved out of the measurement region.

The phase-average axial and radial velocity are shown in Figure 8. At $\varphi = 0°$ the right synthetic jet starts coming out but, differently from the SSJ case, at $\varphi = 24°$ the vortex ring is not fully formed because the starting suction phase of the left synthetic jet delays such a formation (only the right vortex ring footprint is observable). At $\varphi = 48°$ the right synthetic jet impinges but it is slightly deflected towards *x/D* equal to 0. Indeed, near the plate the negative radial velocity is higher and more extended (in the axial direction) than the positive one. At $\varphi = 72°$ the axial distribution is not symmetric with respect to the right synthetic jet centreline (located at *x/D* equal to 0.55) but the maximum is attained along *x/D* equal to 0. This is due to the interaction of the two jet which is stronger near the *x/D* equal to 0 axis. The radial velocity still presents two maxima: the first near the plate where the jet has its complete rotation and the second between the plate and the sweeping vortex ring. At $\varphi = 96°$ the right synthetic jet seems to be less stretched by the left one because the interaction starts decreasing. For $\varphi \geq 120°$ the values become lower.

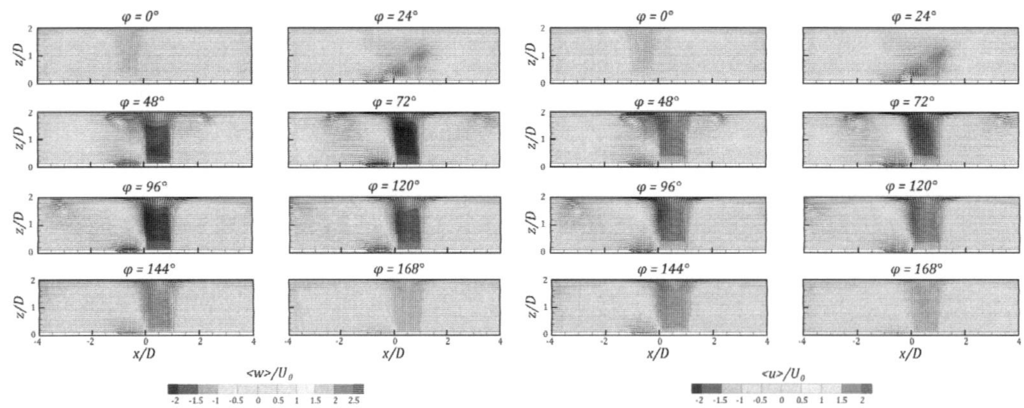

Figure 8. Phase-average axial (left) and radial (right) velocity maps at *H/D* equal to 2.

In Figure 9 the phase-average mean-squared axial and radial turbulent velocity are shown respectively. At $\varphi = 0°$ no zero values of axial and radial turbulence are still present due to the trailing part of the left synthetic jet. At $\varphi = 24°$ the axial turbulence attains high value on the front of the vortex ring while the radial one is located in the vortex ring core and between the two jets. At $\varphi = 48°$ the axial turbulence shows high values along the internal shear layer which are higher than the external one. This is due to the interaction of the two 180° phase-shift jets. Indeed, the extension of the turbulence of the internal shear layer arrives until the nozzle exit, differently from the external one. Moreover, the axial turbulence of the left vortex ring footprint is lower than the right one because of the suction phase of the left synthetic jet. The same considerations can be drawn for the radial turbulence. Moreover, a region of high turbulence, connecting the left nozzle exit and the internal shear layer, is present (above all for the radial component). At $\varphi = 72°$ the turbulence along the internal shear layer increases and is still higher than the external one. The axial turbulence attains also high value in the vortex core and in the zone near the plate between the vortex ring and the shear layer. The radial component acts as the axial one apart from the region near the vortex ring. Indeed, here the maximum is attained near the plate where the counter rotating vortex is located. At $\varphi = 96°$ the behaviour of turbulence is the same, indeed, maxima are located along the shear layer, in the ring vortex centre (on the vortex ring left side near the impinging plate for the radial component) and near the impinging plate. For $\varphi > 96°$ the values start decreasing. In all the phases, it is possible to see that, as previously described, the right synthetic jet is deflected towards the *x/D* equal to 0 axis. Furthermore, even in this TSJ configuration a potential core-like region is detectable.

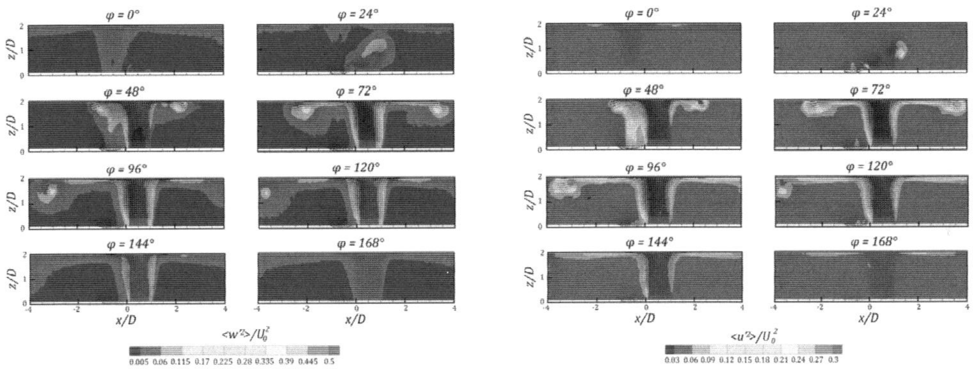

Figure 9. Phase-average mean–squared axial (left) and radial (right) velocity maps at *H/D* equal to 2.

In Figure 10 the phase-average axial and radial velocity components are depicted. At $\varphi = 0°$ the footprint of the trailing part of the left synthetic jet is still present. At $\varphi = 24°$ the right synthetic jet is issued and the behaviour is similar to what shown in the shortest nozzle-to-plate case. At $\varphi = 48°$ the jet is going to approach the impinging plate but, as it is possible to observe, the ring vortex is deflected because of the interaction with the left synthetic jet which is acting like a sink. At $\varphi = 72°$ the jet is impinging and the axial profile is bell-shaped, differently from the *H/D* equal to 2 case.

Moreover, the jet is considerably deflected toward the *x/D* equal to 0 axis. Focusing our attention to the radial component it is possible to see that the negative velocity is higher than the positive one because of the previously described deflection. For $\varphi = 96°$ the left radial value does not show a high peak (as in the previous phase) because the vortex ring is sweeping and for $\varphi \geq 120°$ the velocity values decrease and the vortex ring goes out of the measurement zone. Furthermore, the radial velocity over the plate, for these phases, does not show a peak where the impinging jet rotates. Indeed, this value has the magnitude of the one created between the sweeping vortex ring and the impinging plate (differently from the *H/D* equal to 2 configuration).

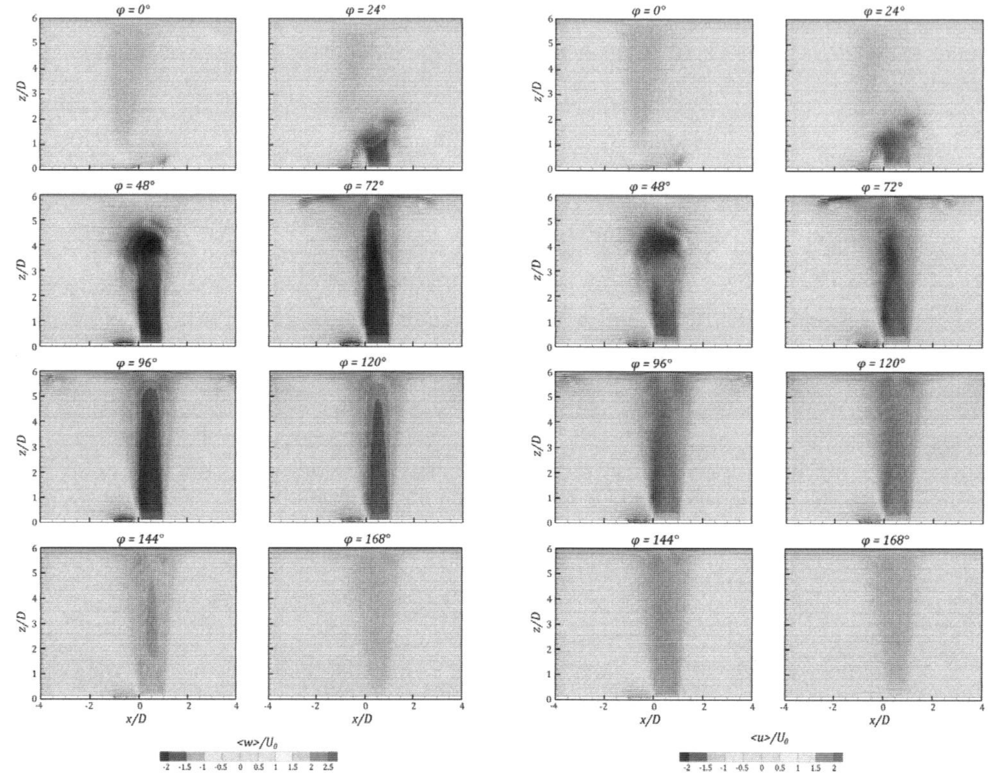

Figure 10. Phase-average axial (left) and radial (right) velocity maps at *H/D* equal to 6.

In Figure 11 the phase-average mean-squared axial and radial turbulent velocity, respectively, are shown. At $\varphi = 0°$ no-zero values of turbulence are still present because of the trailing part of the impinging left synthetic jet. At $\varphi = 24°$ the axial turbulence shows values different from zero along the shear layer and on the front boundary of the jet. The radial turbulence has a peak in the vortex ring core. At $\varphi = 48°$ the axial turbulence along the internal shear layer has increased and is thicker than the external one. Furthermore, the two shear layers start merging at z/D equal to about 3.8 and coalesce with the axial turbulent developed on the front boundary of the convecting vortex ring. Also the radial turbulence seems to start merging but still shows its peak in the vortex ring core. It is necessary to point out that the axial value, as the radial one, of the right vortex ring footprint is higher than the left one. This could be related to the fact that the left part of the vortex ring is affected by the suction phase of the left synthetic jet which damps partially the oscillations. At $\varphi = 72°$ the turbulent axial component shows higher value along the internal shear (than the external one) and also a maximum located in the vortex ring core (with a greater value in the right footprint). The turbulent radial component acts as the axial one along the shear layer but, near the impinging plate, shows a maximum between the plate itself and the convecting vortex ring (where the counter rotating is placed). Furthermore, both components start merging at an axial position of about 3.8 diameters from the nozzle exit. Such a zone is the so-called potential core-like region. At $\varphi = 96°$ the vortex ring continues sweeping the impinging plate, featured by a decreased axial turbulence maximum in the core. Moreover, a high value zone of radial turbulence, which spans from the vortex ring core until the impinging shear layer, is observable. For $\varphi \geq 120°$ the values start decreasing. The deflection and the wider internal shear layer (with high values) presented in these turbulent components is the reason why in the time-average measurement the extension of the potential core-like region seems to be lower (up to 2 diameters).

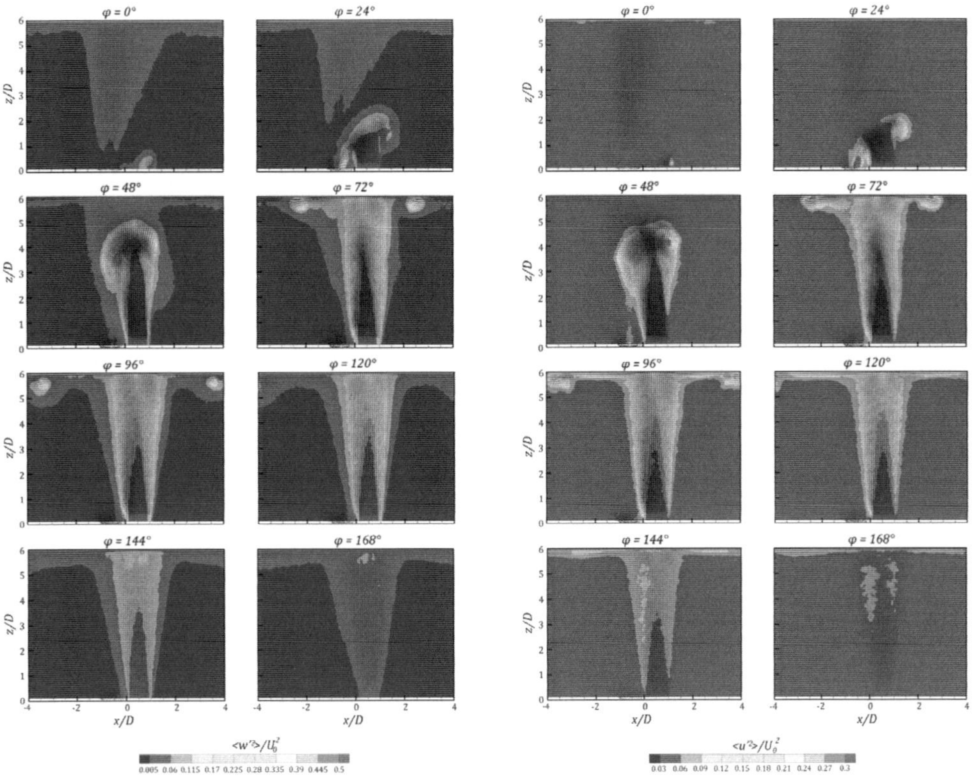

Figure 11. Phase-average mean–squared axial (left) and radial (right) velocity maps at *H*/*D* equal to 6.

Regarding the heat transfer behaviour of this device, Figure 12 shows the phase average Nusselt number maps for the SSJ at *H*/*D* = 2 and 6, starting from $\varphi = 0°$ up to $\varphi = 168°$.

At *H*/*D*=2, the jet reaches the target plate at $\varphi = 0°$ and starts spreading over the foil, as visible at $\varphi = 24°$. At $\varphi = 48°$ the jet is sweeping the surface and the inner ring shaped region arises at radial distance (*r*) from the stagnation point of 0.5 *D* from the centre of impingement. This result agrees with DNS simulation of a continuous circular impinging jet by Rohlfs et al. [40]. This inner ring shaped region is caused by the radial wall acceleration [40, 41].

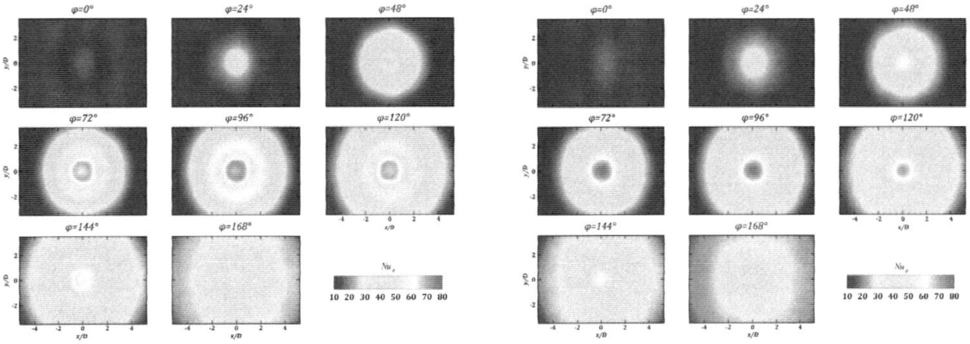

Figure 12. Phase-average Nusselt number for SSJ at *H*/*D* equal to 2 (left) and 6 (right).

According to Rohlfs et al. [40] this radial acceleration generates an increase of the spatial gradient of axial velocity near the impinging plate (for the continuity equation) which leads fresh air near the plate itself. This agrees with the no bell-shaped profile of the axial velocity near the impinging plate which can be inferred from Figure 6. At $\varphi = 72°$ the Nu_φ value at inner ring location increases its value, likely because synthetic jet continues impinging over the foil causing an increase of the wall radial acceleration, and an outer ring shaped region can be detected at a radial distance of 1.9 *D*. The outer ring shaped region can be ascribed to unsteady separation and later reattachment due to the formation of the secondary counter rotating vortex on the wall, generated by the passage of the primary vortex ring and subsequently vortex rings caused by the Kelvin-Helmholtz instability. Hence the delay between the appearances of the two ring shaped regions can be ascribed to the travelling time necessary to the primary vortex ring to assume the requested vorticity in order to generate, for the first time, the secondary counter rotating vortex. Indeed, such ring structures is already present at $\varphi = 48°$ but is not so clearly visible as in the next phases. At $\varphi = 48°$ the outer ring is the one formed by the passage of the sweeping vortex ring which creates the counter rotating one (as seen in PIV experiments). After the ring vortex is passed (indeed the zone featured by high heat transfer is widening) the outer ring shaped region is still present. This one could be ascribed to the column of fluid which follows the ring vortex (as described for a L_0/D higher than 4 in McGuinn et al. [16]). The higher is the dimensionless stroke length (L_0/D), the longer is the following column of fluid. Hence in this case ($L_0/D \approx 42$) this column of fluid acts like a turbulent continuous jet generating vortex rings, because of the Kelvin-Helmholtz instability. Such vortex rings are the cause of the counter rotating vortex rings near the wall and, consequently, the cause of the continuous presence of the outer ring shaped region of Nusselt number maxima. The outer ring shaped region moves from *r/D* equal to about 1.8 to *r/D* equal to approximately 2. Such a motion is caused because the synthetic jet is still impinging, indeed vortex ring is still travelling, so the location of the secondary vortex ring separation is moved towards higher radial position. At $\varphi = 96°$ the Nusselt number acquires its maximum value. After $\varphi = 96°$, phase-average Nusselt number values in this region decrease and the two ring shaped regions weaken probably because the biggest part of the incoming flow (for the cycle) has already impinged. The map for *H/D* equal to 4 (not shown herein for the sake of brevity but present in the attached movie to Greco et al. [28]) presents result similar to those at *H/D*=2.

At *H/*D=6 (Figure 12), differently from the case at *H/D*=2, the two ring shaped regions are not present. The absence of the inner ring shaped region is caused by two main features: the value of *H/D* which is greater than the synthetic jet potential core-like region (shown in PIV measurements); the ring vortex does not have enough vorticity to generate a strong enough secondary counter rotating vortex (as depicted in Figures 9 and 11). It is possible also to note that the maximum value of Nu_φ is attained around $\varphi = 72°$ for *H/D* equal to 6 while it is attained at around $\varphi = 96°$ for *H/D* equal to 2. Moreover, the heat transfer rate is more spatially concentrated near the stagnation point for *H/D* = 6 than the case with *H/D* = 2 where such a cooling occurs inside the area enclosed by the outer ring shaped region. For all the *H/D* values, the phase-average Nusselt number maps for $\varphi > 168°$ (not shown herein for the sake of brevity but present in the attached movie 1 to Greco et al. [28]) present a decreasing value of the heat transfer, with respect to the map at $\varphi = 168°$ and have all a similar spatial distribution.

The phase-average Nusselt number maps for the TSJ with $\sum$ equal to 1.1 at *H/D* equal to 2 and 6 are reported in Figure 13. The phase-average Nusselt number maps for the configuration with $\sum = 3$ and 5 have not been reported because their behaviour is very similar to that of two superimposed synthetic jets. (All phase-average Nusselt number maps are attached to Greco et al. [28] as video sequences in movie 2, 3 and 4 for $\sum = 1.1$, 3 and 5, respectively). The phase-average Nu_φ maps (Figure 12) of TSJ for $\sum = 1.1$ at *H/D* equal to 2 show a different behaviour with respect to SSJ. During the ejection phase, at $\varphi = 0°$, a higher heat transfer (Nu_φ in the stagnation point is about double with respect to the single synthetic jet), is shown mainly due to the strong jet interaction. At $\varphi = 24°$ the phase-average Nu_φ map of TSJ already shows, with respect to SSJ, a strong inner ring shaped region; at $\varphi = 48°$ such a map shows a higher value of Nu_φ at the inner ring shaped region and a strong presence of the outer ring shaped region differently from SSJ. These phenomena are caused by the higher centreline velocity of the TSJ with respect to SSJ (as reported PIV measurements). In fact, such a higher velocity generates a faster spreading of the impinging synthetic jet over the foil, hence a greater wall radial acceleration. Furthermore, at $\varphi = 72°$ the TSJ Nu_φ map shows a higher value at inner and outer ring regions with respect to SSJ. In the following phases the value acquired by the phase-average Nusselt number for TSJ are comparable (at $\varphi = 96°$) or slightly lower (at $\varphi = 120°$) than those of SSJ in the same phase. Finally, the inner and outer ring shaped regions disappear approximately at the same phase for both configurations. For *H/D* equal to 6 the outer and inner ring shaped regions for both configurations (Figures 12 and 13) are not detected because, as in the case of the single synthetic jet, the nozzle-to-plate distance is much higher than the potential core-like region of the synthetic jet. Moreover, it is possible to highlight that also for this value of *H/D* the phase-average Nusselt number map of TSJ, for the first four phases, attain a higher value with respect to the SSJ due to the higher impinging axial velocity and axial fluctuations. On the other hand, the behaviour, in the last four phases, is the same for both configurations.

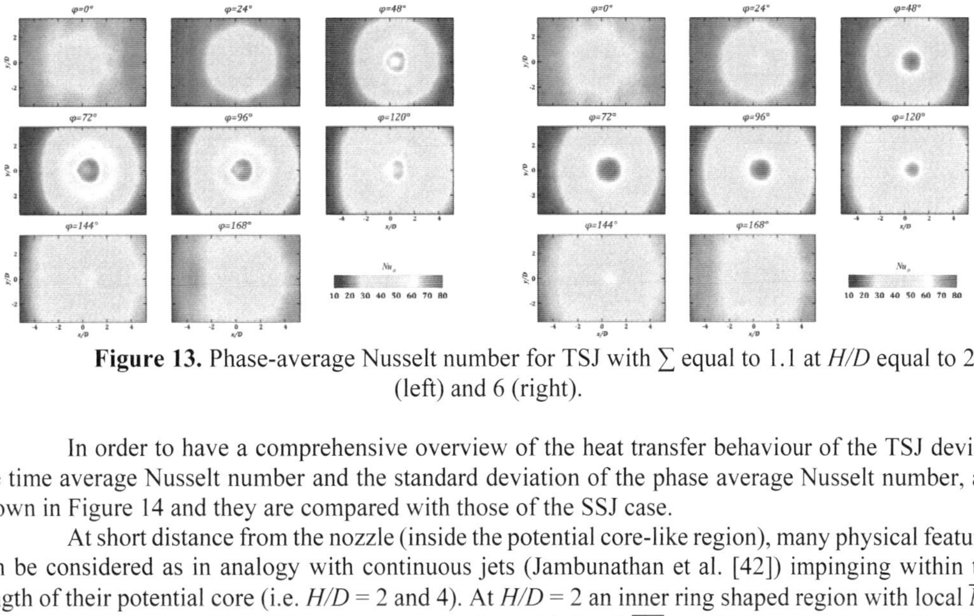

Figure 13. Phase-average Nusselt number for TSJ with $\sum$ equal to 1.1 at *H/D* equal to 2 (left) and 6 (right).

In order to have a comprehensive overview of the heat transfer behaviour of the TSJ device, the time average Nusselt number and the standard deviation of the phase average Nusselt number, are shown in Figure 14 and they are compared with those of the SSJ case.

At short distance from the nozzle (inside the potential core-like region), many physical features can be considered as in analogy with continuous jets (Jambunathan et al. [42]) impinging within the length of their potential core (i.e. $H/D = 2$ and 4). At $H/D = 2$ an inner ring shaped region with local $\overline{Nu}$ maximum appears at $x/D \approx 0.5$ and an outer region with local $\overline{Nu}$ maximum appears at $x/D \approx 1.9$. Maximum values of Nu' are also obtained approximately in the same regions where inner and outer ring shaped structures are located as easily detectable in Nu' map. The $\overline{Nu}$ values, measured in the first ring shaped region, are slightly higher than the values obtained in the stagnation region. This can be related to the behaviour of the time-average axial velocity component and the mean-squared phase-correlated organized contribution to the velocity component. Indeed, a peak of these parameters, for this nozzle-to-plate configuration, is observable at *x/D* equal to 0.5. This ring shaped region is not present for nozzle-to-plate distances higher than 4 (after the end of the potential core-like region), as detectable from $\overline{Nu}$ and Nu' profiles.

The second peak (outer ring shaped region) is present for *H/D* lower than 4 while for *H/D* equal to 6 only a change in the $\overline{Nu}$ slope is visible. For *H/D* greater than 6 such a peak disappears. The presence of this peak, as previously stated, can be ascribed to the formation of the secondary counter rotating vortex on the wall which is generated by the passage of the ring vortex. In these experiments the formation of the counter rotating vortex is visible until *H/D* equal to 6. The size and vorticity of this counter rotating vortex decreases as *H/D* increases. This behaviour agrees with the heat transfer profiles. It has to be point out that such a second heat transfer peak is not only related to the phase-average measured counter rotating vortex caused by the primary vortex ring. Indeed, others counter rotating vortex are generated by the vortex ring caused by the Kelvin-Helmholtz instabilities of the column of the fluid following the primary ring vortex (because the Strouhal number is equal to 0.024). At a nozzle-to-plate distance equal to 6 diameters the maximum heat transfer in the stagnation point is detected, in agreement with what happens for continuous jets.

Nu' resembles the $\overline{Nu}$ profile and is strictly related to the axial fluctuating velocity profile (phase-correlated and turbulent components). Moreover, the value of *Nu'* map becomes more uniform and its maximum value decreases at high *H/D* because, as the distance from the nozzle increases (Shuster and Smith [43]), synthetic jets act like turbulent continuous jets. As matter of fact all the fluctuating velocity values decrease for *H/D* higher than 6.

The $\overline{Nu}$ profile for twin synthetic jets with $\sum = 1.1$ show two distinct stagnation points only for $H/D = 2$. As a matter of fact, the relative $\overline{Nu}$ profile shows a first peak at $x/D = 0$ and a symmetric second peak at approximately $x/D \approx 1.05$. The first and the second peak are related to the existence of the inner ring shaped region. Such a ring occurs around the two stagnation points, located at $x/D \approx 0.55$, and shows a diameter equal to 1 *D*. The superimposition of the two inner ring shaped regions at $x/D = 0$ is the reason why the first peak is higher than the second one. This can be related to the behaviour of the time-average axial velocity component and the time-average phase-correlated organized contribution to the velocity axial component (not presented here).

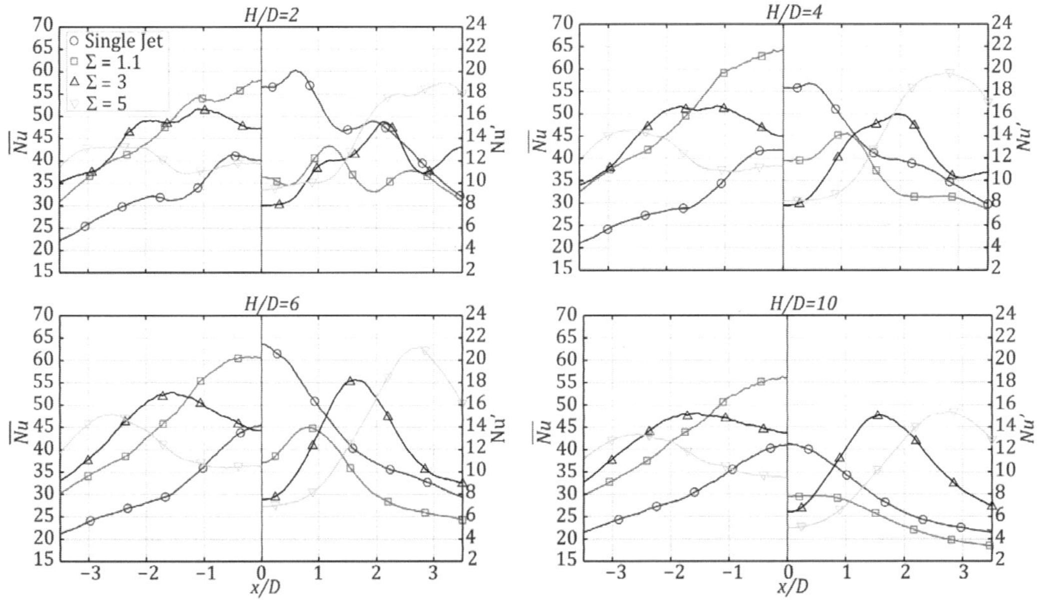

Figure 14. Phase-average Nusselt number for TSJ with $\sum$ equal to 1.1 at *H/D* equal to 2 (left) and 6 (right).

The *Nu'* maps and profile clearly show the effect of the unsteady passage of the vortex ring. Due to the superimposition of the two inner ring shaped regions, the value of *Nu'* acquired at $x/D = 0$ is not equal to the one attained at $x/D \approx 1.05$. As *H/D* increases the strong interaction between the two adjacent jets produces a different heat transfer behaviour characterized by a maximum $\overline{Nu}$ value at $H/D = 4$ as visible in Figure 14. For $\sum = 1.1$ and $4 \leq H/D \leq 8$ *Nu'* presents a different behaviour from the $\overline{Nu}$ profile. In fact, at *H/D* equal to 4 the *Nu'* profile shows the peak relative to the inner ring shaped region but the second peak is strongly reduced. Such a second peak disappears for *H/D* higher than 4 differently from the first one. This could be related to the presence of the second wall jet which passes over the same zone of the plate (already swept by the vortex ring) causing a change in the variables value, differently form the SSJ configuration. Indeed, only small later oscillations, approximately in the same radial position, can be barely seen. The first *Nu'* peak shifts toward the centre as *H/D* increases because the jet, as previously explained, starts acting as a turbulent continuous jet. At *H/D* equal to 10 the *Nu'* profile is very similar to the one of a single synthetic jet. At all the nozzle-to-plate-distances the values of $\overline{Nu}$ measured are higher than the SSJ case. Moreover, it is worth nothing that the values attained by *Nu'* for such a twin synthetic jets configuration are lower than the one acquired by the single synthetic jet. This is related to the presence of two impinging synthetic jets. Indeed, in this configuration there is always a heat transfer caused by the impingement. This condition leads to a decreasing variation of the Nusselt number (*Nu'*) as also occurs for the phase-correlated velocity contributions in the TSJ case (with respect to SSJ case).

Regarding the twin synthetic jets with $\Sigma = 3$ the Nusselt number maps show two clearly distinct stagnation regions, approximately at $x/D \approx 1.5$ with its inner and outer ring shaped regions. In this configuration, for $H/D = 2$, the peak located approximately at $x/D \approx 1$ is higher than the peak at $x/D \approx 2$. This effect is related to the fact that the peak which is closer to the centre is more affected by the presence of the other jet and a beneficial effect is attained. Such a phenomenon decreases with *H/D* increase, disappearing already for $H/D = 4$. Also in this case *Nu'* and $\overline{Nu}$ maps and profiles present a similar behavior. As matter of fact at $H/D = 2$ the *Nu'* profile shows two peaks at approximately $x/D \approx 2.1$ and $x/D \approx 3.5$ showing the presence of the inner and outer shaped regions. The relationship between the velocity component and the heat transfer behaviour is the same previously explained for the SSJ case. As *H/D* increases, the inner and outer structures disappear. Moreover the maxima in *Nu'* profile merge in a unique maximum located at approximately $x/D \approx 1.5$, where the synthetic jet impinges. The behaviour of twin synthetic circular air jets with $\Sigma = 5$ is very similar to the one shown for $\Sigma = 3$.

3.2. *Quadruple Synthetic Jets device: free flow field and heat transfer*

The quadruple synthetic jet flow field is deeply characterized by the formation and convection of large-scale coherent vortex structures (CVSs). These structures originate during the ejection stroke after the separation of the flow at the edges of the jet outlet and then move downstream under their self-induced velocity. As the CVSs propagate, they develop inertial instabilities that cause their break down and, finally, they decay to turbulence. Therefore, two different flow field region can be defined: near and far (velocity) fields. The near (velocity) field is that part of the field characterized by the passage of the CVSs, while the far (velocity) field is the region where the CVSs have completely dissipated and the synthetic jet has turned into a steady turbulent jet flow. The QSJ is characterized by $L_0/D = 5$, so no significant trailing jet is expected. Therefore, the near field of the QSJ is dominated essentially by the dynamics of the CVSs. These CVSs show a complex evolution since they result from the interaction of fours vortex rings issued by the four nozzles. As expected, this interaction strongly depends on the initial phases of the single synthetic jets and, thus, on the considered configuration. In order to explain the behaviour of such structures in the four investigated configurations, the phase-averaged fields of the out-of-plane vorticity are considered and analyzed.

In Figure 15 (left), contours of the non-dimensional phase-averaged $\zeta_z D/2U_0$ in the N/S plane for the MP configuration are reported. Four different phases are presented, corresponding to the beginning and the middle of the ejection part of the cycle (phases $\varphi=0°$ and $\varphi=90°$, respectively) and to the beginning and the middle of the suction part ($\varphi=180°$ and $\varphi=270°$, respectively).

At the phase $\varphi=0°$, two CVSs can be detected at $x/D \approx 8$ and $x/D \approx 4$, named CVS1 and CVS2, respectively. No coherent structures can be observed in the downstream region of the flow field, while the phase-averaged vorticity field shows irregular small- scale variations that is a usual characteristic of unsteady turbulent motions. From $\varphi=90°$, another structure (CVS3) is forming at the beginning of the field of view and, then, it continues its propagation. Indeed, CVS1 and CVS2 can be considered as the evolution of CVS3. It is important to remark that these CVSs are generated by the coalescence of the four separated vortex rings, which are simultaneously issued from the four nozzles. Figure 15 clearly shows that the original vortex rings merge causing a larger structure whose diameter (defined as the distance between the two points with the highest ζ_z value) is approximately 2.5 times the nozzle diameter. Such a structure slowly weakens moving downstream, and, at $x/D \approx 12$, it completely dissipates (CVS1 definitively loses its coherence between φ=180° and φ=270°).

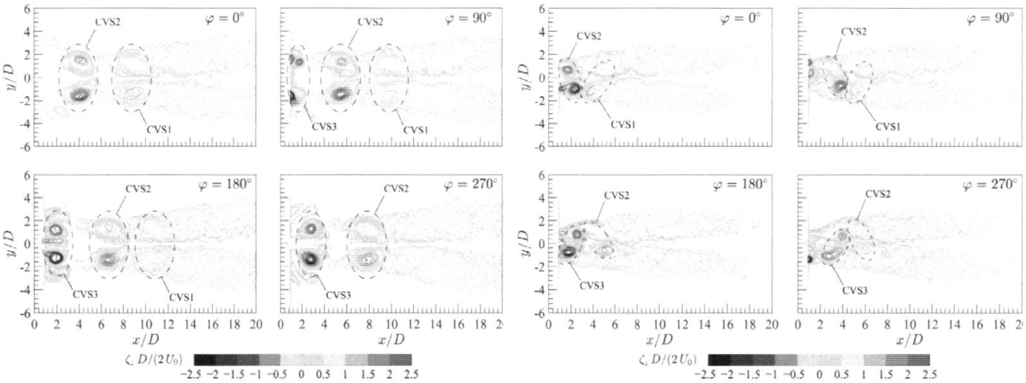

Figure 15. Contours of the non-dimensional phase-averaged out-of-plane vorticity $\zeta_z D/2U_0$ in the N/S plane for the MP configuration (left) and in the NE/SW plane for the DP (right) configuration. *Re* 4000, and *Sr* 0.200.

To better analyse the behaviour of the DP configuration, the phase evolution of the out-of-plane vorticity field in the NE/SW plane is discussed (Figure 15 - right). At φ=0°, the ejection strokes of the north and east jets begin and two CVSs can be observed in the near region ($x/D<6$). These structures seem to have an elongated shape and they are tilted toward the centreline. This particular shape is ascribed to the interaction between the two vortex rings generated by the two nozzles (in the blowing phase) and the influence of the suction phase of the other two nozzles. North and east vortex rings merge in the CVS1 which is distorted by the suction forces of the south and west nozzles. Indeed, the structure denoted as CVS3, observed within the field at φ=180° and already detectable at φ=90°, represents the early stage of the evolution of CVS1. Analogously, the merging of the vortex rings, ejected by the south and west nozzles and affected by the suction phases of the north and east nozzles, generate the CVS2 structure. Its evolution can be followed in a straightforward way as the time phase increases in Figure 15. Anyway, it is important to note that the phase evolution, in the NE/SW plane, during the second half of the period, is specular with respect to the x axis to that in the first half. From Figure 15, it is clear that these CVSs propagate along directions inclined with respect to the x axis (CVS1 and CVS3 toward the negative y direction and CVS2 toward the positive y direction). This oblique propagation of the CVSs is the cause of the greater spreading of the DP jet in this plane than in the other investigated ones. The CVSs preserve their coherence up to $x/D \approx 6$, differently from MP configuration.

The phase-averaged of the out-of-plane vorticity fields of the QP configuration in the N/S plane is illustrated in Figure 16 (left). At φ=0°, the ejection stroke of the north and south jets begin. For this configuration, two different CVSs in the near region, denoted as CVS1 and CVS2, can be observed. As in the DP case, such structures are ascribed to the coalescence of the two vortex rings formed at the exit of the ejection nozzles (which are opposite to each other in this configuration), and their evolution is affected by the suction phase of the other two nozzles. Passing from φ=0° and 180°, it is possible to observe that the CVS1 firstly widens and then decays. On the opposite, CVS2 appears to shrink between φ=180° and 270° and then loses its coherence. Indeed, at φ=270°, such a coherent vortex structure is observed at $x/D \approx 4$, while at φ=0°, no coherent structures are detected downstream of this position. Anyway, these structures lose coherence and dissipate very quickly leaving no evidence of them downstream of $x/D\approx5$.

Finally, in Figure 16 (right), the phase-averaged fields of the out-of-plane vorticity component in the N/S plane for the CSP case are shown. The phase $\varphi=0°$ represents the beginning of the ejection stroke of the north jet. At this phase, two CVSs (CVS1 and CVS2) are observed in the near field region ($x/D<6$). Following the phase evolution, four more CVSs appear in the field of view. CVS3, CVS4, CVS5, and CVS6 are the vortex rings ejected from the north, west, south, and east nozzles, respectively, while CVS1 and CVS2 are advanced stages of the evolution of CVS5 and CVS6, respectively. Indeed, in the CSP configuration, no merging of the vortex rings occurs because the ejection strokes of the four jets are shifted by 90° (one-quarter of period) with respect to each other. Nevertheless, a mutual interaction occurs between these structures because of their closeness. Each vortex moves substantially along the axis of the corresponding nozzle and dissipates completely within eight diameters from the nozzle exits (for instance, CVS1 disappear between $\varphi=90$ and 180° approximately at this location).

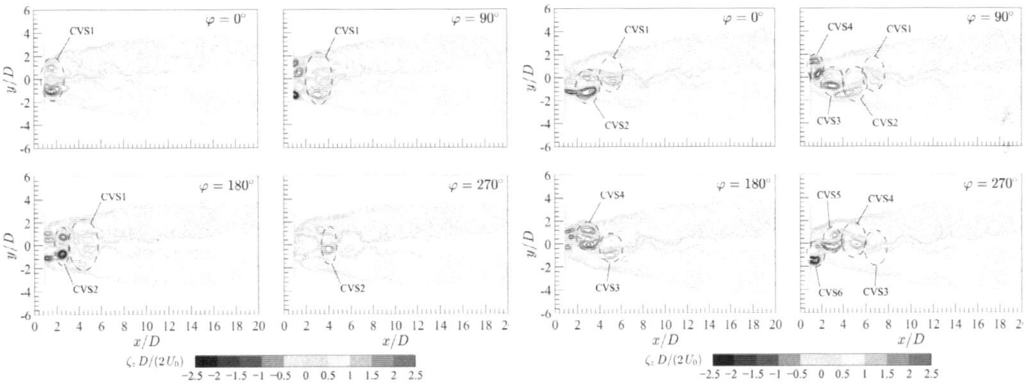

Figure 16. Contours of the non-dimensional phase-averaged out-of-plane vorticity $\zeta_z D/2U_0$ in the N/S plane for the QP configuration (left) and for the CSP (right) configuration. *Re* 4000, and *Sr* 0.200.

The evolution of such coherent vortex structures deeply influence the heat transfer behaviour of the QSJ. In order to have a clearer overview of how such structures evolve, their behaviour is schematically shown in Paolillo et al. [45]. Considering these structures and their evolution, the heat transfer results can be analysed and discussed in an easier way. Figure 17 shows the distribution of the time-averaged Nusselt number for all the impingement distances in all the investigated configurations.

At $H/D = 2.5$ the MP jet attains the lowest values of the time-averaged Nusselt number. This could be due to the frequency of generation of the CVSs in the MP configuration that is smaller than that of the other cases. Interestingly, the Nusselt number spatial distribution reflects the shape of the impinging CVSs. Indeed, the pattern of the inner region, characterized by the highest values of the Nusselt number, has the same shape of the footprint of the impinging CVS. In the MP case at $H/D = 2.5$ such a region has a four-lobe shape with the major axes aligned with the diagonal directions and the minor axes aligned with the longitudinal Y and Z directions. Outside of the high-Nu region, it is possible to observe four regions with relatively high *Nu* values, which form a cross shaped pattern. Such a distribution evidently suggests that the wall jet develops mainly along the Y and Z directions. Moving from the stagnation region along the Y - and Z-axes, it is possible to note a minimum and a subsequent maximum of *Nu* located at 2*D* and 2.8*D* from the centre respectively. Such features are typical of impinging jets at short *H/D* and their origin has been widely investigated and documented in literature (see Carlomagno and Ianiro [44] for instance): local minima are caused with vortex-induced separation, while subsequent local maxima are related to the following reattachment of the wall jet.

At *H/D* = 5 the MP jet exhibits a completely different *Nu* distribution. In particular, the Nusselt number increases and the four lobe pattern has its major and minor axes switched. In this case, the wall jet likely develops mainly in the direction of the minor axes of the four-lobe shape (i.e. Y = Z and Y = −Z,) since *Nu* shows high values over a longer distance when moving away from the plate centre along these directions.

At *H/D* = 7.5 the four-lobe shape disappears and an ovoid pattern is observed, while at higher *H/D* pattern becomes circular. Such a behaviour is ascribed to the CVS shape oscillations that ceases at these nozzle-to-plate distances assuming a circular shape. Indeed, for $H/D \geq 12.5$ the CVSs do not reach the wall because they dissipate before. Therefore, at these *H/D* values, a turbulent jet, generated by the breakdown of the CVSs, reaches the impingement plate. Such a flow is self-preserving in its downstream development and characterized by a Gaussian crosswise velocity profile. Consequently, the *Nu* distribution is bell-shaped, with the maximum value in the plate centre. At higher *H/D,* the bell-shaped distribution is smoothed out and becomes more uniform because of the jet spreading and deceleration.

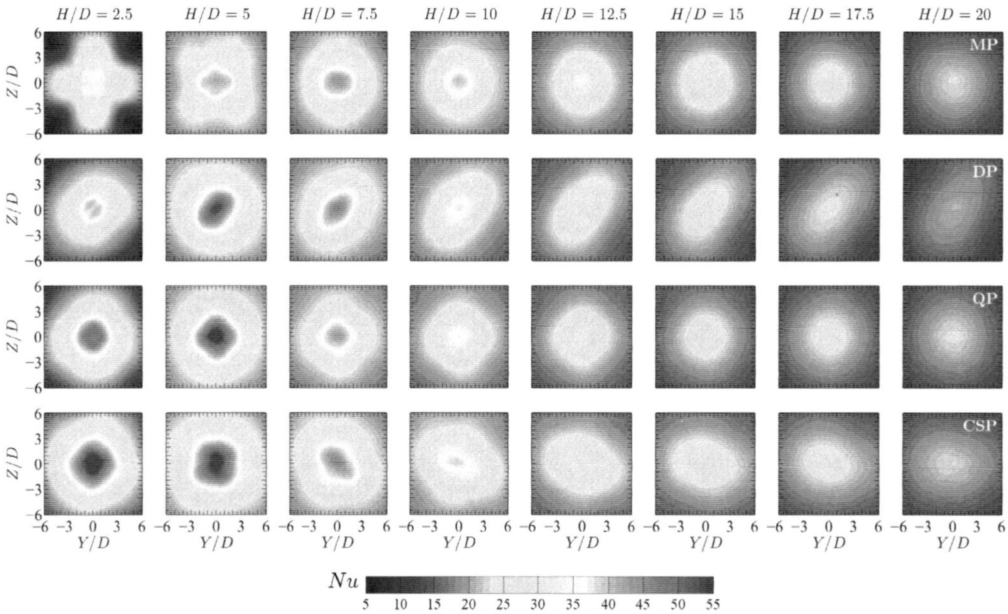

Figure 17. Evolution of the coherent vortex structures in the free flow field of the four investigated configurations. Re = 4000, St = 0.200.

The DP configuration shows, at any impingement distance, an elliptical Nusselt number pattern because of the elongated shape of the CVSs and their oblique propagation. Although the CVSs lose their coherence within the first eight diameters (Figure 15), the *Nu* map preserves an elliptical shape at high impingement distances, opposite to the MP configuration behaviour. This is due to the fact that, despite the decay of the CVSs, the DP jet continues spreading in the streamwise plane containing the trajectories of the CVSs at a faster rate than in the other planes. At *H/D* = 2.5 minimum *Nu* values are observed on the symmetry line (Y=Z) in the central part of the plate between two regions of maximum heat transfer. Such a distribution is due to the particular shape of the impinging CVSs, which are, at this stage, two-lobe shaped. Moving far enough from the centre, the Nusselt number decreases along the direction normal to the symmetry line. At higher *H/D*, the heat transfer attains high values on the symmetry line and in the direction normal to it. Furthermore, the highest value of *Nu* is always detected in the centre of the plate because in its proximity the CVSs impinge and then spread around keeping approximately their shape. The highest *Nu* values are reached at $H/D = 5$; for greater distances, the heat transfer values monotonically decrease.

In the QP configuration, the *Nu* distribution attains a rhomboidal pattern for $H/D \leq 10$ and a circular pattern for greater distances. The rhomboidal pattern is ascribed to the CVSs behaviour: before the vortex reconnection and splitting, the CVSs have a structure elongated along Y or Z direction. Even after the splitting, the two new formed vortex rings move away from each other along one of these two directions. Only after the dissipation of the CVSs, the Nusselt number map shows a more axisymmetric distribution, similar to the MP configuration. As in the DP configuration, the maximum *Nu* values are attained at $H/D = 5$.

The CSP configuration achieved the highest values of Nusselt number and the widest high-*Nu* region at $H/D = 2.5$. At larger impingement distances, the *Nu* distribution seems to continuously rotate in the counterclockwise direction. Such a rotation continues until $H/D = 12.5$, then no further rotation is observed. At this nozzle-to-plate distance (i.e. *H/D*), the Nusselt number pattern is ovoid shaped with its major axis slightly rotated counterclockwise with respect to the diagonal line $Y = -Z$. It is worth noting that the *Nu* distribution resembles that obtained by impinging swirling jets at low swirl numbers (Ianiro and Cardone [46]). In particular, at $H/D = 7.5$, the Nusselt number distribution of the CSP configuration looks very similar to the lozenge shape with rounded edges observed by Ianiro and Cardone [46] at $H/D = 6$ in the case of swirling jets (swirling number equal to 0.4). Clearly, these similar patterns are caused by very different flow topology. Swirling jets are characterized by multiple precessing spiral vortex structures (Cala et al. [47]), while the CSP jet consists of a train of vortex rings arranged in a helicoidal structure along the jet centreline. In the CSP jet case, the significant swirling velocities are generated by the successive vortex rings in their intermediate regions, because of their closeness. Consequently, at the arrival of the vortex rings to the wall, the incoming flow has an azimuthal velocity component in addition to axial and radial velocities. On the other hand, the rotation of such a pattern up to $H/D = 12.5$ is surely related to the helicoidal structure of the vortex train. At greater *H/D* the pattern does not rotate because the CVSs have already dissipated. Finally, it can be noted that the *Nu* values, differently from the other configurations, monotonically decreases with *H* in each point of the plate.

4. Conclusions

The synthetic jet technology has been widely applied in several fields of application because of their main advantages: easy miniaturization and low-cost manufacturing. The scientific community has employed this new technology also in the electronic cooling field because of the continuous request of higher heat power dissipation and miniaturization of the cooling devices. Therefore, the heat transfer capability of impinging synthetic jets has been experimentally and numerically investigated in these last decades. These devices have found to be promising from the heat transfer viewpoints. Many works have underlined the benefits and improvements of synthetic jets, compared with a classic electronic cooling as fan, such as: better efficiency, design-friendly, intrinsic higher reliability, easier miniaturisation, simple noise cancellation.

Early studies on synthetic jet, in the heat transfer field, were focused on the assessment of the effects of all the characteristics parameters on their heat transfer behaviour. After that their heat transfer capabilities were proved and the influence of the characteristic parameters were understood, the researchers paid their attention on the design of innovative configurations in order to further enhance the synthetic jet cooling rate. At the University of Naples Federico II, two new synthetic jet devices were designed in order to achieve a heat transfer enhancement and also a noise reduction, which is one of the main drawbacks of the synthetic jet technology. The heat transfer performances of these two devices have been evaluated through the IR thermography. Obviously, the heat transfer performances of both the innovative devices are strictly related to their fluid dynamic behaviour. For this reason, also the flow field of these two devices has been experimentally investigated through the Particle Image Velocimetry technique (planar and stereo configurations).

The twin synthetic jets device has shown, in the side-by-side condition, higher values of axial velocity and turbulent kinetic energy than the single classical configuration. The other two cases (jet-axes-distances equal to 3 and 5) have shown a behaviour similar to the single case. Moreover, a potential core-like region of low turbulence has been defined and the evolution of the synthetic jet flow field has been described through the phase-average measurements.

Under these results the heat transfer behaviour of this device, always compared to the single case, has been evaluated. The twin synthetic jet device (in its side-by-side configuration) has shown a heat transfer enhancement for all the nozzle-to-plate distances. The time-average heat transfer behaviour of the single and twin synthetic jets has been explained through the results obtained in the impinging flow field study. A strong correlation between the axial velocity components and the heat transfer distribution has been found. Some aspects of the heat transfer behaviour (as the inner and outer region shaped of heat transfer maxima) have been explained through the potential core-like region and the vortex ring behaviour near the plate. The evolution of the heat transfer over the impinging plate has been described through the phase-average measurements.

The quadruple synthetic jet device has shown a complex flow field where the behaviour and the evolution of the coherent vortex structures is fundamental for the explanation of the heat transfer process

The footprint of the CVSs is clearly recognizable in the heat transfer maps at low values of H/D. Therefore, in the MP configuration a four-lobe pattern is detected, which exhibits an axis switching before assuming a circular shape. In the DP configuration, the heat transfer distribution is elongated in the direction of the major dimension of the impinging elliptical CVSs and preserves such a shape also at higher impingement distances. The QP heat transfer map is rhomboidal at small H/D and turns into a circular distribution as H/D is increased. The CSP configuration is characterized by a swirling jet-like heat transfer distribution (consisting in a pattern that rotates with H/D up to a H/D = 12.5).

In particular, the MP configuration exhibits the lowest values of heat transfer rate at small impingement distances and slightly better performance at larger distances. The CSP configuration offers very high heat transfer rates at H/D = 2.5. Intermediate values are attained with the DP and QP configurations, with the latter slightly outperforming the former. The obtained findings and previous acoustic investigation suggest that the CSP configuration is the most performing one for the purpose of electronics cooling, because it combines the lowest noise level with a thermal performance equivalent to that of the other configurations (even better at low impingement distances).

References

[1] Moore G.E., 1965, Cramming more components onto integrated circuits, *Proceedings of the IEEE*, **86(1)**, 114-117.
[2] Coe D.J., Allen M.G., Trautman M.A. and Glezer A., 1994, Micromachined jets for manipulation of macro flows, *Solid-State Sensor and Actuator Workshop*, 243–247.
[3] Allen M.G. and Glezer A., 1995, Jet vectoring using zero mass flux control jets, *AFOSR contractor and grantee meeting on turbulence and internal flows.*
[4] Holman R., Utturkar Y., Mittal R., Smith B.L. and Cattafesta L., 2005, Formation criterion for synthetic jets, *AIAA J.*, **43(10),** 2110-2116.
[5] Smith B.L. and Glezer A., 1998, The formation end evolution of synthetic jets, *Phys. Fluids*, **10(9)**, 2281-2297.
[6] Buchmann N.A., Atkinson C. and Soria J., 2013, Influence of ZNMF jet flow control on the spatio-temporal flow structure over a NACA-0015 airfoil. *Exp. Fluids*, **54(3)**, 1485.
[7] Smith B.L. and Glezer A., 2002, Jet vectoring using synthetic jets. *J. Fluid Mech.*, **458**, 1-34.
[8] Wang H. and Menon S., 2001, Fuel-air mixing enhancement by synthetic microjets. *AIAA J.*, 39(12), 2308-2319.

[9] Müller M.O., Washabaugh P.D., Bernal L.P., Moran R.P., Parviz B.A. and Najafi K., Micromachined acoustic resonators for micro jet propulsion. *AIAA Paper*, 2000, 200-2404.
[10] Gutmark E., Yassour Y. and Wolfshtein M., 1982, Acoustic enhancement of heat transfer in plane channels, *Proceedings of Seventh International Heat Transfer Conference*, 441–445, Munich, Germany.
[11] Mahalingam R. and Glezer A., 2005, Design and thermal characteristic of a synthetic jet ejector heat sink, J. *Electron. Packaging*, 127, 172-177.
[12] Chaudhari M., Puranik B. and Agrawal A., 2010, Heat transfer characteristics of synthetic jet impingement cooling, *Int. J. Heat Mass Tran.*, **53**, 1057-1069.
[13] Valiorgue P., Persoons T., McGuinn A. and Murray D.B., 2009, Heat transfer mechanisms in an impinging synthetic jet for small jet-to-surface spacing, *Exp. Therm. Fluid Sci.*, **33**, 597-603.
[14] Arik M., Icoz T., 2012, Predicting heat transfer from unsteady synthetic jets, J. Heat Transfer, **134**, 1-8.
[15] Persoons T., McGuinn A. and Murray D.B., 2011, A general correlation for the stagnation point Nusselt number of an axisymmetric impinging synthetic jet, *Int. J. Heat Mass Tran.*, **54**, 3900-3908.
[16] McGuinn A., Farrelly R., Persoons T. and Murray D.B., 2013, Flow regime characterization of an impinging axisymmetric synthetic jet, *Exp. Therm. Fluid Sci.*, **47**, 241–251.
[17] Rylatt D.I. and O'Donovan T.S., 2013, Heat transfer enhancement to a confined impinging synthetic air jet, *Appl. Therm. Eng.*, **51**, 468-475.
[18] Chaudhari M., Puranik B. and Agrawal A., 2010b, Effect of orifice shape in synthetic jet based impingement cooling, *Exp. Therm. Fluid Sci.*, **34**, 246-256.
[19] Luo Z.B., Xia Z.X. and Bing L., 2006, New generation of synthetic jet actuators, *AIAA J.*, **44(10)**, 2418-2419.
[20] Luo Z.B., Deng X., Wang L. and Xia Z.X., 2011, Experimental technique based on delay phase angle and PIV measurements of a dual synthetic jets actuator, *Proceedings of the 2011 Symposium on Piezoelectricity, Acoustic Waves and Device Applications,* 1-5, Shenzhen, China.
[21] Persoons T., O'Donovan T.S. and Murray D.B., 2009, Heat transfer in adjacent interacting impinging synthetic jets, *Proceedings of 2009 ASME summer heat transfer conference*, 1-8, San Francisco, California.
[22] Lasance C.J.M. and Aarts R.M., 2008, Synthetic Jet Cooling Part I:Overview of Heat Transfer and Acoustic, *24th Annual IEEE Semiconductor Thermal Measurement and Management Symposium*, 20-35. IEEE.
[23] Russell D.A., Titlow J.P. and Bemmen Y.J., 1999, Acoustic monopoles, dipoles, and quadrupoles: An experiment revisited, *Am. J. Phys*, **67**, 660-664.
[24] Lasance C.J.M., Aarts R.M. and Ouweltjes O., 2008, Synthetic jet cooling part II: experimental results of an acoustic dipole cooler, *24th Annual IEEE Semiconductor Thermal Measurement and Management Symposium*, 26-31. IEEE.
[25] Bhapkar U.S., Srivastava A. and Agrawal A., 2014, Acoustic and heat transfer characteristics of an impinging elliptical synthetic jet generated by acoustic actuator, *Int. J. Heat Mass Tran.*, **79**, 12-23.
[26] Jeyalingam J. and Jabbal M., 2016, Optimization of synthetic jet actuator design for noise reduction and velocity enhancement". *In Proceedingd of 8th AIAA Flow Control Conference. Washington, D.C., United States*, AIAA paper 2016-4236.
[27] Greco C.S., Ianiro A., Astarita T. and Cardone G., 2013, On the near field of single and twin circular synthetic air jets, *Int. J. Heat Fluid Fl.*, **44**, 41-55.
[28] Greco C.S., Ianiro A. and Cardone G., 2014, Time and phase average heat transfer in single and twin circular synthetic impinging air jets, *Int. J. Heat Mass Tran.*, **73**, 776-788.
[29] Greco C.S., Castrillo G., Crispo C.M., Astarita T. and Cardone G., 2016. Investigation of impinging single and twin circular synthetic jets flow field. *Exp Therm Fluid Sci.*, **74**, 354-

367.
[30] Paolillo G., Greco C.S. and Cardone G., 2017, Novel Quadruple Synthetic Jet Device: Flowfield and Acoustic Behavior, *AIAA J.*, **55(7)**, 2241-2253.
[31] Paolillo G., Greco C:S: and Cardone G. 2017, Time-averaged heat transfer in confined impinging quadruple synthetic jets, *9th World Conference on Experimental Heat Transfer, Fluid Mechanics and Thermodynamics*, 12-15 June, 2017, Iguazu Falls, Brazil.
[32] Astarita T. and Cardone G., 2005, Analysis of interpolation schemes for image deformation methods in PIV, *Exp. Fluids*, **38**, 233-243.
[33] Astarita T., 2006, Analysis of interpolation schemes for image deformation methods in PIV: effect of noise on the accuracy and spatial resolution, *Exp. Fluids*, **40**, 977-987.
[34] Astarita T., 2007, Analysis of weighting windows for image deformation methods in PIV. *Exp. Fluids*, **43**, 859-872.
[35] Astarita T., 2008, Analysis of velocity interpolation schemes for image deformation methods in PIV, *Exp. Fluids*, **45**, 257-266.
[36] Hussain A.K.M.F. and Reynolds W.C., 1970, The mechanics of an organized wave in turbulent shear flow, *J. Fluid Mech.*, **41(02)**, 241-258.
[37] Greco C. S., Cardone G. and Soria J., 2017, On the behaviour of impinging zero-net-mass-flux jets, *J. Fluid Mech.*, **810**, 25-59.
[38] Golobic I., Petkovsek J. and Kenning D.B.R., 2012, Bubble growth and horizontal coalescence in saturated pool boiling on a titanium foil, investigated by high-speed IR thermography, *Int. J. Heat Mass Tran.*, **55(4)**, 1385-1402.
[39] Carlomagno G.M. and Cardone G., 2010, Infrared thermography for convective heat transfer measurements, *Exp. Fluids*, **49**, 1187-1218.
[40] Rohlfs W., Haustein H.D., Garbrecht O. and Kneer R., 2012, Insights into local heat transfer of a submerged impinging jet: Influence of local flow acceleration and vortex-wall interaction, *Int. J. Heat Mass Tran.*, **55**, 7728-7736
[41] Gardon R. and Akfirat J., 1965, The role of turbulence in determining the heat-transfer characteristics of impinging jets, *Int. J. Heat Mass Tran.*, **8 (10)**, 1261–1272.
[42] Jambunathan K., Lai E., Mossand M.A. and Button B.L., 1992, A review of heat transfer data for single circular jet impingement, *Int. J. Heat Fluid Fl.*, **13**, 106-115.
[43] Shuster J.M. and Smith D.R., 2007, Experimental study of the formation and scaling of a round synthetic jet, *Phys. Fluids*, **19**, 045109.
[44] Carlomagno G.M. and Ianiro A., 2014, Thermo-fluid-dynamics of submerged jets impinging at short nozzle-to-plate distance: A review, *Exp. Therm. Fluid Sci*, **58**, 15-35.
[45] Paolillo G., Greco C.S. and Cardone G., 2017, The evolution of quadruple synthetic jets, *Exp. Therm. Fluid Sci*, **89**, 259-275.
[46] Ianiro A. and Cardone G., Heat transfer rate and uniformity in multichannel swirling impinging jets. *Applied Thermal Engineering*, 2012, **49**, 89–98.
[47] Cala C., Fernandes E., Heitor M. and Shtork S., Coherent structures in unsteady swirling jet flow, *Experiments in Fluids*, 2006, **40 (2)**, 267–276.

IOP Conf. Series: Materials Science and Engineering **249** (2017) 012002 doi:10.1088/1757-899X/249/1/012002

Improvement of a Pneumatic Control Valve with Self-Holding Function

Shujiro Dohta[1], Tetsuya Akagi[1], Wataru Kobayashi[1] , So Shimooka[1] and Yusuke Masago[2]

[1] Department of Intelligent Mechanical Engineering, Okayama University of Science, 1-1 Ridai-cho, Kita-ku, Okayama 700-0005, Japan
[2] Product Development Division-4, SMC Corporation, 4-2-2 Kinunodai, Tsukubamirai, Ibaraki 300-2493, Japan

Abstract. The purpose of this study is to develop a small-sized, lightweight and low-cost control valve with low energy consumption and to apply it to the assistive system. We have developed some control valves; a tiny on/off valve using a vibration motor, and an on/off valve with self-holding function. We have also proposed and tested the digital servo valve with self-holding function using permanent magnets and a small-sized servo motor. In this paper, in order to improve the valve, an analytical model of the digital servo valve is proposed. And the simulated results by using the analytical model and identified parameters were compared with the experimental results. Then, the improved digital servo valve was designed based on the calculated results and tested. As a result, we realized the digital servo valve that can control the flow rate more precisely while maintaining its volume and weight compared with the previous valve. As an application of the improved valve, a position control system of rubber artificial muscle was built and the position control was performed successfully.

1. Introduction

Recently, it has been strongly desired to develop a system to aid in nursing care and to support the activities of daily life for the elderly and the disabled [1],[2],[3]. In such a control system, an actuator and a driving device such as a control valve are mounted on the human body [4],[5],[6]. When we consider to develop a wearable control valve that can drive pneumatic actuators to support the multi-degrees of freedom of human motion, the size and weight of the valve become serious problems. The typical electro-magnetic solenoid valve drives its spool with a large-sized solenoid. The solenoid valve has a complex construction to keep a seal while the spool is moving. This complexity makes its miniaturization and the fabrication of a low-cost valve more difficult. The purpose of our study is to develop a small-sized, lightweight, inexpensive control valve with lower energy consumption which can be mounted on the human body. We thought that the small sliding force to the spool can be useful to decrease the required force to open the valve compared with the longitudinal directional force as shown in Fig.1. In addition, it is possible to save the electrical power by utilizing the permanent magnet to drive the valve. Thus, our key concepts are summarized as follows:

1) The force from the side is applied to the ball to make a driving force smaller as shown in Fig.1.
2) The ball is set in the tube without any mechanical connecting part to realize “no sealing”.
3) The self-holding function is added by using permanent magnets to save the energy.

Figure 1(a) shows a construction of a simple check valve which consists of a ball, an orifice and a tube. For example, in a valve using ball diameter D of 2 mm and orifice diameter d of 0.5 mm, the horizontal force *f* to open the valve is about a quarter of the longitudinal force *F* as shown in Fig.1(b). Based on these concepts, we have developed two types of small-sized on/off control valve. One is the valve which can be driven by a small vibration motor [7]. Another is the valve which can be driven by an electromagnetic solenoid and a permanent magnet [8], [9]. Furthermore, we developed a digital servo valve which can adjust the output flow rate and pressure [10] described in the next section. The valve consists of some check valves, a cylindrical magnet and a tiny servo motor. This valve has an inherent problem: In order to control the flow rate more precisely, it is required to increase the number of check valve. However, it might make the valve larger and heavier. In this paper, therefore, we try to solve this problem and improve the valve. In addition, as an application of the improved digital servo valve, a position control system of rubber artificial muscle is built and tested. On the other hand, we have also developed an analogue type control valve using a buckled tube [11],[12]. This servo valve has some advantages; inexpensive, small-sized and applicable for any fluid.

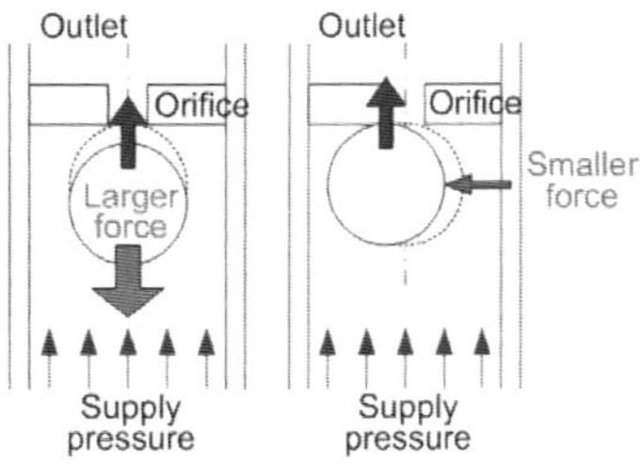

(a) Construction of a check valve

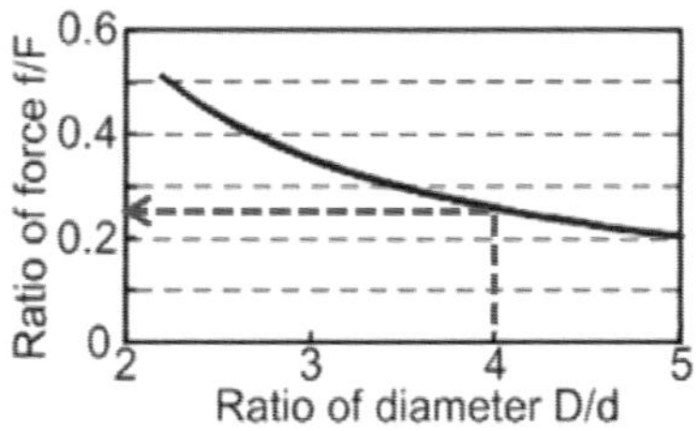

(b) Required force to open a valve

Figure 1. Construction of a check valve and required force

2. Digital servo valve with self-holding function

Figure 2 shows the digital servo valve that includes four valves with self-holding function. The valve consists of a servo motor(GWS Co. Ltd., PICO/STD/F, speed: 500 deg./s), a disk with a cylindrical magnet and four check valves which consist of an orifice in the flexible tube with inner diameter of 2.5 mm and a magnetic ball with outer diameter of 2 mm. The center of the disk is connected with the shaft of the servo motor. The four check valves are set on parallel to the surface of the disk. The rotational angle of the disk is easily controlled by a micro-computer with PWM timer function. The mass of the tested valve is 22.5 g and the size is 24 x 36 x 50 mm. The operating principle of the valve is as follows. First, a cylindrical magnet stays at the neutral position as shown in Fig.2(b). To increase the output flow rate from the valve, the servo motor rotates the disk from the neutral position to 40 degrees counterclockwise. The cylindrical magnet attracts the ball in the check valve. Then, the check valve 1 in Fig.2 opens. By rotating the disk an additional 20 degrees (60 degrees from the neutral position) counterclockwise using the servo motor, the check valve 2 opens. At the same time, the ball in the check valve 1 loses the attracting magnetic force and the ball automatically moves toward the orifice by the momentum of the flow and closes the check valve 1. Then, the larger flow rate at the output port can be obtained according to the sectional area change from 1 to 2. In the same manner of increasing the output flow rate, the flow rate can be decreased by rotating the disk clockwise. The tested valve can realize five states that include holding the pressure and two types of flow rate in supply and exhaust.

Figure 3 shows the relation between the valve opening (state) and output flow rate from the valve. The valve opening is defined as the percentage to the maximum sectional area in the orifice diameter of 0.6 mm. The output flow rate was measured by a digital flow meter (SMC Corp., PF2A750-01-27). In the measurement of the supply flow rate, the supply pressure of 500 kPa is applied to the supply port in the condition that the output port is open to the atmosphere through the flow meter. In the case of measuring the exhaust flow rate, the output port is pressurized to 500 kPa in the condition that the supply ports are closed. From Fig.3, it can be observed that the relation between the output flow rate and the valve opening is proportional in both cases of supply and exhaust states. In order to control the flow rate more precisely (multi-stage), it is required to increase the number of check valve; the number of orifice tube. However, the more the tube increases parallel on the surface of the disk, the larger the

IOP Conf. Series: Materials Science and Engineering **249** (2017) 012002 doi:10.1088/1757-899X/249/1/012002

valve size becomes. Therefore, we try to locate the tube around the disk rotated by a servo motor three-dimensionally as shown in Fig.4.

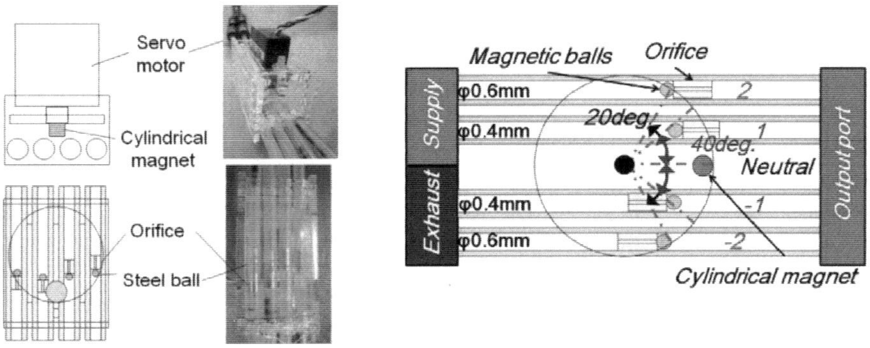

(a) Construction (b) Schematic diagram

Figure 2. Digital servo valve with self-holding function

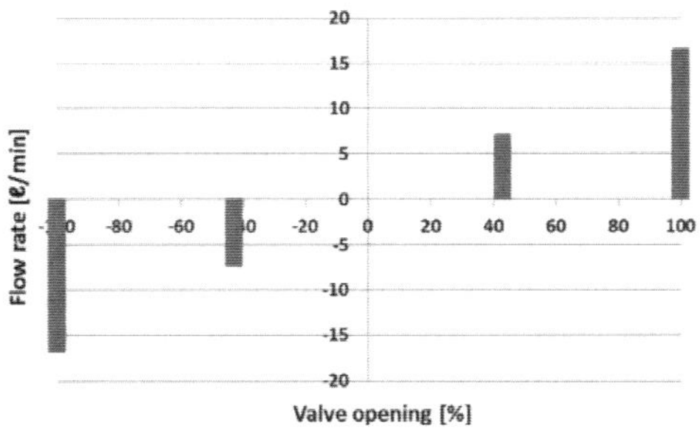

Figure 3. Output flow rate

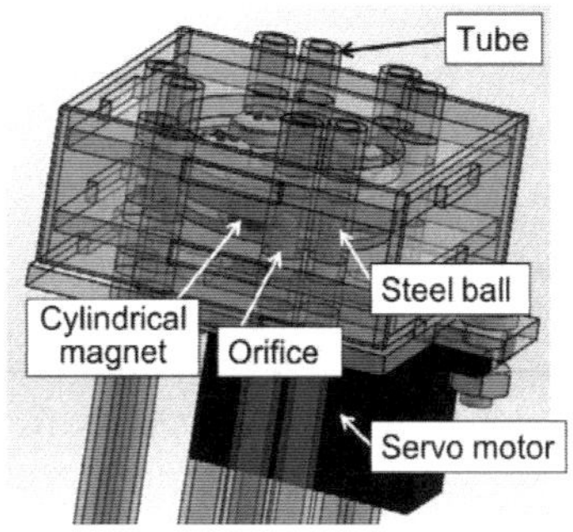

Figure 4. 3-dimensional arrangement

3. Analysis for improvement of the valve

3.1. Analytical model of the digital servo valve

In order to design the valve which can control the flow rate more effectively with limited number of check valve, we build an analytical model of the tested valve and investigate the optimal combination of the orifice diameter by simulation. Figure 5 shows an analytical model of the digital valve. The rotational angle of a servo motor θ_S is expressed as follows.

$$\frac{d\theta_S}{dt} = k_a sign\{\theta_r(t-L)-\theta_S\} \tag{1}$$

Where, k_a and θ_r mean the rated angular velocity (=6.28 rad/s) and the reference angle, respectively. The dead time of the motor L of 0.035 s is determined by experiment.

The following equations are concerned with single check valve and the same equations can be applied to eight check valves. The relations between cylindrical magnet and steel ball can be obtained from the geometry in Figs.5 and 6 , and they are given as follows.

$$\ell = \sqrt{r_1^2 + r_2^2 - 2r_1r_2\cos\theta_s} \tag{2}$$

$$r_1 = r_{10} - 0.5\{d_0 - D_b\sin(\theta_0-\theta)\} \tag{3}$$

Where, ℓ, r_1, r_2, r_{10}, d_0, D_b, θ_0, θ represent the distance between cylindrical magnet and steel ball, distance between rotation axis and steel ball, distance between rotation axis and cylindrical magnet, initial distance, inner diameter of orifice, diameter of steel ball, initial angle of steel ball and rotational angle of steel ball, respectively.

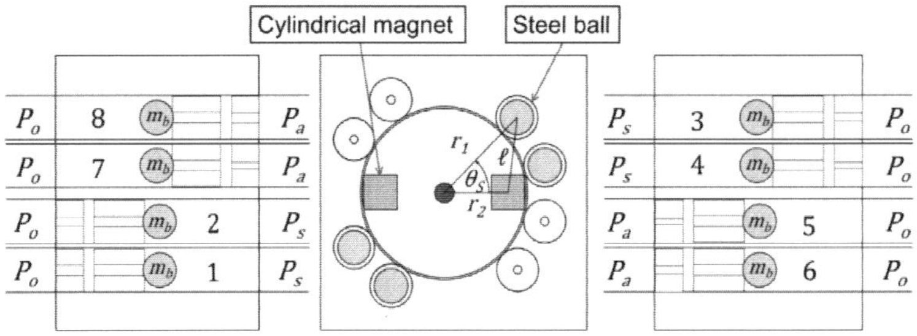

Figure 5. Analytical model for digital servo valve with eight check valves

The force acted on the steel ball by the cylindrical magnet can be expressed as follows.

$$\begin{aligned} &F_b = 0 && (\ell > \ell_{\max}) \\ &F_b = F_{b0} && (\ell \le \ell_{\max}\ \&\ F_b = 0 \rightarrow F_{b0}) \\ &F_b = F_{b0}(t - L_b) && (\ell \le \ell_{\max}\ \&\ F_b = F_{b0} \rightarrow 0) \end{aligned} \tag{4}$$

Where, ℓ_{max} (=5.3 mm) is a critical distance which a steel ball and a magnet attract each other and can be determined experimentally. It is assumed that there is a dead time L_b when the attractive force F_b becomes from F_{b0} to zero. The dead time L_b of 10 ms and the constant force F_{b0} of 0.14 N are determined by simulation and experiment, respectively.

Figure 6 shows an analytical model for the opening of the valve. In the model, the steel ball rotates at the fulcrum C as shown in Fig.6. In the initial state of the ball (dotted line as shown in Fig.6), the ball has plugged up the orifice. The initial angle θ_0 is given by the following equation from geometric relation.

$$\theta_0 = \sin^{-1}(d_o/D_b) \tag{5}$$

The total torque τ (a clockwise torque is positive) acted on the steel ball is the sum of the torques generated by the magnet, the differential pressure and the momentum of the fluid flow, and can be expressed as follows.

$$\tau = \frac{D_b}{2} F_b \cos(\theta_0 - \theta) - \frac{\pi}{4} d_o^2 \frac{d_o}{2}(P_s - P_o) - \frac{4}{\rho\pi} \frac{D_b^2}{d_t^4} \cdot Q_o^2 \frac{d_o}{2} \tag{6}$$

The equation of rotational motion of the steel ball is given as follows.

$$\frac{7}{5} m_b \left(\frac{D_b}{2}\right)^2 \frac{d^2\theta}{dt^2} + C_b \frac{d\theta}{dt} = \tau \tag{7}$$

Where, m_b and C_b mean the mass of the steel ball and the viscous drag coefficient of the ball.

The opening area S_o of the valve changes with the angle θ of the ball. In order to express this area simply, S_o is approximately given by the difference between the sectional area of the inner bore of the tube and the ball, and is expressed by the following equation.

$$S_o = \frac{\pi}{4}\{d_o^2 - D_b^2 \sin^2(\theta_0 - \theta)\} \tag{8}$$

The mass flow rates Q_o, Q_a as shown in Fig.7 are expressed as follows.

$$Q_o = S_o P_s \sqrt{\frac{2}{RT}} f\left(\frac{P_o}{P_s}\right) \tag{9}$$

$$Q_a = S_a P_o \sqrt{\frac{2}{RT}} f\left(\frac{P_a}{P_o}\right) \tag{10}$$

Where, P_s , P_o , P_a , R and T represent the supply pressure, output pressure, atmospheric pressure, gas constant and absolute temperature, respectively. S_a can be expressed by same relation as Eq.(8). The function $f(z)$ that expresses the state of flow state is given as follows.

$$\begin{aligned} f(z) &= \sqrt{\frac{\kappa}{\kappa - 1}\left(z^{2/\kappa} - z^{(\kappa+1)/\kappa}\right)} \quad (0.528 \le z \le 1) \\ f(z) &= \sqrt{\frac{\kappa}{\kappa + 1}\left(\frac{2}{\kappa+1}\right)^{2/(\kappa-1)}} \quad (0 \le z < 0.528) \end{aligned} \tag{11}$$

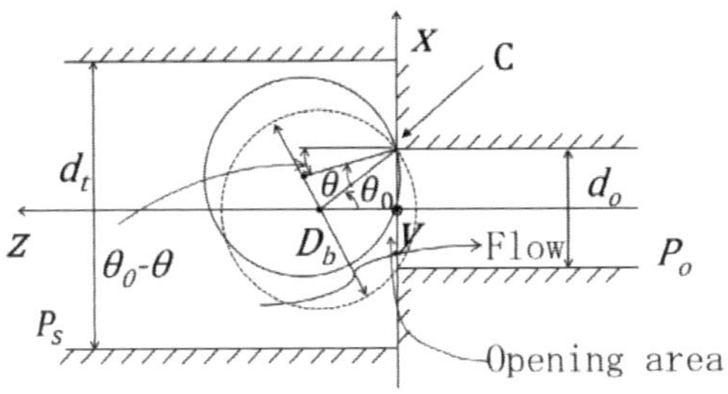

Figure 6. Analytical model for valve opening

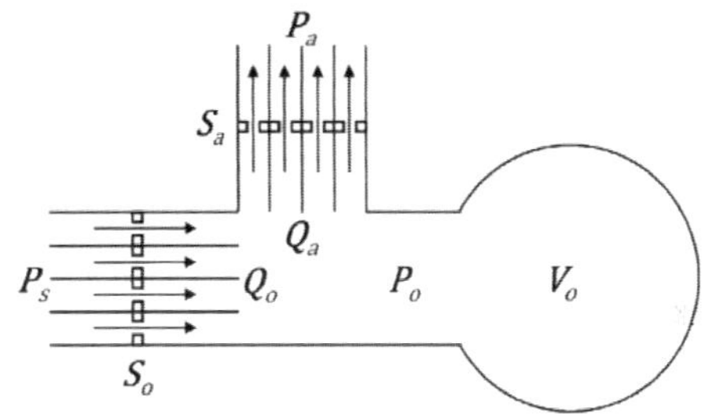

Figure 7. Model for the valve with 8 check valves

The pressure change in a tank connected to the valve according to the difference between the supply mass flow rate Q_o and the exhaust flow rate Q_a is given by following equation.

$$\frac{dP_o}{dt} = \frac{\kappa RT}{V_o}(Q_o - Q_a) \tag{12}$$

Where, κ and V_o mean the specific heat ratio (=1.4) and the volume of tank, respectively. By using equations from (1) to (12), we can calculate and predict the dynamics of the valve.

3.2 Identified system parameters and simulation

Table 1 shows the values of main system parameters. The optimal diameters of each orifice were obtained by simulation so that the pressure control by using the valve shows higher response and smaller steady state error. As a result, the following diameters were selected; d_1=d_8=0.6 mm, d_2=d_3=d_6=d_7=0.3 mm and d_4=d_5=0.2 mm. The numbering of the orifice is shown in Fig.5.

Table 1. System parameters

Parameter	Value	Unit
Viscous drag coefficient C_b	$0.6\text{x}10^{-6}$	$kg \cdot m^2/s$
Diameter of steel ball D_b	$2.0\text{x}10^{-3}$	m
Inner diameter of orifice d_0	$6.0\text{x}10^{-4}$	m
Inner diameter of tube d_t	$2.5\text{x}10^{-3}$	m
Mass of steel ball m_b	$3.3\text{x}10^{-5}$	kg
Initial distance r_{10}	$13.25\text{x}10^{-3}$	m
Distance r_2	$8.75\text{x}10^{-3}$	m
Volume of tank V_o	$3.8\text{x}10^{-5}$	m^3

Figure 8(a) and (b) show examples of step response and frequency response of the output pressure, respectively. P control scheme is basically used. In both figures, chain line, solid line and broken line show the desired, simulated and experimental pressure, respectively. From Fig.8, it can be found that the pressure control can be realized and the simulated result agrees with the experimental result. We can confirm the validity of the proposed analytical model and the identified system parameters.

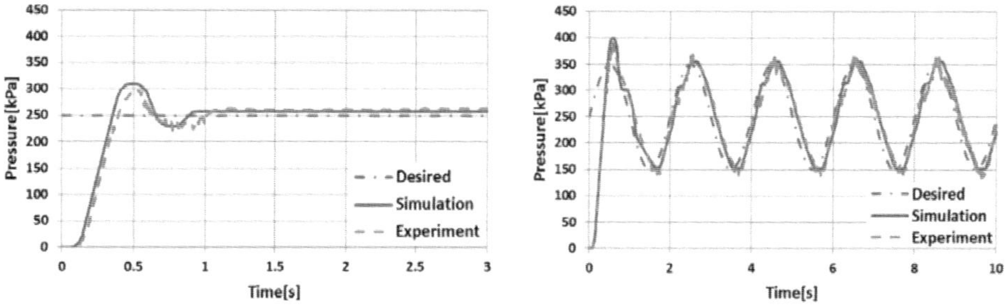

(a) Step response (b) Frequency response

Figure 8. Comparison of simulated result with experimental

4. Improved digital servo valve and its application

4.1 Improved digital valve

Figure 9 shows the improved digital servo valve. The valve consists of a servo motor, a disk with two cylindrical neodymium magnets (diameter of 4 mm, length of 5 mm) and eight check valves located parallel around the disk. The location angles of the tube are decided so that the total size of the valve keeps small and the valve does not open during a holding state. The size of the valve is 42 x 38 x 42 mm and the mass is 24 g. Compared with the previous valve, the improved valve has 4 more states although the volume and mass are almost same. In addition, the maximum flow rate of the improved valve is increased because of two cylindrical magnets. Figure 10 shows the relation between valve opening and output flow rate. The supply pressure is 500 kPa. It can be found that the output flow rate changes linearly and precisely with the opening in small openings. It can be also found that the maximum flow rate is increased to 25 lit/min compared with 16.5 lit/min of the previous valve.

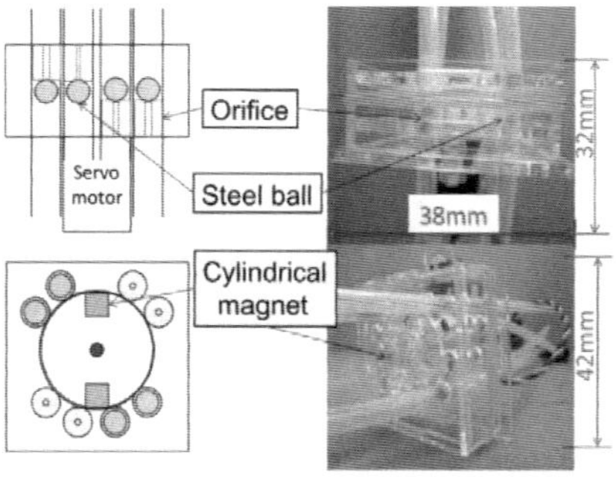

Figure 9. Overview of improved valve

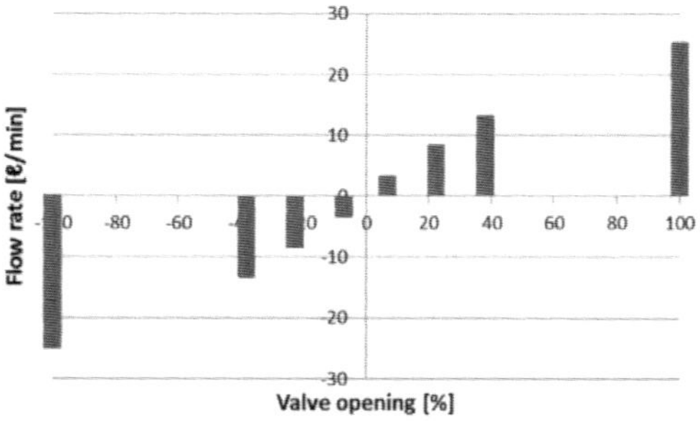

Figure 10. Flow rate of improved valve

4.2 Positon control of a rubber artificial muscle using the tested valve

As an application of the improved valve, we applied it to the position control of the rubber artificial muscle. Figure 11 shows a schematic diagram of the control system. The system consists of a tested valve, a rubber artificial muscle, two potentiometers and a micro-computer. The rubber muscle has a stroke of 63 mm and an inner diameter of 10 mm. The position control is done as follows. The micro-computer gets output voltages from potentiometers; desired and measured value and calculates the deviation from the desired position. Then, suitable check valves for supply and exhaust are selected to be opened based on the magnitude of the deviation. This means a quasi P control scheme. Figure 12 shows the transient response of axial displacement of the muscle. The blue chain line shows the desired position of 22 mm and red solid line shows the measured displacement. And the broken line shows the response using the previous valve. It can be seen that in the improved case, the dead time is a little smaller than the previous case, and the displacement is settled down the final position after 2 seconds. It can be concluded that the position control using the improved valve can give much better control performance than that using the previous valve.

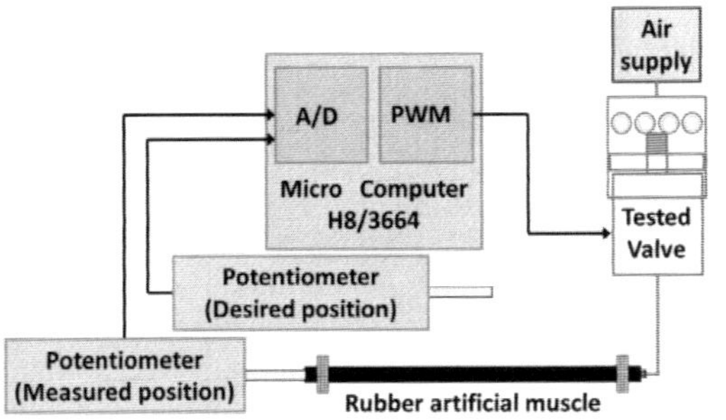

Figure 11. Position control system

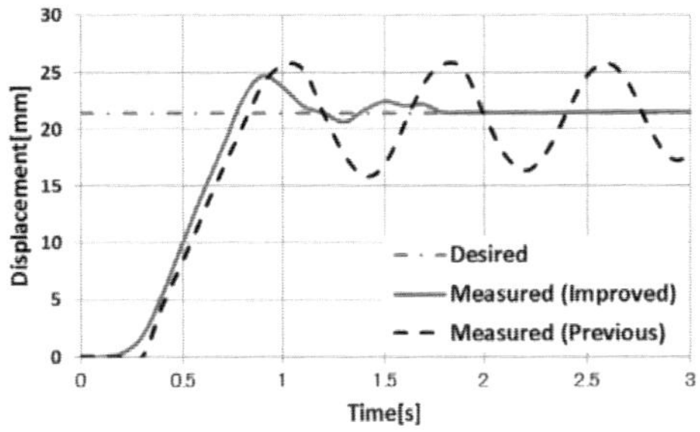

Figure 12. Transient response of displacement

In order to apply various control schemes to the tested valve, the relation between the valve opening (digital amount) and the valve opening (analogue amount; control input) should be described. Figure 13 shows the relation between both valve openings. For example, when the control input calculated by a control scheme is between 15 and 30%, the valve opening becomes 23 % and the rotational angle of a servo motor becomes 28 degrees. Based on the relationship, the following PID scheme was applied to the position control of the muscle.

$$u(i) = K_P e(i) + K_I \sum e(i)\Delta t + K_D \frac{e(i) - e(i-1)}{\Delta t} \qquad (13)$$

Where, $u(i)$, $e(i)$, K_p , K_I , K_D and Δt represent the control input, deviation of displacement, proportional gain, integral gain, derivative gain and sampling period for control, respectively.
Figure 14 shows the transient response of the displacement using above PID control scheme for several kinds of desired position. Each broken line shows the desired position and solid line shows the measured displacement. PID gains are tuned for the desired position of 20 mm ; $K_p = 5$ %/mm, $K_I = 175$ %/(mm·s) and $K_D = 10^{-5}$ %·s/mm. The sampling period Δt is 2 ms. It is found that the position control for different desired positions can be realized well.

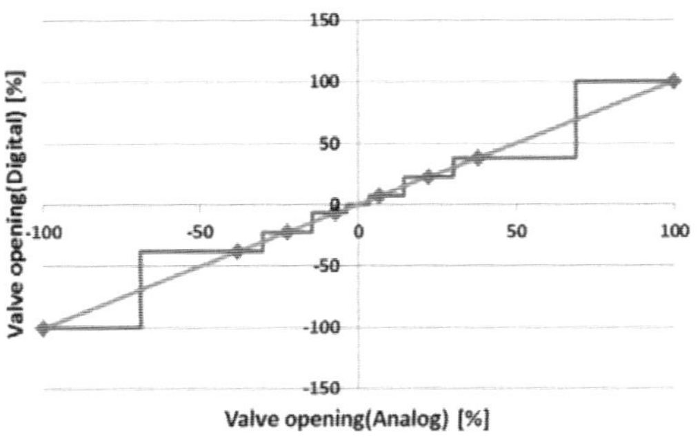

Figure 13. Relation between valve openings

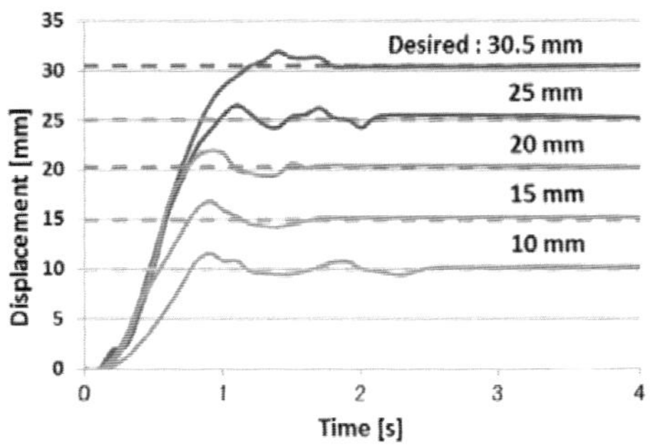

Figure 14. Transient response of displacement

5. Conclusions

This study on the pneumatic digital servo valve can be summarized as follows.

1) To improve the previous digital servo valve, an analytical model of the valve is proposed. And the optimal combination of the orifice diameter is investigated by the simulation. As a result, the validity of the proposed model is confirmed by comparing the calculated results with the experimental. And the optimal combination of orifice was obtained.

2) Based on the results of simulation, the improved valve was produced and tested. And it was applied to the position control of the rubber artificial muscle. As a result, it was confirmed that the improved valve can control the flow rate and pressure more precisely than the previous valve while keeping its volume and mass. And the position control using the improved valve can give much better control performance than that using the previous valve.

Acknowledgements

We express our thanks that this work was supported in part by the MEXT of Japan through a QOL Innovative Research Program (2012-2016).

References

[1] Ishii M, Yamamoto K, and Hyodo K, 2005, Stand-Alone Wearable Power Assist Suit -Development and Availability-, *Journal of Robotics and Mehatronics,* vol 17, no 5, pp 575-583
[2] Kaisumi A, Hirata Y and Kosuge K, 2013, Investigation of User Load and Evaluation of Power Assistive Control on Cycling Wheelchair, *Journal of Robotics and Mehatronics,* vol 25, no 6, pp 959-965
[3] Li X, Noritsugu T, Takaiwa M and Sasaki D, 2013, Design of Wearable Power Assist Wear for Low Back Support Using Pneumatic Actuators, *International Journal of Automation Technology*, vol 7, no 2, pp 228-236
[4] Naruoka Y, Hiramitsu N, and Mitsuya M, 2016, A Study of Power-Assist Technology to Reduce Body Burden During Loading and Unloading Operations by Support of Knee Joint Motion, *Journal of Robotics and Mehatronics,* vol 28, no 6, pp 949-957
[5] Kobayashi H, Aida T, and Hashimoto T, 2009, Muscle Suit Development and Factory Application, *International Journal of Automation Technology,* vol 3, no 6, pp 709-715
[6] Sasaki D, Noritsugu T and Takaiwa M, 2011, Development and Application of High Contractile Pneumatic Artificial Muscle, *International Journal of Automation Technology*, vol 5, no 4, pp 531-537
[7] Akagi T, Dohta S and Katayama S, 2010, Development of Small-sized Flexible Pneumatic Valve using Vibration Motor and Its Application for Wearable Actuator, *Journal of Computer Applications in Technology*, vol 39, issue 1/2/3, pp 86-92
[8] Akagi T, Dohta S and Ueda H, 2010, Development of Small-sized Fluid Control Valve with Self-holding Function Using Permanent Magnet, *Journal of System Design and Dynamics*, vol 4, no 4, pp 552-563
[9] Akagi T, Dohta S and Zhang Y, 2015, Theoretical and Experimental Analysis of Wearable Control Valve with Self-holding Function using Permanent Magnets, *International Journal of Fluid Power System*, vol 8, no 1, pp 113-118
[10] Dohta S, Akagi T, Masago Y, Matsushita H and Zhang Y, 2012, Development of small-sized digital servo valve for wearable pneumatic actuator, *Journal of Procedia Engineering*, vol 41, pp 97-104
[11] Nasir A, Akagi T, Dohta S and Ono A, 2013, Development of Small-Sized Servo Valve Controlled by Using Buckled Tube and Its Application, *Journal of System Design and Dynamics*, vol 7, no 4, pp 516-527
[12] Ono A, Akagi T, Dohta S, Nasir A, 2015, Improvement of Low-cost Wearable Servo Valve Using Buckled Tube, *International Journal of Fluid Power System*, vol 8, no 2, pp 119-125

IOP Conf. Series: Materials Science and Engineering **249** (2017) 012003 doi:10.1088/1757-899X/249/1/012003

Pressure Response of Various Gases in a Pneumatic Resistance Capacitance System and Pipe

J Peng[1], C Youn[2], K Tadano[1], T Kagawa[1]

[1]Laboratory for Future Interdisciplinary Research of Science and Technology, Tokyo Institute of Technology, 4259 Nagatsuta-cho, Midori-ku, Yokohama, Japan
[2]Azbil Corporation, Fujisawa Technology Center, 1-12-2 Kawana, Fujisawa-shi, Kanagawa-ken, Japan
E-mail: pengjiehongjjj@yahoo.co.jp

Abstract: City gas, such as propane and methane, is widely used as a fuel in households and factories. Recently, hydrogen as a clean and efficient fuel has been proposed for fuel cell vehicles. However, few studies have investigated pressure control and response of gases considering their properties. This study investigated the static flow rate characteristics in an orifice with four gases—air, propane, methane, and hydrogen. Then, a pressure response experiment was performed using a pneumatic resistance capacitance system comprising an isothermal chamber and a nozzle flapper, and the time constant of the pressure response with various gases was analysed with a mathematical model. The simulation results agreed with the experimental data. Finally, the differences in pressure propagation in a pipe with various gases were explicated by a pressure response experiment. The results showed that the pressure response speed of hydrogen is faster than that of the other three gases because of its small molecular weight. Therefore, the pressure control equipment of hydrogen needs a high response speed.

1. Introduction

Compressible fluid is widely used in industry as a transmission medium or an energy source. Compressed air has been applied in pneumatic products that are driven by pressure differences, such as air cylinders and rotary actuators. Pneumatic systems have numerous advantages, such as high effectiveness, high durability and reliability, and environmental friendliness. The most city energy supply systems use methane as a city gas because it is a cleaner energy source than coal. Propane, which can be liquefied, is usually employed in areas that lack the necessary infrastructure, such as gas pipelines, because of its high energy density. Recently, hydrogen has been proposed as an environmentally friendly fuel that produces only water as a combustion product [1][2]. The most compelling application is the hydrogen fuel cell vehicle [3], the pressure is needed to control within about 10 kPa in the fuel cell from the hydrogen storage which owns the ultra-high pressure about 70 MPa [4]. Pressure control technology is indispensable for utilizing different gases. Air and propane are generally controlled by a pressure regulator called the pressure reducer valve and city gas is controlled by a gas governor [5]. Recently, many manufacturers and researchers are developing a new type of pressure regulator and bombe for the application of hydrogen [6][7].

It is known that the pneumatic resistance and the capacity, are the most important constituent of pneumatic and gas pressure control systems. And the pressure response experiment is an important method to evaluate the property of the gas pressure control equipment. However, city gases or hydrogen are all flammable and dangerous. Therefore, engineers generally use air in their pressure response research and experiments for developing new pressure control equipment for the flammable gases. Because air source can be obtained conveniently and safely, and its properties are similar to those of city gas. Kagawa performed several studies on the dynamic characteristics and transient response of air in a chamber, nozzle flapper, and pipe 30 years ago [8-10]. Many researchers investigated and resolved the pressure transient response [11] of the compressible fluid in a pipe using the characteristic method [12-14]. Recently, Takeuchi [15] developed a pressure frequency response test to evaluate the stability of gas pressure regulator. Rahman [16] presented a pressure and temperature response within control volume of a pneumatic system with thermal consideration. Li [17] investigated the pressure control performance of the solenoid valve under PWM-controlled conditions by CFD simulation. Lee [18][19] performed numerical modeling of fluid pressure transients with air entrainment and proposed an improved numerical method and computational procedure for implementing typical air vessel responses and their influence on the pressure transient of unsteady flow in a pipeline system. However, these studies are all based on the properties of air. Shin [20] investigated the characteristics of transient flow and the possibility of freezing in a pressure regulator and the rear connecting pipe of the pressure regulator during the closing process of the pressure control valve by using the properties of nature gas in the calculation. However, this study only provided a theoretical calculation without sufficient supporting experiment data. Up to now, it is difficult to find some literatures for pressure response considering the different physical properties of various gases.

Therefore, this study adopted experimental method to investigate the effect of different gas properties on the pressure response in a pneumatic resistance capacitance system and pipe. Four gases—air, hydrogen, methane, and propane—were chosen to perform a static flow rate characteristic experiment and two dynamic pressure response experiments. The flow rate equation for an orifice with the four gases was confirmed based on the experimental data, and the pressure response speed with various gases in the pneumatic resistance capacitance system was discussed. Finally, changes in the propagation speed of pressure wave and pressure vibration through a pipe with the various gases were studied. The research findings can provide the reference and help in the design and manufacture of different gas pressure control systems and equipment.

2. Laminar flow rate meter

In our experiment, we used a quick flow sensor (QFS; Figure 1) based on the laminar flow theory to measure the flow rate of the various gases [21]. The flow rate calculation of the QFS uses the Hagen–Poiseuille equation:

$$Q = \frac{\pi r^4 \Delta P}{8\mu L} \tag{1}$$

where Q is the volumetric flow rate, r is the pipe radius, L is the pipe length, μ is the dynamic viscosity, and ΔP is the differential pressure.

Table 1. Properties [a] of various gases

Gas	Density [kg/m^3]	Relative density	Viscosity [10^{-7} Pa·S]	Relative viscosity	Gas constant [J/(kg·K)]	Specific heat ratio
Air	1.205	1.000	181	1.000	287	1.40
Methane	0.668	0.554	108	0.597	518.3	1.32
Hydrogen	0.083	0.068	88	0.486	4128.5	1.41
Propane	1.882	1.562	80	0.442	189	1.13

[a] Temperature = 20 °C at atmospheric pressure

Figure 1. Laminar flow rate meter (QFS)

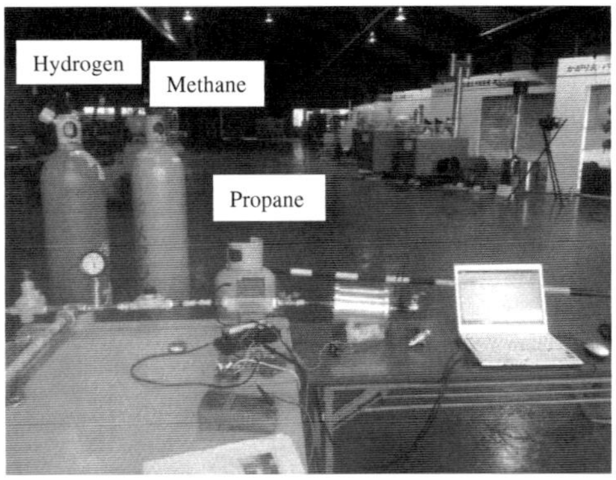

Figure 2. Experimental setup

A flow rate meter is usually designed for air because of its universality. Therefore, the parameters of the QFS can only be used for air measurements; that is, the dynamic viscosity μ is of air. The parameters of the QFS must be modified to enable measurements for other gases. Table 1 presents the properties of the four gases. Based on the experimental data Q_m obtained with the QFS, the real flow rate Q_{real} of various gases can be obtained with equation (2), which uses the relative viscosity μ^*.

$$Q_{real} = \frac{Q_m}{\mu^*} \tag{2}$$

3. *P-Q* characteristics of an orifice with various gases

An orifice as a restriction is the most widely used pneumatic resistance. The relationship between the pressure and the flow rate G is summarized by the well-known flow rate equation [22]:

when $\frac{P_D}{P_U} \leq (\frac{2}{\kappa+1})^{\frac{\kappa}{\kappa-1}}$,

$$G(P_U, P_D, S_e) = S_e P_U \sqrt{\frac{\kappa}{R\theta_1}(\frac{2}{\kappa+1})^{\frac{\kappa+1}{\kappa-1}}} \tag{3.1}$$

and $\frac{P_D}{P_U} > (\frac{2}{\kappa+1})^{\frac{\kappa}{\kappa-1}}$

$$G(P_U, P_D, S_e) = S_e P_U \sqrt{\frac{2\kappa}{\kappa-1} \cdot \frac{1}{R\theta_1}\left[\left(\frac{P_D}{P_U}\right)^{\frac{2}{\kappa}} - \left(\frac{P_D}{P_U}\right)^{\frac{\kappa+1}{\kappa}}\right]} \tag{3.2}$$

IOP Conf. Series: Materials Science and Engineering **249** (2017) 012003 doi:10.1088/1757-899X/249/1/012003

$$S_e = C_D \frac{\pi d^2}{4} \tag{4}$$

where S_e is the effective area, d is the orifice diameter, C_D is the discharge coefficient, R is the gas constant, P_U is the upstream pressure, P_D is the downstream pressure, θ_1 is the flow temperature, and κ is the specific heat ratio.

Generally, the discharge coefficient C_D is mostly dependent on the structure and dimensions of the orifice, while at times, it is affected by the Reynolds number. In this experiment, the orifice was not changed and the Reynolds number was large most of the time. At large Reynolds numbers, changes in C_D are very small. Therefore, to compare the primary factor of influence, C_D was set to a constant equal to the C_D of air. Curve fitting of the experiment data revealed that the discharge coefficient is 0.89. On the other hand, the specific heat ratio κ of the different gases is close to each other; therefore, it is not a primary influencing factor compared with the gas constant. If the specific heat ratios of all gases are assumed to be the same as that of air, that is 1.4, it is not difficult to determine that the mass flow rate of the various gases is only dependent on the gas constant under a certain pressure difference. Therefore, the calculation results indicate that the volume flow rate of hydrogen, propane, and methane were 3.80, 0.79, and 1.34 those of air, respectively, under the same pressure difference. Note that we are talking about the volume flow rate, and not the mass flow rate.

A static experiment using a ϕ1 mm orifice was conducted to confirm the *P-Q* characteristic curve with various gases. The experimental location and sources of the various gases were provided by KATSURA Co., Ltd (Figure 2). Figure 3 shows the pneumatic circuit of the static experiment. Here, the upstream pressure was set constant at 0.1 MPaG and the downstream pressure was changed from 0.1 MPaG to atmospheric pressure through the flow rate control valve.

Figure 4 shows the static flow rate characteristic curve, that is, the *P-Q* characteristic curve with the various gases. The dashed curve represents the calculation results without considering the influence of κ, that is, κ of all gases set to 1.4. The solid curve represents the calculation results considering the different values of κ of the various gases. The two curves of the same gas were generally coinciding, and the experimental data were consistent with these calculations. Therefore, it is believed that the flow rate equation for the orifice with various gases can neglect the effect of the specific heat ratio when its value is from 1.13 to 1.4.

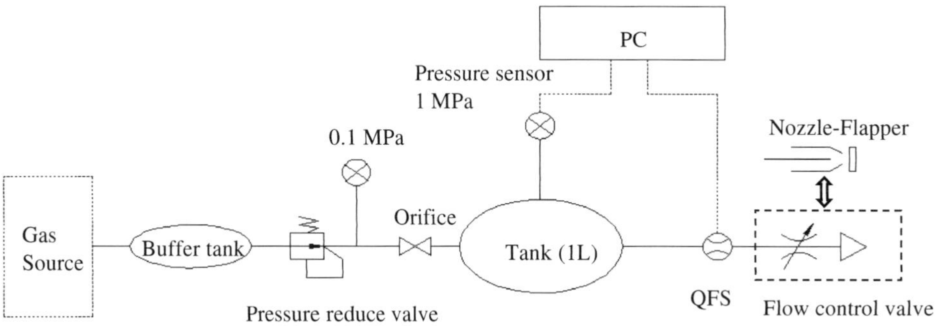

Figure 3. Pneumatic circuit of experiment

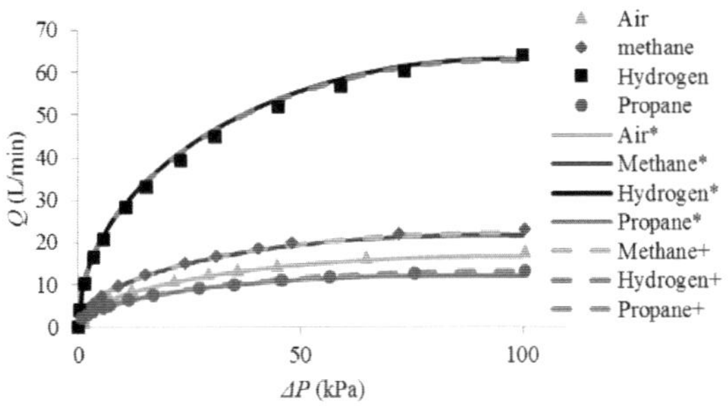

(*represents change in κ with various gases, + represents κ =1.4)

Figure 4. *P-Q* characteristic curve with various gases

4. Pressure response in a pneumatic resistance capacitance system

The gas flow through a valve or a pressure regulator can sometimes be seem as though a nozzle flapper. Next, the pressure response with the various properties of gases in an isothermal chamber (ITC) [23], which is a tank filled with copper wire, with an orifice and the step action of a nozzle flapper was investigated.

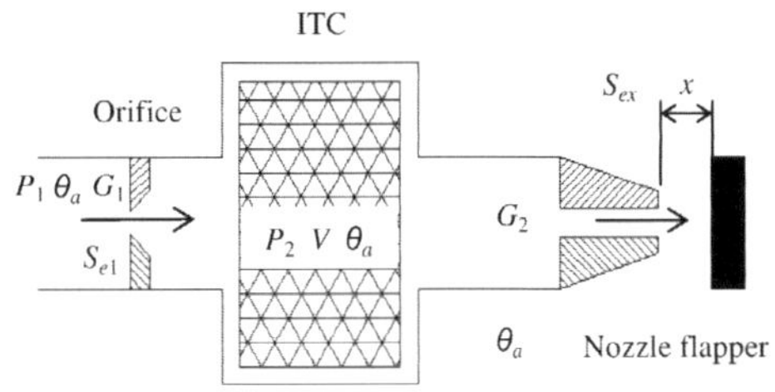

Figure 5. Isothermal chamber with nozzle flapper model

4.1. Calculation

The mathematical model of the ITC with an orifice and nozzle flapper is shown in Figure 5. The ideal gas equation for the ITC was differentiated with respect to time:

$$P_2 V = mR\theta \tag{5}$$

$$\frac{VdP_2}{dt} = R\theta\frac{dm}{dt} + mR\frac{d\theta}{dt} \tag{6}$$

where P_2 is the pressure in the chamber, V is the volume, m is the mass of air in the chamber, θ is the temperature, and R is the gas constant. Because the chamber is isothermal, $d\theta / dt = 0$ and $\theta = \theta_a$. Furthermore, $dm / dt = G$, where G is the mass flow rate. Therefore, equation (6) was changed to

$$\frac{dP_2}{dt} = \frac{R\theta_a}{V}(G_1 - G_2) \tag{7}$$

$$P_2 = \int \frac{R\theta_a(G_1 - G_2)}{V} dt \tag{8}$$

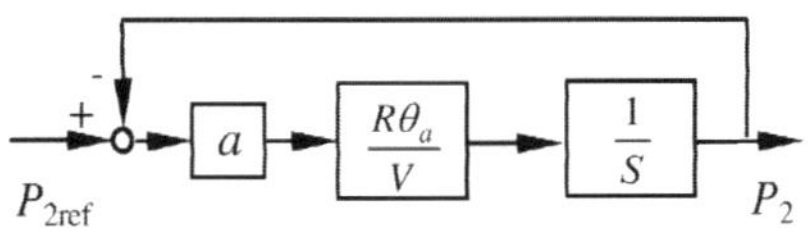

Figure 6. Block diagram

The flow rate through the upstream orifice G_1 and the downstream nozzle flapper G_2 are also determined by equations (3) and (4). The specific heat ratios of all gases are set to 1.4 according to the conclusion of Section 3. So, we propose the gas constant and simply note:

$$G_1 = \frac{1}{\sqrt{R}}\varphi_1(P_1, P_2, S_{e1}), G_2 = \frac{1}{\sqrt{R}}\varphi_2(P_2, P_a, S_{ex}) \tag{9}$$

The flow rate G_1 and G_2 are linearized at the pressure balance point $P_{2\mathrm{ref}}$ and the flow rate gain a is defined as follows:

$$a = \frac{\Delta G_1 - \Delta G_2}{\Delta P_2} \tag{10}$$

The block diagram is shown in Figure 6.

From this block diagram, the transfer function is obtained:

$$F(s) = \frac{1}{1 + T_p s} \tag{11}$$

Here, T_p is the time constant [9] of the system:

$$T_p = \frac{V}{aR\theta_a} \tag{12}$$

Using equations (9) and (10), we have

$$a = \frac{1}{\sqrt{R}}\left(\frac{\Delta\varphi_1(P_1, P_2, S_{e1}) - \Delta\varphi_2(P_2, P_a, S_{ex})}{\Delta P_2}\right) \tag{13}$$

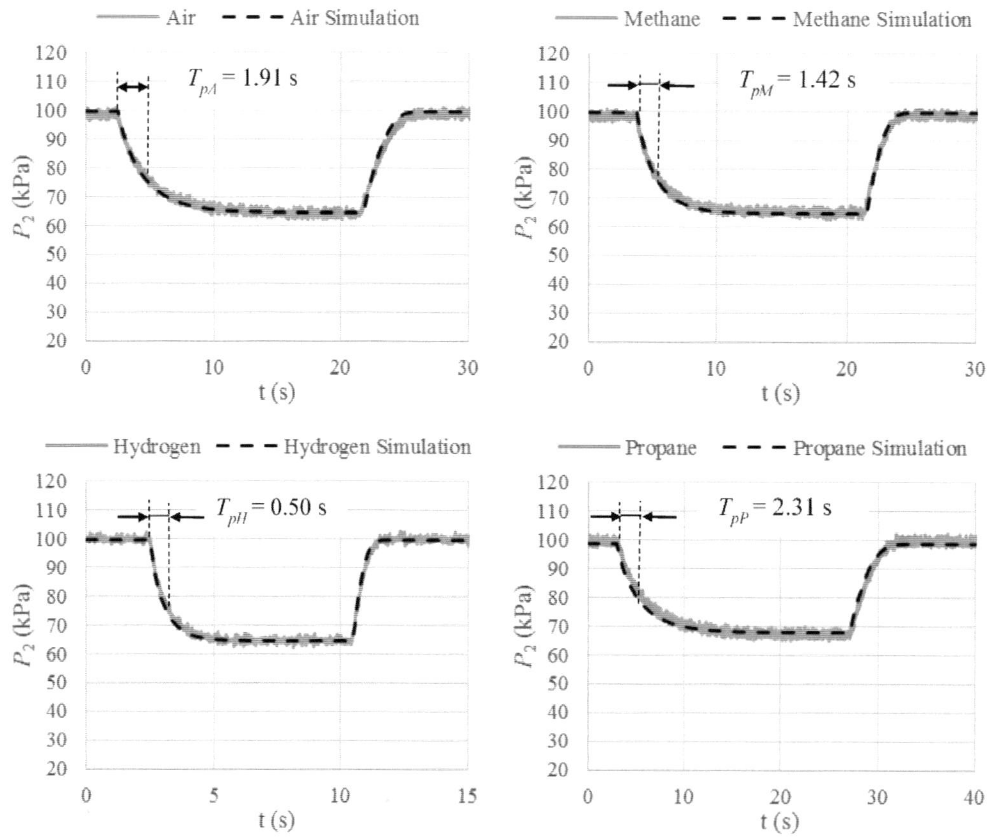

Figure 7. Pressure response of P_2

Defining

$$a^* = \frac{\Delta\varphi_1(P_1, P_2, S_{e1}) - \Delta\varphi_2(P_2, P_a, S_{ex})}{\Delta P_2} \tag{14}$$

we get

$$a = \frac{1}{\sqrt{R}} a^*, \text{ and } T_p = \frac{1}{\sqrt{R}} \frac{V}{a^* \theta_a} \tag{15}$$

From equation (15), it can be seen that the time constant is inversely proportional to the square root of the gas constant of the various gases when the other experimental conditions are fixed.

4.2. Pressure response experiment

The pneumatic circuit of the pressure response experiment is also shown in Figure 3, but the flow control valve is changed to a nozzle flapper. In the discharge process, the effective area of the downstream restriction was changed from 0.15 mm^2 to 0.85 mm^2 by controlling the displacement between the nozzle and flapper with propane, while the effective area of the downstream restriction was changed from 0.1 mm^2 to 0.9 mm^2 with the other gases. Of course, the action is opposite in the charge process. The upstream pressure was set to 0.1 MPa.

The pressure response of P_2 in the ITC is shown in Figure 7. The black dashed curve represents the simulation results obtained with MATLAB, and the orange curve is the experimental sampling data. The results obviously show that the pressure response speed of hydrogen is the fastest, while that of propane is the slowest. The time taken to reach an 63.2% drop in the response pressure is defined as the time constant of the various gases. Note that T_{pA}, T_{pH}, T_{pM}, and T_{pP} represent the time constant of air, hydrogen, methane, and propane, respectively. The calculation results show that T_{pA}= 1.91 s, T_{pH}= 0.50 s, T_{pM}= 1.42 s, and T_{pP}= 2.31 s. Therefore, the time constant of air is 3.82 times that of hydrogen, which is precisely equal to the reciprocal of the square root of the gas constant times. The experiment data also show agreement with the calculations, that T_{pA} *= 2.12 s, T_{pH} *= 0.55 s, T_{pM}* = 1.48 s, and T_{pP} *= 2.61 s. Therefore, the results verify the conclusion of equation (15). As we know, the gas constant refers to the molecular weight of the gas. Thus, it is certainly believed that the high response speed of hydrogen is owing to its small molecular weight. Finally, it must be noted that the pressure control system needs a high response speed when hydrogen is used. These results can help engineers in designing gas pressure control equipment with various gases, particularly, hydrogen pressure control systems.

5. Pressure propagation through a pipe with various gases

Pipe is another important constituent of pneumatic and gas pressure control systems. We performed a pressure response experiment using an 20 m pipe, and the dynamic characteristics of the pressure propagation with various gases were explicated; the pipe's inner diameter was 6.5 mm.

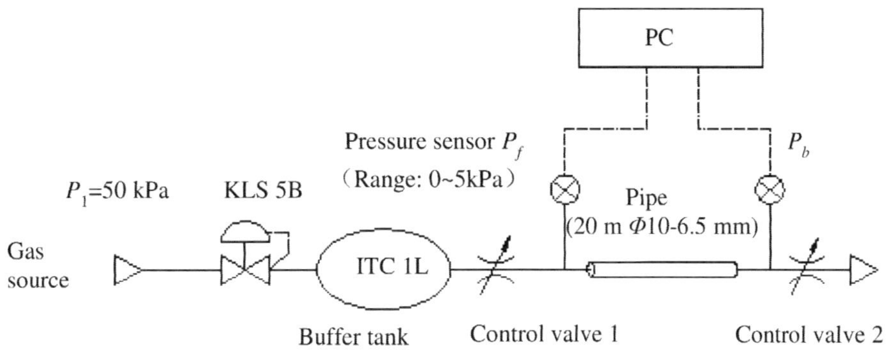

Figure 8. Pneumatic circuit of pressure response experiment with a pipe

5.1. Experiment

Figure 8 shows the pneumatic circuit of the dynamic pressure response experiment. In this experiment, the upstream pressure was set to 50 kPaG, and the downstream pressure was maintained at 2.8 kPaG using a second-stage regulator KLS-5B. The pressure in front of the pipe was recorded with a pressure sensor P_f and that at the back of the pipe was recorded with P_b. The steps of the experiment are illustrated as follows:

(1) Discharge process:

a. Slowly shut control valve 2, hold the downstream pressure at 2.8 kPaG.

b. Shut control valve 1.

c. Open control valve 2 quickly and record the data until the pressure is stable.

(2) Charge process:

a. Shut control valve 1, wait till the downstream pressure decreases to the atmospheric pressure and stabilizes.

b. Shut control valve 2.

c. Open control valve 1 quickly and record the data until the pressure is stable.

5.2. Results and discussion

Figures 9 and 10 show the results of the pressure response in the discharge and charge processes, respectively, with various gases. The black curve represents the pressure response P_f in front of the pipe, and the blue curve represents the pressure response P_b at the back of the pipe.

In the discharge process, it can be observed that there is some time delay in the pressure response curve of P_f with air, methane, and propane, but the time delay with hydrogen is very small. This is thought to be because the time taken for the pressure wave to travel from the back to the front of the pipe with hydrogen is way lesser than that with the other gases. The speed of sound is determined by $c = \sqrt{\kappa RT}$. Therefore, the speed of sound in hydrogen is way faster than that in the other three gases because of their large gas constants. On the other hand, the large gas constant means that the molecular weight of hydrogen is very small, and that of propane is large. That is, propane is very heavy and hence

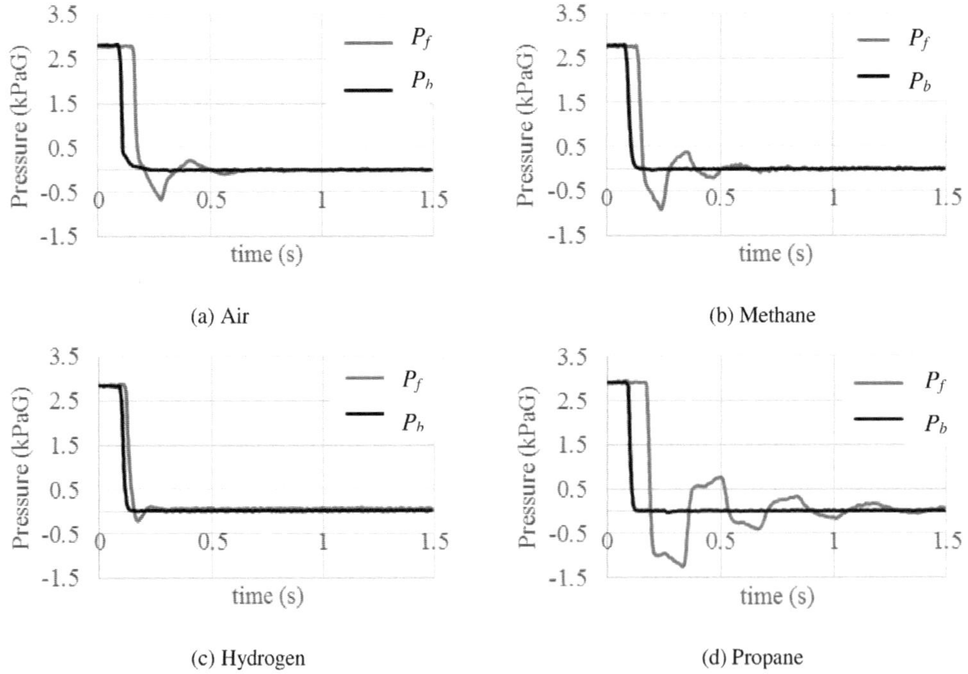

Figure 9. Pressure response in pipe (Discharge)

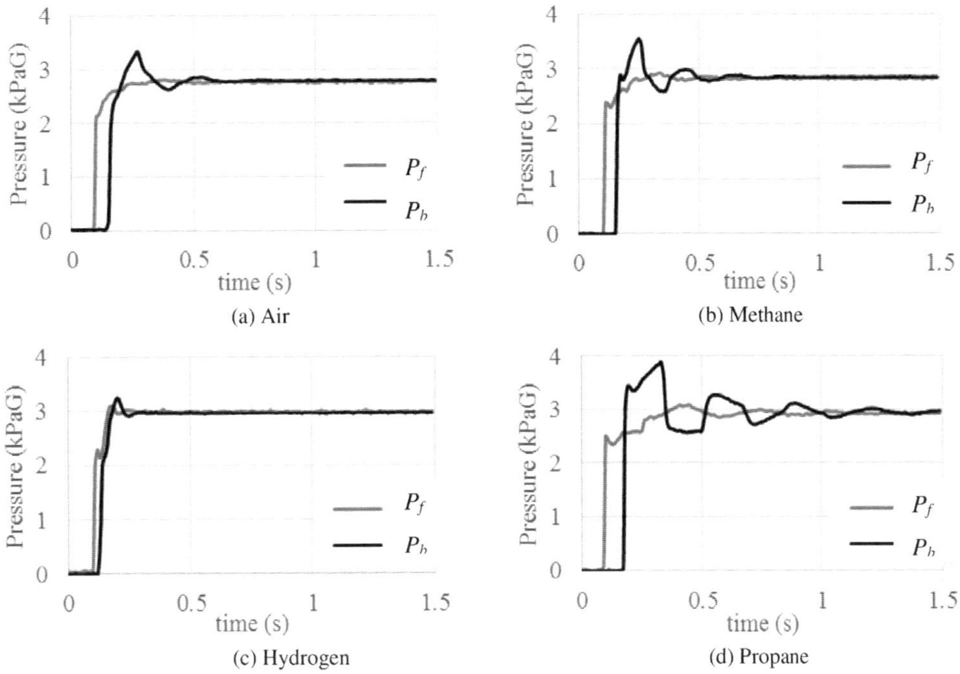

Figure 10. Pressure response in pipe (Charge)

has a large inertia. In Figure 9(d), it seems that there is a large vibration in P_f with propane when the pressure reaches the atmospheric pressure, while a slight vibration is observed with hydrogen in figure 9(c).

The time delay phenomenon also occurs in the charge process, but the delayed pressure is P_b because of the reverse flow direction. A pressure vibration which likes the water hammer is occurred at the back of the pipe because of the propagation of the pressure wave. The period of the pressure vibration is determined by $T_t = 4L/c$. Here, L is the length of the pipe. The pressure vibration periods for air, methane, propane and hydrogen are 0.23 s, 0.17 s, 0.29 s, and 0.06 s, respectively. From the form of the pressure wave in Figure 10, the results roughly accord with the calculations. The amplitude of the pressure wave demonstrates the vibration intensity. In figure 10(d), when the propagation medium is propane, the amplitude of the pressure wave at the back of the pipe is larger than that of the other three gases. Meanwhile, the pressure wave owns the smallest amplitude in hydrogen, and the vibration time is extreme short that no more than one period. Similar to the phenomenon of the water hammer, the maximum amplitude A_P of the pressure vibration at the back of the pipe is determined by

$$A_P = \rho c u \tag{16}$$

Therefore, the maximum amplitude of pressure wave in hydrogen is certainly small, because the density of hydrogen is only 0.068 times that of air, even though its sound speed is larger than the other gases. On the contrary, the pressure vibration owns a large amplitude in propane due to its large density. Of course, many other factors, such as friction at the pipe's inner wall and the boundary condition affect the pressure response and the pressure wave propagation. Therefore, a high-precision simulation needs to be performed in a future study.

6. Conclusions

In this study, a static flow rate characteristic experiment using an orifice was performed with various gases. The effect of the viscosity and gas constant of the gases on the flow rate measurement and P-Q

characteristic curve was understood. Two dynamic pressure response experiments with a pneumatic resistance capacitance system and a pipe were performed to verify the effect of the gas constant of various gases on the pressure response speed. Finally, the following conclusions were obtained:

(1) The flow rates can be calculated by only considering the gas constant of the various gases in an orifice with a constant area under a certain pressure difference because the influence of κ is negligible when its value is from 1.13 to 1.4.

(2) The time constant of the various gases is equal to the reciprocal of the square root of the gas constant. That is, the pressure response speed of hydrogen is 3.82 times that of air in a pneumatic resistance capacitance system.

(3) The pressure response speed in a pipe with hydrogen is much faster than that with the other three gases, and a large pressure vibration does not easily occur because of its small density. In contrast, propane has a large pressure vibration.

References

[1] Zuttel A, Remhof A, Borgschulte A and Friedrichs O 2010 Hydrogen: the future energy carrier *Philosophical Transactions of the Royal Society A* **368** 3329–3342.

[2] Kler A M, Tyurina E A, Potanina Y M and Mednikov A S 2015 Estimation of efficiency of using hydrogen and aluminum as environmentally-friendly energy carriers *International Journal of Hydrogen Energy* **40**(43) 14775-14783.

[3] Hwang Y G 2016 Technical trends of hydrogen manufacture, storage and transportation system for fuel cell vehicle *Journal of The Korean Institute of Resources Recycling* **25**(1) 48-59.

[4] Wong J and Gambone L, 2006 70 MPa fueling station for hydrogen vehicles *16th World Hydrogen Energy Conference* Lyon France.

[5] Takeuchi T and Kagawa T, 2013 Applicability of frequency response test for stability evaluation of gas pressure regulator *Transactions of the Society of Instrument and Control Engineers* **49**(8) 747-754.

[6] Nakano A, Maeda T, Ito H, Masuda M, Kawakami Y, Tange M, Takahashi T and Nishida K 2012 Study on absorption/desorption characteristics of a metal hydride tank for boil-off gas from liquid hydrogen *International Journal of Hydrogen Energy* **37**(6) 5056-5062.

[7] Mizuno T, Youn C, Nakamura Y and Kagawa T 2013 A simulation study of radial slits pressure regulator for hydrogen gas *AsiaSim 2013: 13th International Conference on Systems Simulation* (Singapore) Tan G, Yeo G K, Turner S J and Teo Y M pp 288-297.

[8] Kagawa T and Shimizu M 1988 Nondimensional pressure responses of pneumatic RC circuits considering heat transfer *Hydraulics & Pneumatics* **19**(4) 308-311.

[9] Kagawa T 1982 Transient pressure response of pneumatic nozzle flapper *Transactions of the Society of Instrument and Control Engineers* **18**(6) 617-621.

[10] Kagawa T, Kitagawa A and Takenaka T 1984 An analysis of transient response of a pneumatic transmission line terminated by volume using characteristics method *Transactions of the Society of Instrument and Control Engineers* **20**(11) 1014-1018.

[11] Brown F T, Margolis D L and Shah R P 1969 Small-amplitude frequency behavior of fluid lines with turbulent flow *ASME, Journal of Basic Engineering* **91**(4) 678-693.

[12] Zielke W 1968 Frequency-dependent friction in transient pipe flow. *Trans. ASME, Journal of Basic Engineering* **90**(1) 109-115.

[13] Kitagawa A, Kagawa T, Sanada K and Okada T 1997 A study on the transient characteristics of the flow in pneumatic transmission line *Transactions of the Society of Instrument and Control Engineers* **33**(4) 227-233.

[14] Yoshida M, Kawato T, Fujita T, Kawashima K and Kagawa T 2003 Modeling of Gas Transmission Systems Considering Heat Transfer *Transactions of the Society of Instrument and Control Engineers* **39**(3) 253-258

[15] Takeuchi T and Kagawa T 2013 Applicability of frequency response test for stability evaluation of gas pressure regulator *Transactions of the Society of Instrument and Control Engineers*

49(8) 747-754.

[16] Rahman F 2015 Pressure and Temperature Response of Pneumatic System with Thermal Consideration *Global Journal of Research In Engineering* **15**(4) 23-27.

[17] Li S, Wu P, Cao L, Wu D and She Y 2017 CFD simulation of dynamic characteristics of a solenoid valve for exhaust gas turbocharger system *Applied Thermal Engineering* **110**(5) 213–222

[18] Lee T S 1994 Numerical modelling and computation of fluid pressure transients with air entrainment in pumping installations *International Journal For Numerical Methods In Fluids* **19** 89–103.

[19] Lee T S 1998 A numerical method for the computation of the effects of an air vessel on the pressure surges in pumping systems with air entrainment *International Journal For Numerical Methods In Fluids* **28** 703-718.

[20] Shin C H 2013 A numerical study on the characteristics of transient flow in a pressure regulator resulting from closure of the pressure control valve *Journal of Mechanical Science and Technology* **27**(2) 443-449.

[21] Kagawa T, Saisu Y, Nishimura R and Youn C, 2009 Development of high speed response laminar flow meter for air conditioning *10th International Conference on Fluid Control, Measurements, and Visualization* (Moscow, Russia)

[22] Oertel H 2004 *Prandtl's Essentials of Fluid Mechanics, 2nd ed* (Springer, New York, USA) pp 190-193.

[23] Kawashima K, Kagawa T and Fujita T 2000 Instantaneous flow rate measurement of ideal gases *Transactions of the ASME Journal of Dynamic Systems, Measurement, and Control* **122** 174-178.

IOP Conf. Series: Materials Science and Engineering **249** (2017) 012004 doi:10.1088/1757-899X/249/1/012004

Improvement of Wearable Power Assist Wear for Low Back Support using Pneumatic Actuator

Feifei Cho[1], **Riku Sugimoto**[1], **Toshiro Noritsugu**[1] , **Xiangpan Li**[2]

[1] National Institute of Technology, Tsuyama College Tsuyama, Japan

[2] Henan University of Science and Technology, Kaiyuan road,Luoyang, Henan,471023,China

Abstract. This study focuses on developing a safe, lightweight, power assist device that can be worn by people who like caregivers during lifting or static holding tasks to prevent low back pain (LBP). Therefore in consideration of their flexibility, light weight, and large force to weight ratio we have developed a Wearable Power Assist Wear for caregivers, two types of pneumatic actuators are employed in assisting low back movement for their safety and comfort. The device can be worn directly on the body like normal clothing. Because there is no rigid exoskeleton frame structure, it is lightweight and user friendly. In this paper, we proposed the new type of the wearable power assist wear and improved the controller of control system.

1. Introduction

Resent years, the number of people in Japan age 65 and over was about 31 million, account for about 25% of the population. With the rapid arrival of an aging society, the demand for professional caregivers has increased drastically. It was said that 70% of caregivers experience back pain reference. So the development of wearable power assist devices and non-wearable lifting assist devices are greatly needed in nursing care fields.

Wearable robots are a mechatronic systems that are designed around the shape and function of the human body. Early designs were mainly exoskeleton types. These robots have multiple degree of freedom (DOF), but are not suitable for use in small spaces and in daily life.

To meet the aging society's needs for power assist robotic technology, and devices that wear like normal clothing, this study proposes a new wearable power assist device for low back support using pneumatic actuators. First, the biomechanical model of the human spine was analyzed to better understand the reason for low back pain (LBP). Since many health caregivers are troubled by low back pain problems, the requirement for low back support is needed. It is critical to understand the cause of LBP when designing power assist wear for low back support. LBP is common in various occupations, its presence being related to activities requiring repetitive lifting, repeated activities in bending forward positions, and lifting high energetic loads. Related results also suggest that bending activities involving higher degrees of trunk flexion are associated with disabling types of LBP in certain working populations. Such work characteristics are common among nursing caregivers. The prevalence of LBP in nursing is high compared to other occupations and in relation to other types of work. Risk factors include physical work such as manual lifting and transferring of patients, working conditions such as working time and rest during the night shift, and the working environment. Among these factors, exposures to frequent manual lifting and transferring of patients are widely recognized factors. Based on this analysis, two types of pneumatic actuators were selected to support the human's back from two aspects: increased support force and related lever arm length. And in this study, for cut the weight of the device, we used the quasi-servo valve to control the air. In order to provide efficient assisted force, control strategies were determined to minimize the interference of human actions and provide assistance effectiveness for the proposed device. The proposed design has been proven through experiments.

2. Wearable power assist wear

2.1. Elongation-type pneumatic rubber artifical muscle

Developing wearable power assist devices that wear like normal clothing requires the development of a new type of pneumatic rubber artificial muscle to meet the requirements of safety and user-friendliness. In this part, an elongation-type of pneumatic rubber artificial muscle referred as actuator A, is employed for the special requirement in assisting low back movement. Figure 1 shows the structure of the elongation-type pneumatic rubber artificial muscle. It is composed of a rubber tube covered with a bellows sleeve and closed with ties at both ends. The outer diameter of the rubber tube is 12 mm; the inner diameter is 10 mm. The bellows sleeve is woven of twisted fiber cord, which has a maximum expansion diameter of 14 mm and a minimum contraction diameter of 4 mm. When first pressurized, the elongation-type pneumatic rubber artificial muscle expands slightly in the radial direction, putting the internal rubber tube into close contact with the external sleeve. As the pressure continues to rise, due to the fact that radial expansion is limited by the external sleeve, the artificial muscle expands in the axial direction, and it becomes longer. The higher the pressure, the longer it becomes. Figure. 2 shows the relation between supplied pressure and L_d. related displacement of elongation-type pneumatic rubber artificial muscle with no external load connected. The maximum length is attained when it is pressurized; and the minimum length is attained when it is not. For this type of muscle (original length L_o=320 mm), the total length can reach 490 mm (L_d = 170 mm) when pressurized to 500 kPa.

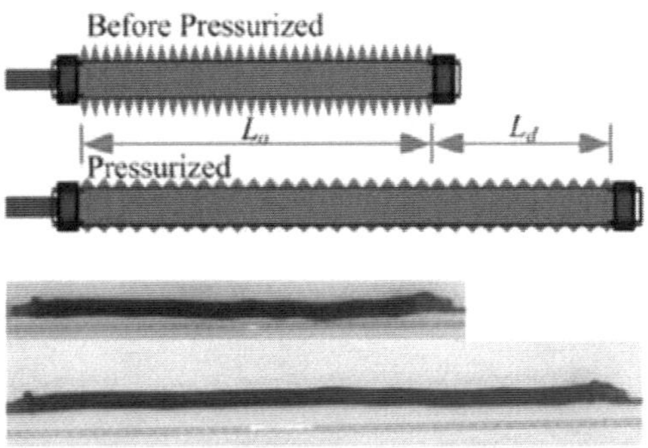

Figure 1. Pneumatic rubber artificial muscle

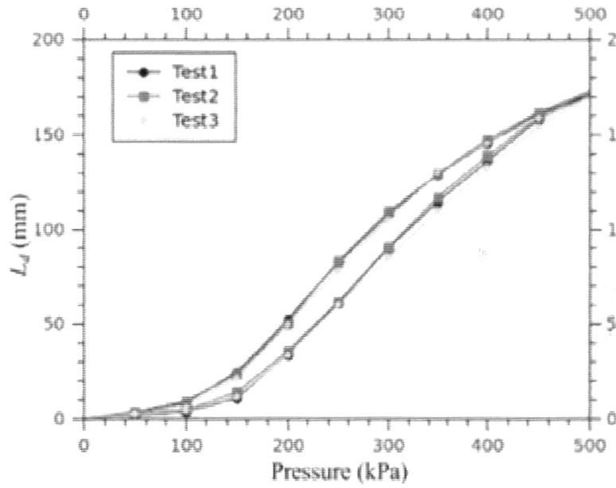

Figure 2. Relation between supplied pressure and Displacement

2.2. Layer-type pneumatic actuator

The layer-type pneumatic actuator called actuator B, is composed of two TPU balloons. The TPU material is a composite material that combines the properties of rubber and plastic. It has excellent weight bearing capacity and impact resistance, and it is widely used in producing massage chairs and airbags. The TPU balloon used in this device is 150 mm long, 100 mm wide, and 2 mm thick, as shown in Fig.3. When the balloon is supplied with compressed air, it will become taller, reaching 50 mm. A TPU balloon can take a maximum air pressure of 250 kPa. The expansion force reaches 450 N at a pressure of 60 kPa. TPU balloons are put inside pockets made with nylon bands. In this device, the actuator is installed in the inner layer of the garment. In order to increase the moment arm A of assistance force, the expansion displacement in height can be adjusted by changing the air pressure applied.

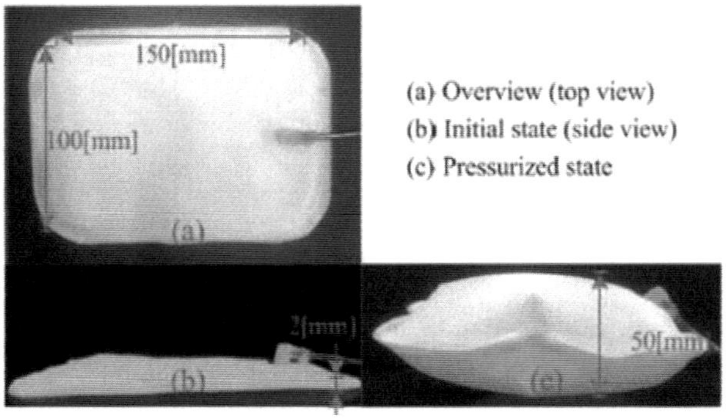

Figure 3. Overview of TPU balloon

2.3. Previous control system

The whole structure of the control system is shown in Figure 4. The system consists of A/D board (PCI-3133, Interface Corporation), D/A board (PCI-3341A, Interface Corporation), pressure sensor (AP-43, KEYENCE Corporation), pneumatic servo valve (EVD-1500-008AN, CKD Corporation), and IMU sensors.

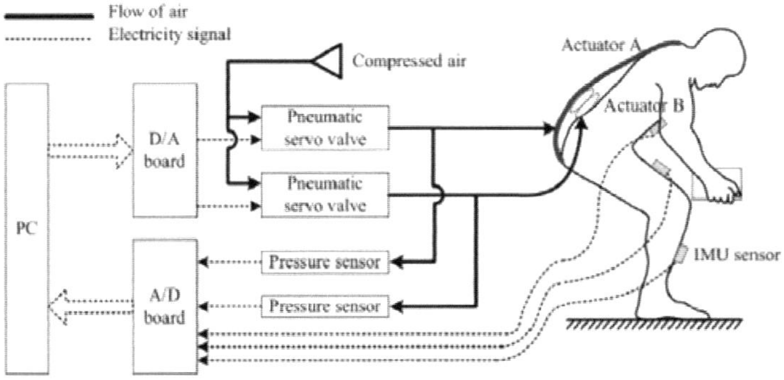

Figure 4. Structure of previous control system

3. Tested assist wear device for waist

3.1. Assist wear device

This research focuses on developing a safe, lightweight, power assist device that can be worn by people during lifting or static holding tasks to prevent LBP. In consideration of their flexibility, light weight, and large force to weight ratio, two types of pneumatic actuators were employed in assisting low back movement for their safety and comfort. The biomechanical model of the human spine was analyzed, and to understand the main causes of LBP: when the human is bending forward and lifting a load. The erector spinae muscles have a small lever arm, causing the spine to bear a large amount of force, several times the body weight. By taking into account the biomechanic structure of the human spine, this device can provide support in two ways. Actuator acts as an external muscle power generators to reduce the force requirement for the erector spinae muscles. Actuator B acts as a moment arm for the contractile force generated by actuator A, and increase the effective torque. The device shown in Fig. 4 can be worn directly on the body like normal clothing. Because there is no rigid exoskeleton frame structure, it is lightweight and user friendly.

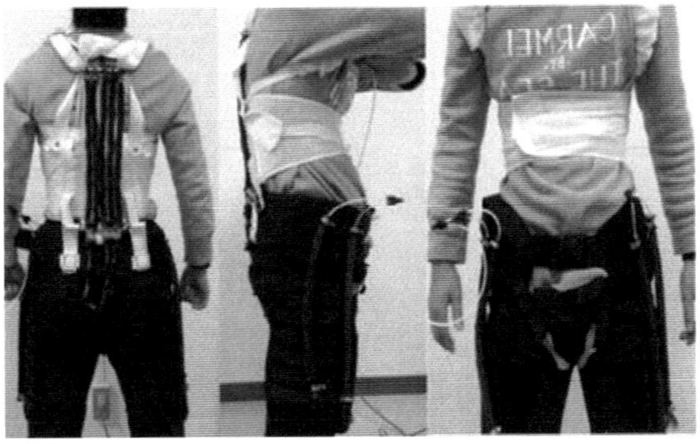

Figure 5. Wearable power assist wear device

3.2. Waterproof actuators
This study proposed actuators for use in environment of high humidity. The basic concept for the waterproof of the actuators are summarized as follows.

[1] The actuators don't use substances of metal. Because it prevents corrosion.
[2] Between the actuator and the pulp is glue with acrylic resin.

The connector portion of the actuators processed acrylic-silicon glue is resistant to water.

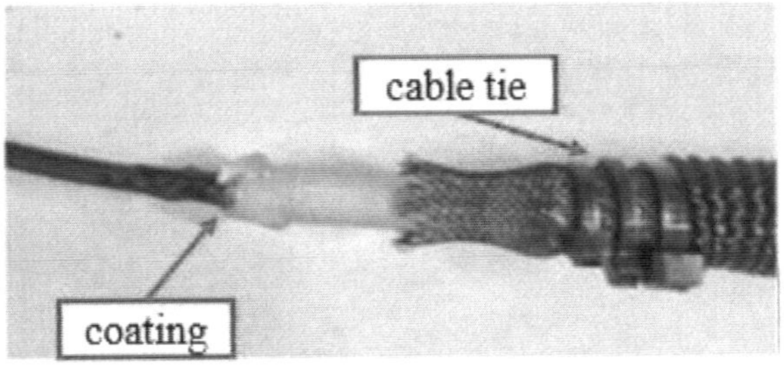

Figure 6. Waterproof actuators

4. Quasi-servo valve
Figure 7 shows the schematic diagram of the tested servo valve. The valve consists of two on/off type control valves (G010HE-1, Koganei Co.Ltd.) whose both output ports are connected each other. The valve connected with the actuator is a two-port valve. The other is a three-port valve that can change the direction of fluid flow such as a supply or exhaust. The two-port valve is driven by pulse width modulation (PWM) method in order to adjust the valve opening per time. It becomes the quasi fluid resistance. Then, the latter valve is called as "PWM valve", the former is called as "switching valve". The size of the valve is 33×19.6×10mm, and the mass is 15.3g.
Each valve is controlled by a micro-computer through the transistors. The maximum supply pressure of the on/off valve is 500kPa, and the maximum output flow rate is 38 liter/min.
In the case of supplying to the actuator, the switching valve is turned on while the PWM valve is being turned on according to the duty ratio. In the case of exhausting, the switching valve is turned off while the PWM valve is being driven. In the case of closing, the PWM valve is turned off.

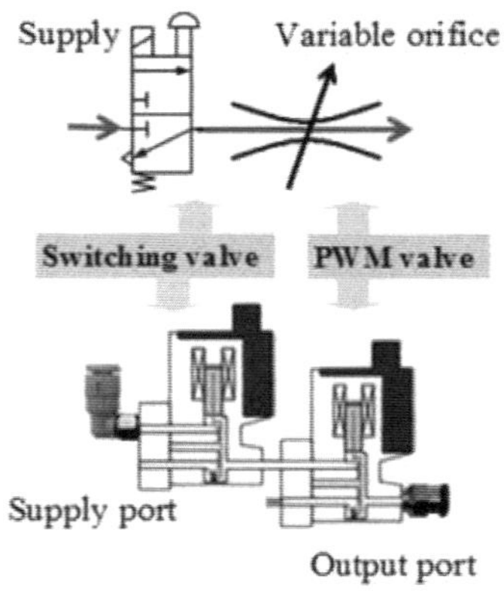

Figure 7. Schematic diagram of quasi-servo valve

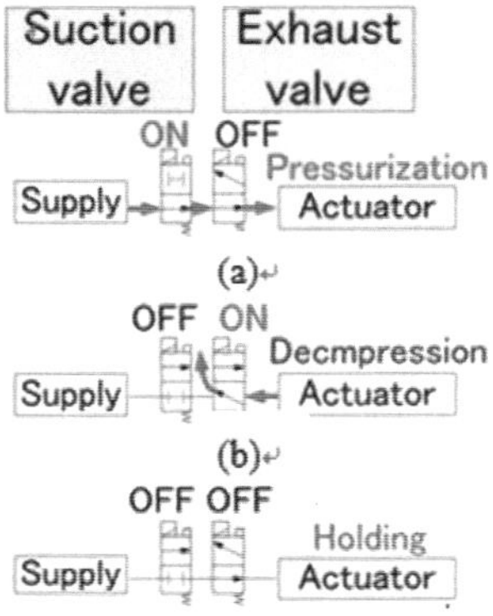

Figure 8. Oprating principle of quasi-servo

5. Control system of wearable power assist wear

5.1. Improved control system

The previous control system used PC and big solenoid valve. Because it is not easy to move freely. In this study we use the microcomputer (Arduino NANO) to control the assister wear. Figure 9 and Figure 10 show the exterior of improved control system and structure of improved control system. The size of the improved control system is 90×100×90mm, and the mass is 250g.

The improved control system used an acceleration sensor (KXM52-1050, Kionix Corporation) and an EMG sensor (101020058, Seeed Corporation) and a pressure sensor(SEU11-4UA-S3, PISCO Corporation), five quasi-servo valves. Each one of the sensors are used to t detect inclination θ of the human body, the force that human output, air pressure value.

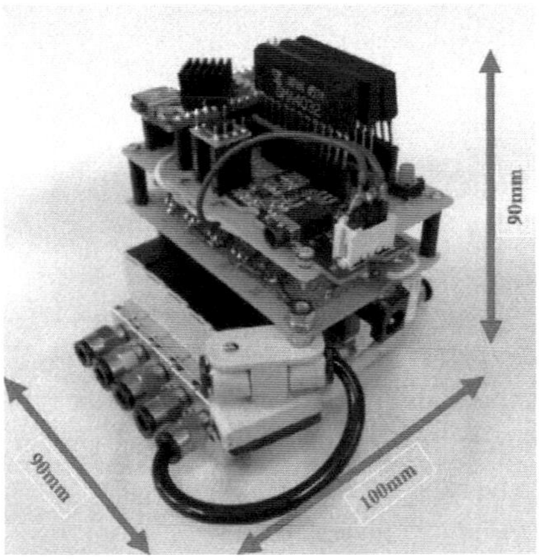

Figure 9. The exterior of improved system

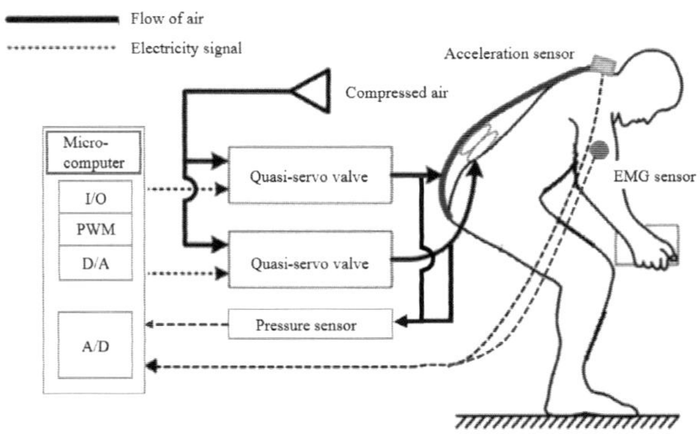

Figure 10. Structure of improved control system

5.2. PID control

In this study, we performed to construct a pneumatic flow controller supplied to the actuators using PID control. The reason for selecting the PID control, it is possible to compensate the steady-state error. Control output U of PID control is expressed by the following equation.

$$U = K_p e + K_i \int edt + K_d \frac{de}{dt} \qquad (1)$$

K_p is proportinal gain, K_i is the intergral gain, K_d is the derivative gain. These control parameters are adjusted based on Ziegler-Nichols' Ultimate Gain method.

5.3. Device control strategy

The block diagram of control system is shown in Figure 11. Through the acceleration sensor mounted on the human body, the human movement can be determined. At the time that the power assistance is needed during the human bends forward and lifts the load, the reference pressure Ps is determined through the posture to pressure algorithm, the controller will output the desired driving voltage on the quasi-servo valve, regulate compressed air, and then the required assistance force can be obtained.
Figure 12 shows flow chart of the control system. First of all, each one of the sensors detect inclination of the human body (θ) and air pressure value (p_ value). And EMG sensor is calculated the difference between the current value and the previous value (de). Next it decided to target air pressure (Pa) of the θ , it is to adjust the air puressure by using PID control. Finally if de is than 30, exhaust the air pressure of the actuators. If not, once again, decide each one of pressure values. Table 1 shows the target pressure of each inclination angle with the upper half part of the humam body.

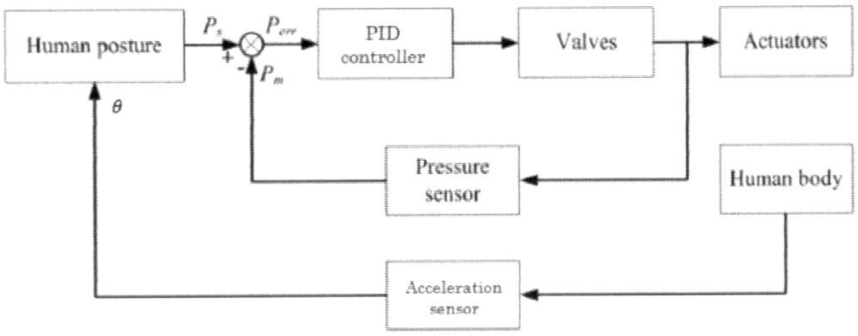

Figure 11. Block diagram of the control system

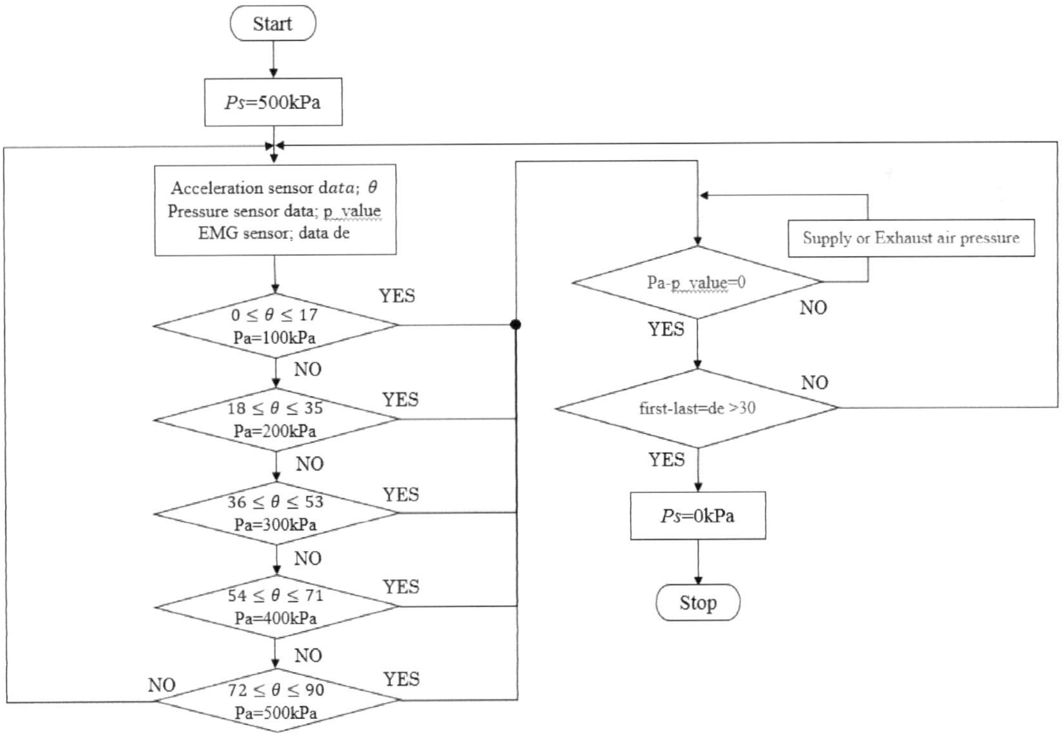

Figure 12. Flow chart of the control system

Table 1. The target pressure of each inclination angle with the upper half part of the body

Inclination angle [deg]	Pressure [kPa]
1≤ θ_T ≤17	100
18≤ θ_T ≤35	200
36≤ θ_T ≤53	300
54≤ θ_T ≤71	400
72≤ θ_T ≤90	500

5.4. Control experiment

Figure 13 shows an experimental setup for pneumatic control of an actuator. In this experiment, the compressed air that is input to the actuator is provided by an air compressor [CHST-25, ANEST IWATA Corporation]. The input air pressure is arranged using the improved control system. The air pressure that is input to the actuator is measured using a pressure sensor. The measured air pressure is input to a PC through an A/D board with the sampling time set to 20ms.

Experiment result is shows in Fig.14 (a), the desired air pressure that is input to the actuator are set to 100kPa, 200kPa, 300kPa, 400kPa, and 500kPa. Each lines shows the represent results of the PID control. From this experimental result, it can be seen that the air pressure reaches the target value. At the maximum pressure (500kPa), the air pressure is stable at about 2.5 seconds.

Experiment result (b) shows response of exhaust air pressure. The desired air pressure that is input to the actuator are set to 400kPa, 300kPa, 200kPa, and 100kPa. At the minimum pressure (100kPa), the air pressure is stable at about 1.7 seconds.

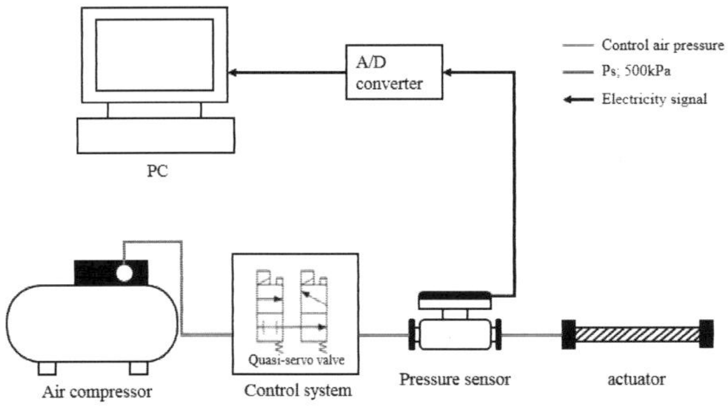

Figure 13. Experimental system

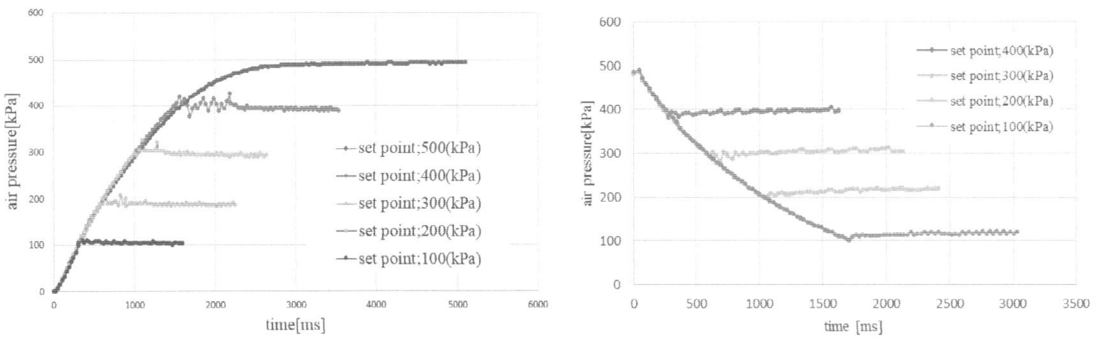

(a) Response of supply air pressure (100kPa~500kPa)

(b) Response of exhaust air pressure (400kPa~100kPa)

Figure 14. Response of PID control

IOP Conf. Series: Materials Science and Engineering **249** (2017) 012004 doi:10.1088/1757-899X/249/1/012004

5.5. Waterproof controller

This study makes control system for use in environment of high humidity. The basic concept for the waterproof of the control system is summarized as follows.

[1] Put into the waterproof case in order to use the control system in a high-humidity environment.

[2] This device was considered for use in such as the bathroom. Therefore, choose a case that not affected by the direct injection of water.

It is possible to [1] by miniaturized controller. For [2], it choose a case that not affected by the direct injection of water (From a distance of 3m, 100L/min, 100kPa).

6. Effect of support for human body with EMG signal

6.1. Movement of lifting with subject

In this study, we put the EMG amplifier on the erector spinae muscle to measure muscle activity. Figure 15 shows the subject movement of lifting. From the start of measurement of the EMG signal to 4[s], the subject maintains a vertical posture. Then the subject bends upper part of the body angle of inclination θ _T from 4[s] to 7.5[s] until it becomes 65 degrees. And let subject maintain the same posture (anteflexion posture) during until 10.5[s] from 7.5[s]. These movement is carried out with or without power assist device. In addition, subject always keep the angles of the knee back with 180 degrees because we have to make sure the EMG signal only from erector muscles of spine.

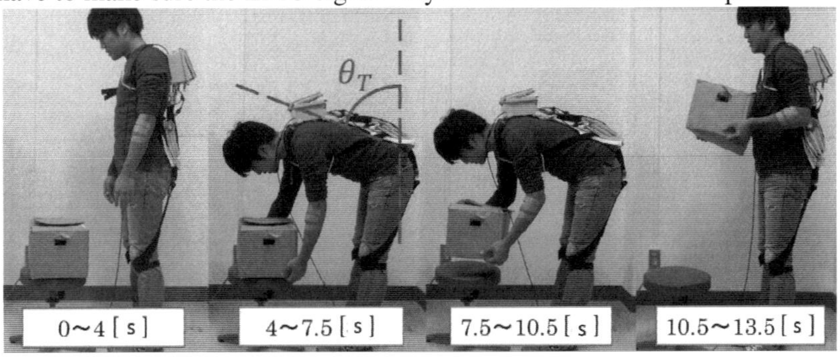

Figure 15. The subject movement of lifting

6.2. Activity amount of muscle comparison

For assess the prototype power assist wear effect, there are 3 subjects to maked the experiment. Subject 1(Height: 180cm, Weight: 82kg), Subject 2(Height: 175cm, Weight: 75kg), Subject 3(Height: 172cm, Weight: 64kg). Figure 16 shows the signal of EMG amplifier without assist wear and Figure 17 shows the signal with power assist wear. From two figures, we can see that the activity amount of muscle with power assist wear is smaller than the other date. Table 2 shows the result of the experiment.

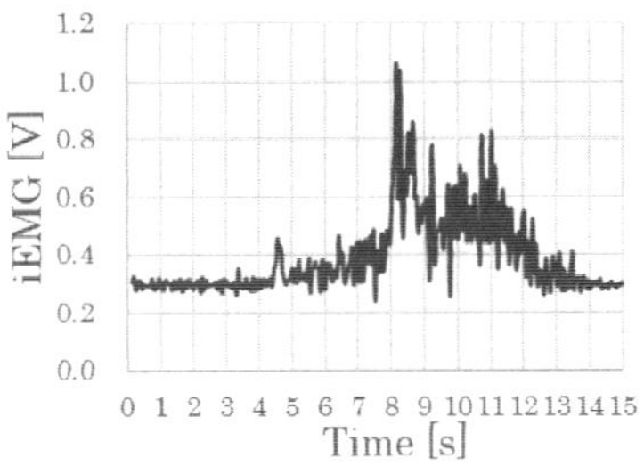

(a) Subject 1

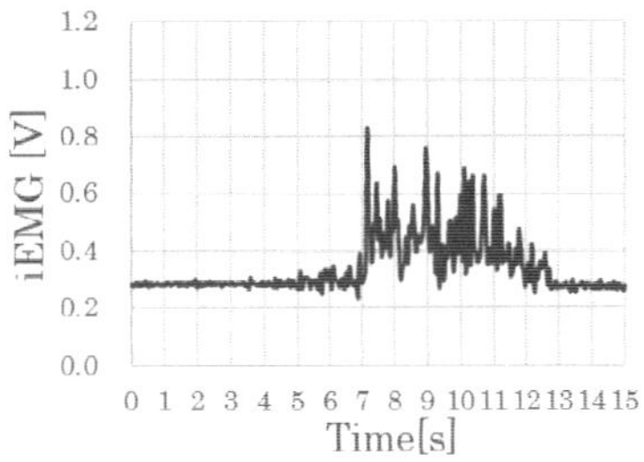

(b) Subject 2

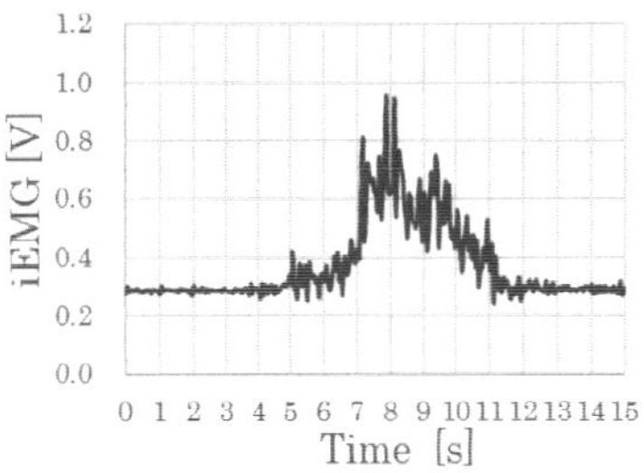

(3) Subject 3

Figure 16. The signal of EMG amplifier without assist wear

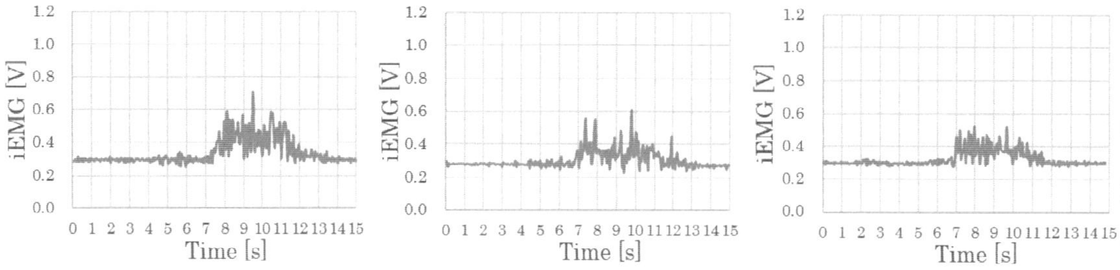

(a) Subject 1 (b) Subject 2 (3) Subject 3

Figure 17. The signal of EMG amplifier with assist wear

Table.2 Decrement of Maximum muscle activity and activity time

	Maximum muscle activity	Activity time [s]
Without assist wear	1.06	9.22(t_n)
With assist wear	0.71	7.06(t_a)
Decrement of Maximum muscle activity [%]	33.13	
$t_n - t_a$ [s]		2.16

7. Conclusions

In this study, we had proposed a power assist wear for low back support, a device using new types of pneumatic actuators. Compared with McKibben-type pneumatic rubber artificial muscle, the contraction rate of elongation-type pneumatic rubber artificial muscle is larger. As it does not use an exoskeleton structure, the device can be wear on the human body just like normal clothing. It can provide assistance force for the low back, reducing the possibility of LBP, and the assistance power of the device can be adjusted by changing the pressure of the compressed air. The effectiveness of the device has been verified through experiments.This study that is aimed to improve the pressure control performance using the quasi servo valve for Wearable Power Assist Wear. For the control system, we improved control system which enable free movement. The control system include quasi-servo valve for adjust the compressed air is light, and the mass is 250g. Because it is possible to equip the human body. Also we performed to construct a pneumatic flow controller supplied to the actuators by PID control method. The PID control parameters are adjusted based on the limit sensitivity method. The effectiveness of the improved control system has been verified through experiments. We also get the result of 3 subjects without and with power assist wear.

References

[1] Mineo Ishii, Keijiro Yamamoto, Kazuhito Hyodo. Stand-Alone Wearable Power Assist Suit–Development and Availability-, Journal of Robotics and Mechatronics Vol.17, No.5, pp. 575-583, 2005.

[2] Josephson M, Vingard E. Workplace factors and care seeking for low-back pain among female nursing personnel. MUSIC-Norrtalje Study Group. Scand J Work Environ Health 24:465-472, 1998.

[3] Eugene K. Wai, Darren M. Roffey, Paul Bishop, Brian K. Kwon, Simon Dagenais. Causal assessment of occupational bending or twisting and low back pain: results of a systematic review Review Article, The Spine Journal, Vol. 10, Issue 1, pp.76-88, 2010.

IOP Conf. Series: Materials Science and Engineering **249** (2017) 012004 doi:10.1088/1757-899X/249/1/012004

[4] Darren M. Roffey, Eugene K. Wai, Paul Bishop, Brian K. Kwon, Simon Dagenais. Causal assessment of workplace manual handling or assisting patients and low back pain: results of a systematic review, Review Article, The Spine Journal, Vol. 10, Issue 7, pp.639-651, 2010.
[5] Eugene K. Wai, Darren M. Roffey, Paul Bishop, Brian K. Kwon, Simon Dagenais. Causal assessment of occupational lifting and low back pain: results of a systematic review Review Article, The Spine Journal, Vol. 10, Issue 6, pp.554-566, 2010.of the paper.
[6] Xiangpan Li, Toshiro Noritsugu, Masahiro Takaiwa, Daisuke Sasaki, "Design of Wearable Power Assist Wear for Low Back Support Using Pneumatic Actuators", International Journal of Automation Technology, Vol.7, No.2, pp. 228-236, 2013.
[7] Toshiro Noritsugu, Masahiro Takaiwa, Daisuke Sasaki, "Development of Power Assist Wear Using Pneumatic Rubber Artificial Muscles", Journal of Robotics and Mechatronics, Vol.21, No.5, pp.607-613, 2009.

Parametric Optimization of Multi-Electrode Plasma Actuator with Serrated Electrode

T Matsuno[1], A Fukuda[1], H Kawazoe[1], K Nakai[2] and H Nishida[3]

[1] Department of Mechanical and Aerospace Engineering, Tottori University, Tottori, Tottori 680-8550, Japan
[2] Department Mechanical Systems Engineering, Tokyo University of Agriculture and Technology, Koganei, Tokyo 184-8588, Japan
[3] Institute of Engineering, Tokyo University of Agriculture and Technology, Tokyo 184-8588, Japan

E-mail: matsuno@mech.tottori-u.ac.jp

Abstract. The thrust performance of the multi-electrode plasma actuator has been investigated and optimized in terms of the configuration of the actuator element and the driving conditions, from the viewpoint of electrical mobility of ions. A trielectrode plasma actuator that has two exposed electrodes with serration in bilateral symmetry was used in this study. The induced body force of the plasma actuator was investigated on quiescent air, by measuring thrust of the induced jet. With the experimental results, the physical factors that affect the growth of thrust is considered, by estimating total momentum of ions from their drift velocity.

1. Introduction

Single dielectric barrier discharge (SDBD) plasma actuator, a flow control device that utilizes dielectric barrier discharge to generate wall-jet flow, demonstrate their potential in delaying separation and significant improvement in aerodynamic performances [1-3]. At the same time, due to the limitation of the physical mechanism of dielectric barrier discharge, it is difficult to increase plasma density to enhance the body force of the SDBD jet. In order to solve this problem, several types of the plasma actuators which derived from standard SDBD type have been proposed and developed.

One promising approach is the use of additional electrode to enhance discharge and assist acceleration of the generated ions from that discharge. Sosa et al. have tested a multi-electrode plasma actuator that has additional electrode to apply DC voltage [4, 5]. This type of the device is called trielectrode (TED) plasma actuator. The basic SDBD plasma actuator would consist of a pair of electrodes isolated by a dielectric material. One of the electrodes is exposed to the air and the other is buried in a dielectric material so never contacts to the air. The electrodes are arranged in a staggered manner that shares a single edge, as shown in figure 1(a). Figure 1(b) shows the general configurations of the trielectrode plasma actuator. We call the additional exposed electrode as "DC electrode" in this paper. The original exposed electrode for applying AC voltage is called as "AC electrode" to prevent confusion.

Characteristics of the TED plasma actuator have been investigated on the electric aspect [6] and improvement of tangential thrust compared to SDBD plasma actuator [7, 8]. These researches clarified that TED plasma actuator could induce a jet with more than ten times larger thrust compared to a SDBD plasma actuator at same AC voltage. The application of DC voltage extends the ionization region and produce counter jet at the edge of the DC electrode. The interaction of these jets results an upward vectored jet. In order to understand the working mechanism of TED plasma actuator with DC high voltage, Nishida et al. conducted numerical simulation for plasma and induced flow. Figure 2 shows the time variation of plasma distribution for the case of positive DC voltage application obtained by numerical simulation [9]. In this case, the electrons produced from the AC discharge are attracted by the DC electrode, then they trigger the ionization to generate positive ions at the DC electrode. The drift motion of the generated positive ions results in the negative body force. These results indicated that the distance between the exposed electrodes and AC voltage frequency should have a strong impact on the body force produced from the TED plasma actuator, since the number of electrons and the drift velocity of ions are determined by these parameters.

In this research, the impact of the frequency of the AC voltage and the distance between their exposed electrodes on the thrust characteristics of the TED plasma actuator has been estimated by using a simple model and investigated by the experiment.

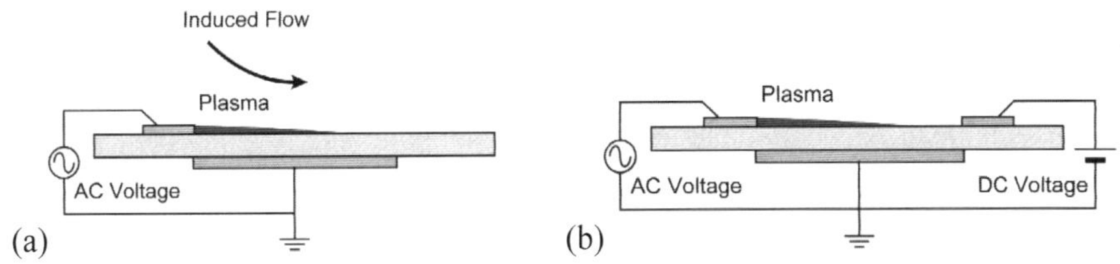

Figure 1. Schematic configurations of SDBD and the trielectrode plasma actuators: (a) SDBD plasma actuator, (b) Trielectrode plasma actuator.

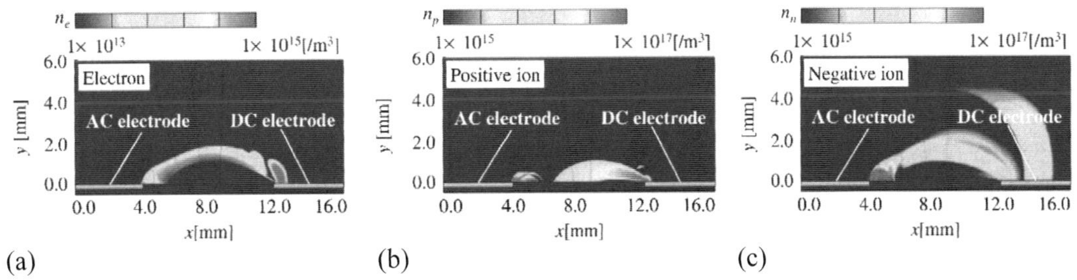

Figure 2. Time variation of the plasma distribution for the case of the DC voltage of 13.2 kV at a negative-going voltage half cycle: (a) electron density, (b) positive ion density, and (c) negative ion density [9].

2. Experimental setup

2.1. Plasma actuator

In this research, a trielectrode plasma actuator with serrated electrodes has been employed for experiment. The example image of the actuator and its serrated electrode are shown in figures 3 and 4. In this research, the distance between vertexes of the triangle, d_s, was fixed at 1mm. Other dimensions and specifications of the plasma actuator employed in this work are summarized in Table 1.

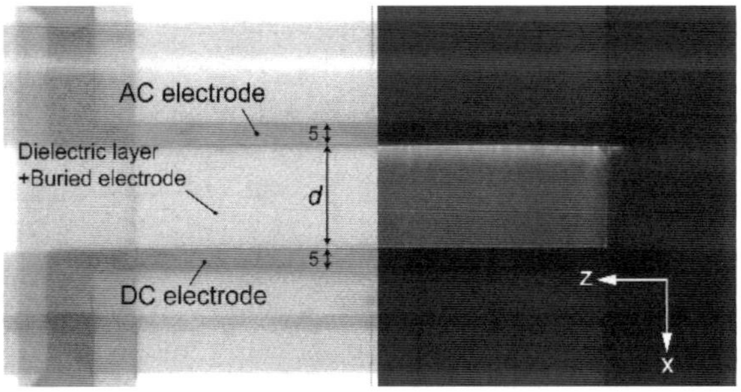

Figure 3. Photograph of the trielectrode plasma actuator element and its discharge.

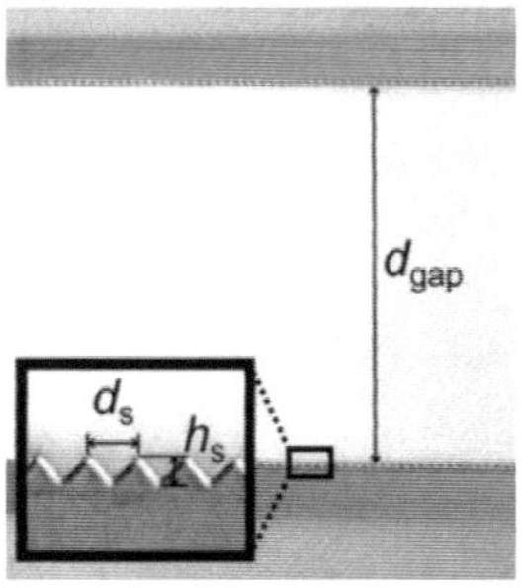

Figure 4. Image of the serrated exposed electrodes (triangular, 1mm spacing).

Table 1. Experimental conditions for thrust test of the trielectrode plasma actuator.

AC frequency: f_{ac}	4 ~ 18 kHz
AC voltage: V_{ac}	15.6 kVpp
DC voltage: V_{dc}	0 ~ 30.0 kV
Electrode material	Copper
Dielectric layer material	PTFE
Dielectric layer thickness	2 mm
Buried electrode width (Gap length): d_{gap}	40 ~ 80 mm
Exposed electrode width	5 mm
Overlap of electrodes	0 mm
Discharge length	200 mm
Distance between vertexes of serration: d_s	1 mm
Height of serration: h_s	0.5 mm

2.2. Power supply
Figure 5 shows a schematic of the power supply of plasma actuator used in this study. A reference waveform of a high-voltage AC input was generated by a function generator and amplified by a solid-state high power amplifier, which increases input power up to 400 W, with the amplitude of 70 V_{pp}. By a high voltage transformer, the AC voltage attains amplitude up to 30 kV at a frequency of 4 ~ 18 kHz. The voltage and current of AC input was monitored by a digital oscilloscope. DC voltage was applied directly from the high voltage power supply (Matsusada Precision HAR 30), which can generate up to 30 kV with 10 mA of output.

2.3. Thrust measurement
In this research, the thrust of the induced jet by the plasma actuator was measured, as a reaction of the aerodynamic force from the actuator, and was evaluated as an index of the flow control performance. The schematic of the apparatus for the thrust measurement is illustrated in figure 6. The thrust from plasma actuator was sensed by an analytical balance (Shimadzu AUW320) with a lever. Since the TED plasma actuator generates vectored jet, we need to measure the magnitude of the thrust and their angle.

In this experiment, two types of mounting base for actuator element were used to measure horizontal (tangential to the actuator surface) and vertical (perpendicular to the surface) thrust component separately, as shown in figure 6(a) and 6(b). The direction of the horizontal thrust was set as same as the thrust generated by SDBD actuator. The reference of the thrust was set at the centerline of exposed electrodes on the dielectric surface.

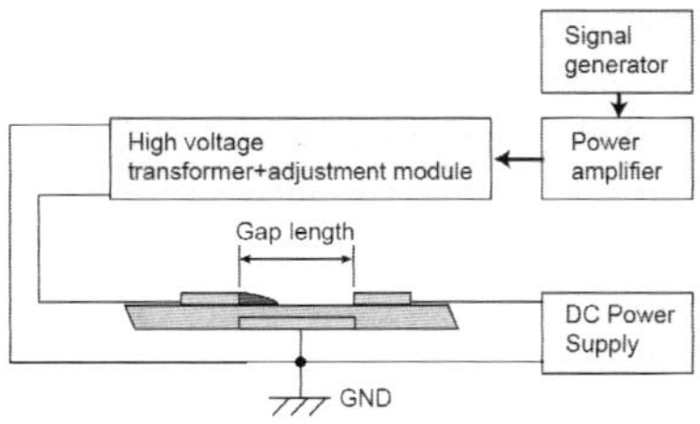

Figure 5. Connection diagram of the power supply for TED plasma actuator.

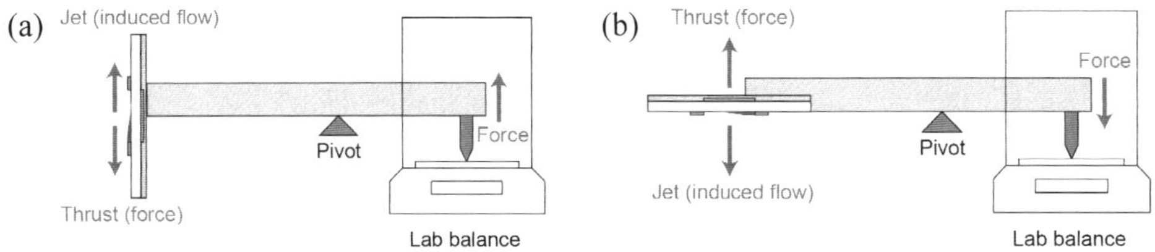

Figure 6. Schematics of thrust measurement system: (a) horizontal thrust measurement, (b) vertical thrust measurement.

3. Results and discussion

3.1. Estimating momentum of ions from drift velocity

The physical factors that affect the body force from plasma actuator are considered. The fundamental physical mechanism to produce induced jet is the drift motion of ions from the discharge as shown in figure 7. The momentum transfer from the ions to the neutral molecules through collisions produces the body force which induces the fluid motion to generate jet. In this process, the body force must have a causal relation to the total momentum of drifting ions. The total momentum is the product of the total mass of the ions and their drift velocity. The total mass of the ions can be represented by a number of ions from discharge, and the number of ions has a linear relation with the number of micro-discharge in each AC cycle. The report pointed [9] that the number of the micro-discharge is a function of a rate of AC voltage change over time, dV_{ac}/dt.

For the drift velocity of ions, it can be calculated with ion mobility and the electric field strength, assuming uniform electric field;

$$v_{\mathrm{i}} = \mu_{\mathrm{i}} E$$

The magnitude of the electric field strength was estimated from the RMS of the electric field between the electrodes;

$$E_{RMS}[\mathrm{V/m}] = \sqrt{\frac{1}{T}\int_0^T \left(\sin \omega t - \frac{V_{\mathrm{dc}}}{V_{\mathrm{ac}}}\right) dt} \cdot \frac{V_{\mathrm{ac}}}{d_{\mathrm{gap}}}$$

The ion mobility is set to be constant; μ_i = 20.0/p m^{2}V^{-1}s^{-1}, where p is the ambient gas pressure (1.013 x 10^5 Pa).

Figure 8 shows the calculated contours of the momentum of ions. The magnitude of momentum was simply estimated from the production of the ion drift velocity, v_i, and the nominal rate of the AC voltage change dV_{ac}/dt. These were normalized by the maximum value in the calculated conditions. The increase in the frequency relates directly to the rate of the AC voltage change. The increase in dV_{ac}/dt causes a growth in the number of micro-discharge, which results the increase of the number of ions. Figure 9 shows the effect of the gap length on the drift velocity of ions at the same conditions. For the gap length, the smaller gap length makes higher electric field strength, which leads higher drift velocity of ions.

Figure 10 shows the estimated drift velocity plotted against the DC voltage for various gap lengths. The trend is as expected; the larger drift velocity can be achieved at the high DC voltage. The effect is primarily same as the reduction of the gap length. This simply interprets the mechanism to enhance the jet on the AC electrode at high DC voltage. The plot also presents the trend that the change in the drift velocity becomes small as the gap length increases, even in the high DC voltage region. This is because the impact of the difference in the gap length on the electric field is diminished at large gap length.

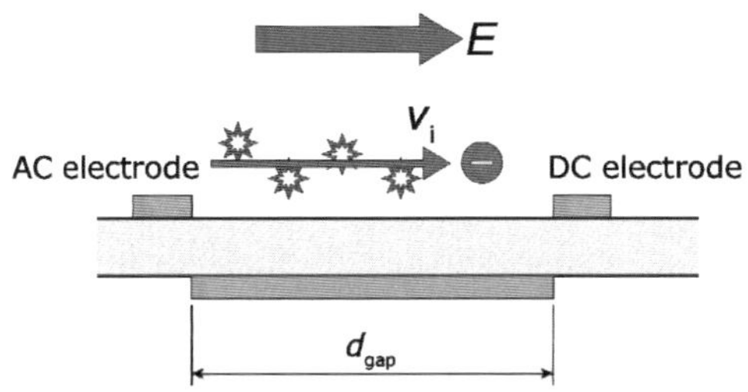

Figure 7. Schematic of ion drift between the exposed electrodes of trielectrode plasma actuator.

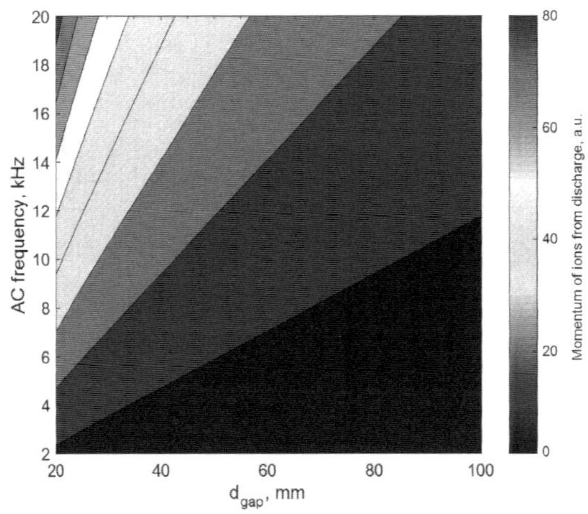

Figure 8. Calculated contours of the momentum of ions estimated from the rate of AC voltage change over time and the drift velocity.

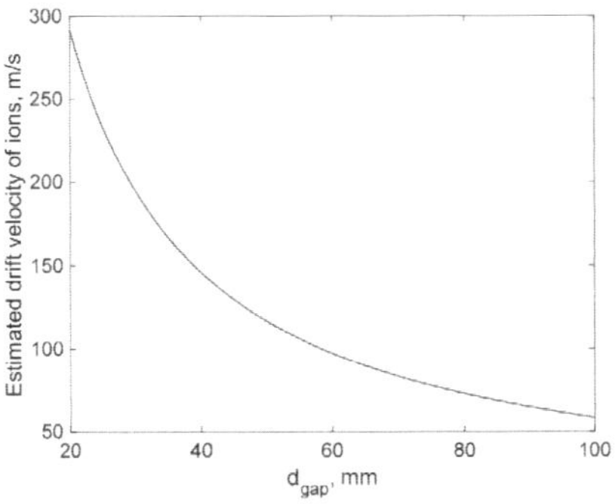

Figure 9. Effect of the gap length on the estimated drift velocity of ions.

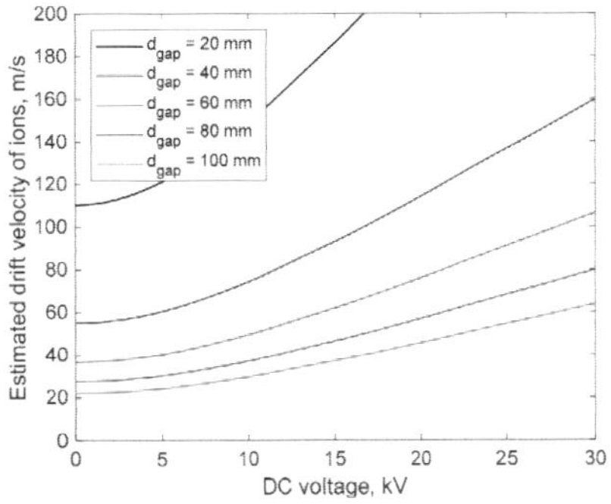

Figure 10. Estimated drift velocity plotted against the DC voltage for various gap lengths, at V_{ac} = 15.6 kV_{pp}.

3.2. Thrust of the induced jet for various AC frequencies

Figure 11 presents the vertical thrust component of the induced jet from the TED plasma actuator for various AC frequencies. Figure 11(a) is the case for the distance between two exposed electrodes (gap length, d_{gap}) was set at 40 mm, and (b) and (c) represent the case for d_{gap} = 60 mm and 80 mm, respectively. To discuss the basic characteristics, we focus on the case of the applied AC voltage was set at V_{ac} = 15.6 kV_{pp} in this paper.

As pointed by the previous research [5], the vertical thrust component starts to rise when the discharge between the electrodes starts at a certain DC voltage. The discharge leads the negative body force at the DC electrode, resulting in the generation of the counter jet to the standard SDBD plasma jet. Once the threshold DC voltage is exceeded, the vertical component of the thrust increases until the maximum DC voltage at which the irregular arc discharge starts between the electrodes. In all cases, the positive vertical thrust, which indicated vertical jet from the surface of the actuator, was observed in the present experimental results.

From figure 11(a) ~ (c), we can observe the effect of the gap length on the thrust characteristics. When d_{gap} is small, the threshold DC voltage to start the counter jet generation is relatively low. For d_{gap} = 40 mm, this threshold voltage corresponds to 14~16 kV, while it becomes >17 kV at d_{gap} = 80 mm. The growth rate of the thrust is relatively large at d_{gap} = 40 mm, then decreases as d_{gap} becomes large.

Then, the effect of the frequency of AC voltage is discussed. From figure 11(a), the difference of the frequency can be observed. At a high AC frequency (f_{ac} = 16 ~ 18 kHz), the threshold DC voltage is small, and the vertical component of the thrust linearly increases at a relatively high rate. At slightly lower (f_{ac} = 13 ~ 15 kHz) frequency, the overall characteristics is same as the high AC frequency cases, however the thrust at the same DC voltage was lowered about 20 %. This is due to the delay of the threshold DC voltage, since the growth rate of the thrust is not changed in this frequency range. On the other hand, at the low frequency (f_{ac} = 4 ~ 7 kHz) the thrust characteristics showed significant change. The threshold voltage did not change largely, however the growth rate of the thrust became small and nonlinear as the frequency was decreased.

The effect of the frequency is moderated as the gap length increases. Even the case d_{gap} = 80 mm, small difference in the thrust curve is observed, and the higher frequency generates higher thrust. However, the growth rate of the thrust is entirely deteriorated, and the difference in the thrust is diminished, as presented in figure 10.

IOP Conf. Series: Materials Science and Engineering **249** (2017) 012005 doi:10.1088/1757-899X/249/1/012005

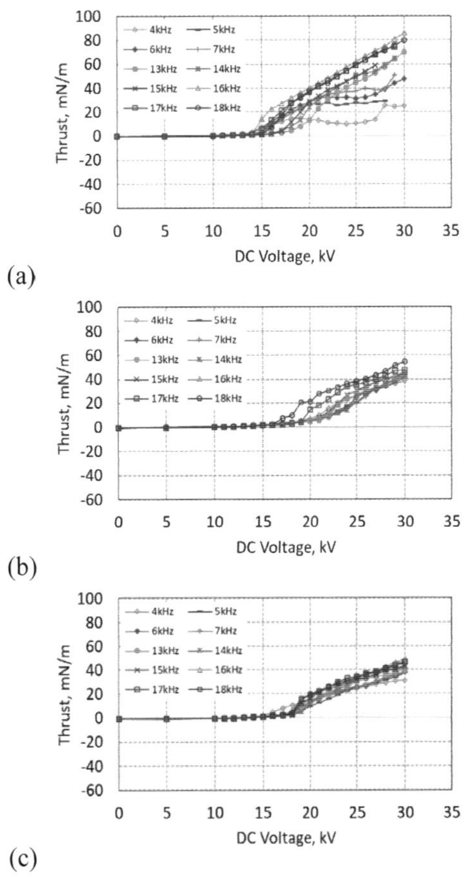

Figure 11. Vertical thrust component versus DC Voltage at V_{ac} = 15.6 kV_{pp} for various AC frequencies: (a) d_{gap} = 40 mm, (b) 60 mm, and (c) 80 mm.

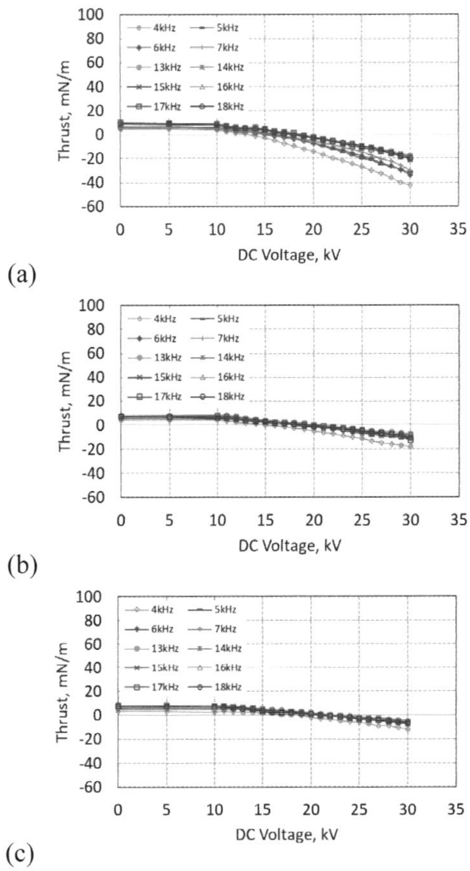

Figure 12. Horizontal thrust component versus DC Voltage at V_{ac} = 15.6 kV_{pp} for various AC frequencies: (a) d_{gap} = 40 mm, (b) 60 mm, and (c) 80 mm.

Figure 12 shows the horizontal component of the thrust of the induced jet from the TED plasma actuator for various AC frequencies. Figure 12(a) ~ (c) represent the case for d_{gap} = 40, 60 and 80 mm, respectively. At V_{dc} = 0, which represents the SDBD condition, the thrust exhibited 7.45 ~ 8.85 mN/m in the current cases. The horizontal thrust decreased and become negative as the counter jet from the DC electrode produced. As expected from the results shown in figure 11, the horizontal thrust component shows that the effect of the gap length is moderated as d_{gap} is increased, as well as for the effect of the frequency.

Figure 13 presents the deflected angle of the thrust vector calculated from figures 11 and 12 versus applied DC voltage. At small DC voltage, the jet is almost parallel to the dielectric surface as SDBD jet. The direction of the induced jet is rapidly changed upward as the DC voltage rise at V_{dc} is from 10 to 20 kV. Once the angle is reached to 90 degrees, the direction shows small change even the DC voltage is increased for the most of the cases.

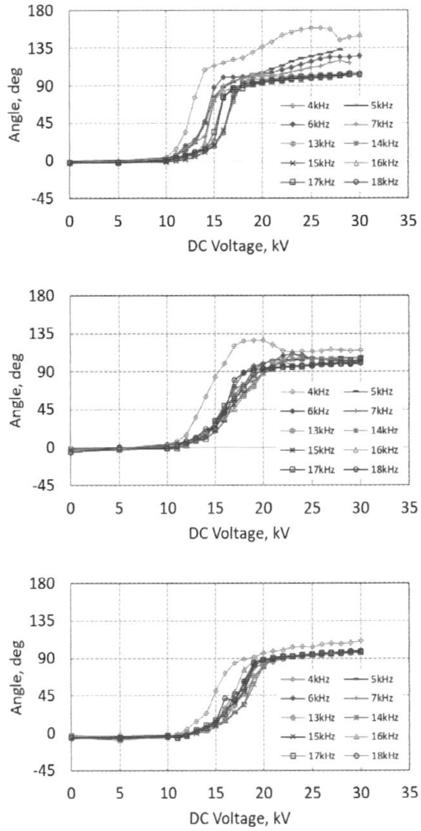

(a)

(b)

(c)

Figure 13. Angle of thrust vector from the wall at V_{ac} = 15.6 kV$_{pp}$: (a) d_{gap} = 40 mm, (b) 60 mm, and (c) 80 mm.

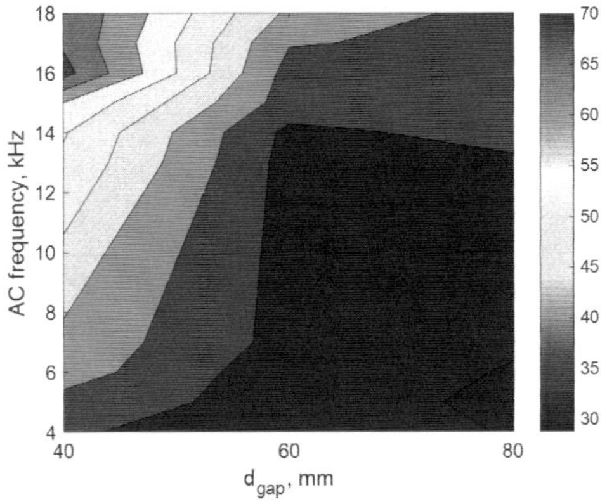

Figure 14. Contour plot of the thrust about d_{gap} and AC frequency at V_{dc} = 27 kV.

The impact of the gap length and the frequency is analyzed with the DC and AC voltage is fixed. Figure 14 shows the same results as in figure 8 as contour map of the thrust at The DC voltage of 27 kV. V_{dc} = 27 kV is the maximum conditions where all of the cases have the data without irregular arc discharge. This again illustrates the effect of the gap length. The impact of the gap length on the thrust becomes significantly small at a large d_{gap}. At a small d_{gap}, the thrust shows rapid increase as the frequency rise. The trend of the contour map of the thrust about these two parameter is quite similar to the estimated momentum shown in figure 8. This analogy indicates that the characteristics of the TED plasma actuator can be predicted from those two physical parameters (the rate of AC voltage change and the drift velocity of ions) and the gap length and the AC voltage frequency affect them as modeled at section 3.1.

Figure 15 shows the total thrust of the induced jet plotted against the power consumption of the driving apparatus for various gap lengths. As expected from the estimation, the input power rises as AC frequency increased, for all gap length cases. Nevertheless, the power efficiency is significantly improved by utilizing the serrated-edge TED plasma actuator, especially when the larger thrust is generated. At this case, the thrust

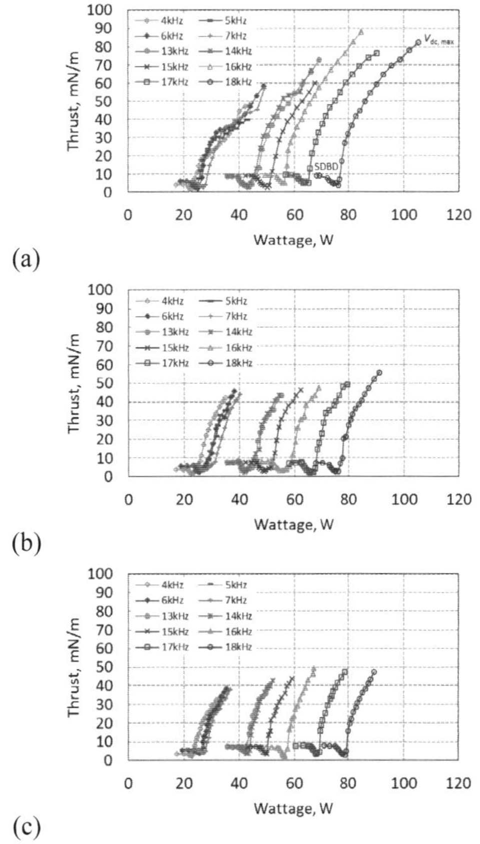

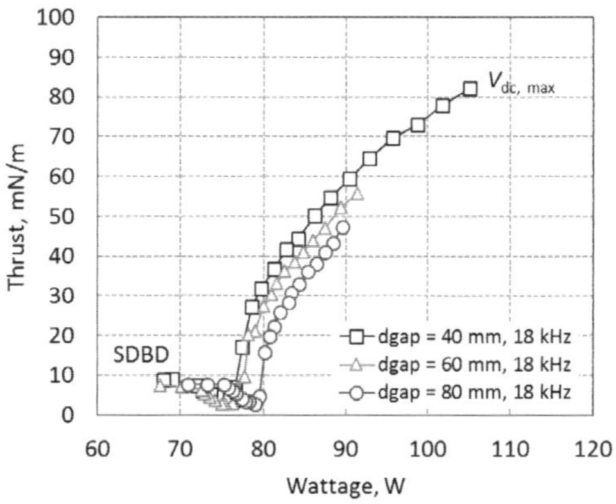

Figure 15. Thrust versus power consumption at $V_{ac} = 15.6$ kV$_{pp}$: (a) $d_{gap} = 40$ mm, (b) 60 mm, and (c) 80 mm.

Figure 16. Total thrust versus power consumption at $f_{ac} = 18$ kHz.

becomes 830 percent larger than that of the SDBD plasma actuator, by additional 55 percent of the input power.

Figure 16 compares the total thrust produced as a function of the power consumption for each gap length. All of the cases shows similar curve, as their thrust rapidly rise at a same rate as the power increases at high-thrust regime, at which the peculiar discharge of the trielectrode plasma actuator occurs. The plot also shows that small gap length could produce larger thrust at the same input power. The wattage at the maximum V_{dc} was largely different even at the same DC voltage ($V_{dc} = 30$ kV), however, this condition still shows the highest thrust per unit input power in all of the conditions tested.

4. Summary

The impact of the frequency of the AC voltage and the distance between their exposed electrodes (gap length) on the thrust characteristics of the TED plasma actuator have been investigated experimentally. The induced body force of the serrated TED plasma actuator is investigated on quiescent air. In addition to that, the physical factors that affect the growth of thrust is considered by estimating total momentum of ions from the drift velocity.

From the results of the thrust measurement, the effect of the gap length and the frequency on the thrust characteristics has been clarified. The impact of the gap length on the thrust becomes significantly small at large gap. At small gap, the thrust shows rapid increase as the frequency rise. The power efficiency is significantly improved. The thrust becomes 830 percent larger than that of the SDBD plasma actuator, by additional 55 percent of the input power. The basic characteristics of the TED plasma actuator can be predicted from two physical parameters, the rate of the AC voltage change and the drift velocity of ions, and they are affected by the gap length and the AC voltage frequency.

Acknowledgments
This research was partially supported by KAKENHI (23760771, 15K05797) from JSPS, JAPAN.

References

[1] Corke T C, Post M L and Orlov D M 2007 SDBD plasma enhanced aerodynamics: concepts, optimization and applications *Prog. Aerosp. Sci.* **43** 193–217
[2] Wang J-J, Choi K-S, Feng L-H, Jukes T N and Whalley R D 2013 Recent developments in DBD plasma flow control *Prog. Aerosp. Sci.* **62** 52–78
[3] Matsuno T, Maeda K, Yamada G, Kawazoe H and Kanazaki M 2013 Improvement of Flow Control Performance of Plasma Actuator Using Wind-Tunnel Test Based Efficient Global Optimization *31st AIAA Applied Aerodynamics Conference* 1–10
[4] Sosa R, Artana G, Moreau E and Touchard G 2007 Stall control at high angle of attack with plasma sheet actuators *Exp. Fluids* **42** 143–67
[5] Sosa R, Arnaud E, Memin E and Artana G 2009 Study of the flow induces by a sliding discharge *IEEE Trans. Dielectrics and Electrical Insulation* **16** 305–11
[6] Lago V, Grondona D, Kelly H, Sosa R, Marquez A and Artana G 2009 Sliding discharge optical emission characteristics *IEEE Trans. Dielectrics and Electrical Insulation* **16** 292–8
[7] Matsuno T, Kawaguchi M, Fujita N, Yamada G and Kawazoe H 2012 Jet Vectoring and Enhancement of Flow Control Performance of Trielectrode Plasma Actuator *6th AIAA Flow Control Conference*
[8] Matsuno T, Sugahara M, Koyama J, Fujita N, Yamada G and Kawazoe H 2016 Vectored Jet Control for Trielectrode Plasma Actuator with Serrated Electrode *Trans. Japan Soc. Aeronaut. Sp. Sci. Aerosp. Technol. Japan* **14** Pe_55-Pe_61
[9] Nishida H, Nakai K and Matsuno T 2017 Physical Mechanism of Tri-Electrode Plasma Actuator with Direct-Current High Voltage *AIAA J.* **55** 1852–61

Plasma-Assisted Control of Mach-2 Flowfield over Ramp Geometry

Yasumasa Watanabe[1,2*], Sergey B. Leonov[2], Alec Houpt[2], Brock E. Hedlund[2], and Skye Elliott[2]

[1]Department of Aeronautics and Astronautics, Graduate School of Engineering, The University of Tokyo, 7-3-1 Hongo, Bunkyo-ku, Tokyo 113-8656, Japan.
[2]Department of Aerospace and Mechanical Engineering, University of Notre Dame, Notre Dame, IN 46556, The United States.

E-mail: watanabe@daedalus.k.u-tokyo.ac.jp

Abstract. This study examined the effect of Reynolds number on plasma-assisted flow control ahead of a compression ramp geometry in Mach-2 supersonic flow. The experiments were conducted in the supersonic wind tunnel SBR-50 at the University of Notre Dame. Stagnation temperature and pressure were varied as T_0=294-500K and P_0=1-3bar to attain Reynolds number ranging from $3.4x10^5$-$2.2x10^6$. Ramp pressure measurements, schlieren visualization, and high-speed camera imaging were used for the evaluation of plasma-assisted flow control effects. A linear dependency was found between the ramp pressure change per averaged plasma power and Reynolds number.

1. Introduction

Plasma-based flow actuators are gaining a good deal of research attention. When plasma-assisted flow actuation is applied to surface flow control on a supersonic vehicle, it can modify the aerodynamic characteristics of said vehicle. It has several advantages over conventional mechanical actuators [1-3], two of which are a short response time due to electrical actuation that will realize quick flow-control operation and the simplicity of the device structure will prevent any mechanical failure by eliminating movable parts. Especially in case of high speed vehicles such as supersonic transports, the quick-response feature will be highly beneficial to the prevention of delay in generating an aerodynamic moment. In case of supersonic transports, a vehicle will travel several kilometers if the aerodynamic control is delayed for one second, and it will deviate from the designated orbit. However, if the delay in aerodynamic control can be reduced by plasma-assisted flow control, it will be highly useful in keeping the desired orbit and thereby minimize fuel consumption for trajectory modification. Therefore, plasma-based aerodynamic control is considered to be promising in realizing efficient and more secure flight.

Flow control techniques in supersonic flow were examined by Leonov et al. and they found that the near-surface electrical discharge was useful for surface flow modification, shock formation, and flow modification purposes [3-5]. They also found great utility in combustion and flameholding inside scramjet engines [6, 7]. Therefore, it is expected that plasma-based flow control is promising both for internal flow control and for flow modification around a supersonic vehicle. In this work, the main research interest is plasma actuation ahead of a body-flap geometry, which has been found to be beneficial to the pitching moment control of high-speed vehicles. Watanabe and Suzuki investigated the Mach-7 hypersonic flow control ahead of 20-degree ramp geometry [8, 9]. A schematic drawing of this plasma-assisted flow control is illustrated in Fig. 1. In these studies, plasma actuation was found to be beneficial when a pair of electrodes were installed ahead of ramp geometry, which is a simplified form of space vehicle and its body flap. As in Fig. 1(a), electrodes are located ahead of flap/ramp geometry to generate plasma that leads to flow separation and subsequently changing the pressure distribution over flap/ramp geometry as illustrated in Fig. 1(b). When plasma actuation was applied to the 20-degree ramp geometry, pitching moment acting on the ramp surface can be controlled by 10-20 percent within 0.09 seconds. Hence, it is clear that plasma-based flow actuation is highly effective in high-speed flow.

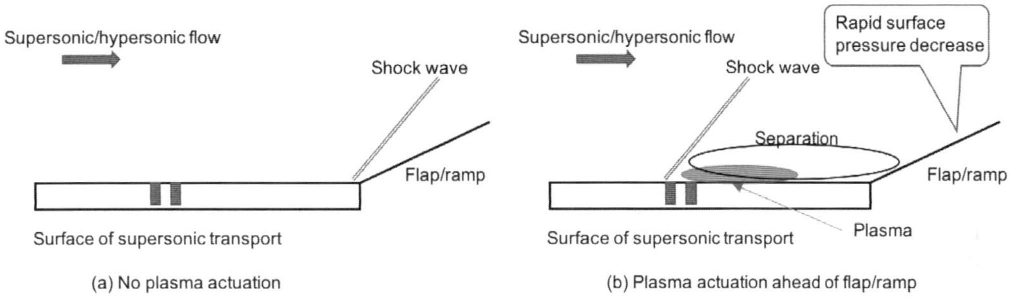

Figure 1. Plasma-assisted aerodynamic control ahead of flaps for supersonic/hypersonic transport.

Although a number of studies have focused on plasma-assisted supersonic flow control, most of them are done on the lab-scale basis, and thereby it is still necessary to clarify the scale effect, i.e. Reynolds number effect, before applying the plasma-assisted flow control method to practical-scale problems.

In light of the background above, experimental investigation was carried out in Mach-2 supersonic flow. The objective of this work is to examine the effect of Reynolds number on plasma-assisted flow control ahead of a ramp by correlating Reynolds number with parameters such as plasma power and ramp pressure change.

2. Experimental facility and instrumentation

2.1. Supersonic wind tunnel

The experiments were carried out in the supersonic blow-down wind tunnel SBR-50 at the University of Notre Dame. A photographic view of the wind tunnel facility is shown in Fig. 2. The test section is connected to a vacuum tank. A Mach-2 supersonic nozzle, plenum chamber with an ohmic heater, and high-pressure reservoir are connected to the upstream side of the test section. The plenum chamber pressure can be adjusted by a wind tunnel control panel, but the actual stagnation pressure is identified based on the static pressure at the nozzle exit. In the present configuration, Mach-2 supersonic flow is realized at the test section, which has 76.2 mm x 76.2 mm rectangular cross section at the nozzle exit and a 1-degree inclination along the upper and bottom walls to avoid choking the flow. The wind tunnel offers the stagnation temperature from room temperature up to 800 K. The stagnation pressure can be controlled ranging from 0.1 to 5 bars. The flow duration is 1 second at maximum. In the present work, the stagnation pressure P_0 was changed from one atmospheric pressure up to 3 bars and the stagnation temperature T_0 was varied from 294 to 500 K. By changing stagnation temperature and pressure, various Reynolds number conditions were achieved with Reynolds number ranging from 3.4 x 10^5 to 2.2 x 10^6 based on the characteristic distance, 55 mm, measured from the nozzle exit to the leading edge of the ramp surface along the flat test-section wall. While the flow at the test section was steady, plasma-assisted flow control was applied for 0.1 second.

Figure 2. Supersonic wind tunnel SBR-50 at the University of Notre Dame.

2.2. Ramp model

Schematics of the ramp model are illustrated in Fig. 3. It has 11 electrodes flush mounted on the wall in the transverse direction, 5 of which are anodes and the others are cathodes. A 20-degree ramp with 12 mm height was attached with the leading edge 30 mm downstream of the electrodes as illustrated in Fig. 3(a). The ramp surface was made of an alumina ceramics plate with 1-mm thickness. It has pressure ports along the center line, with their location indicated in Fig. 3(a). The ramp angle was set to 20 degrees to match up with a previous experiment in reference [9] that was tested at higher Mach number.

The ramp model was fixed in the test section as shown in Fig. 3(b). Both side walls are made of quartz and visualization was conducted through these windows. There are several pressure ports over the surface of the wall and ramp.

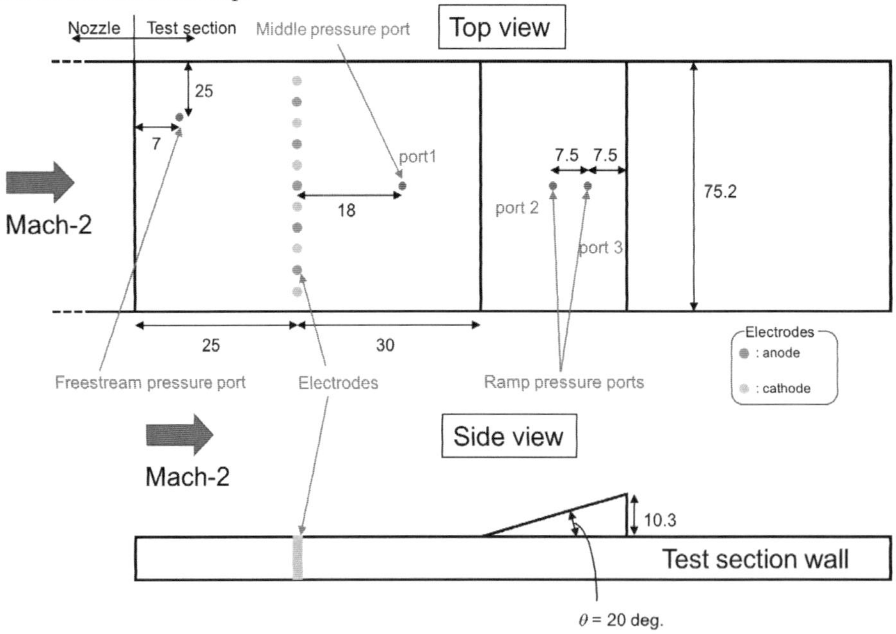

(a) Geometry of ramp model (units are in millimeters).

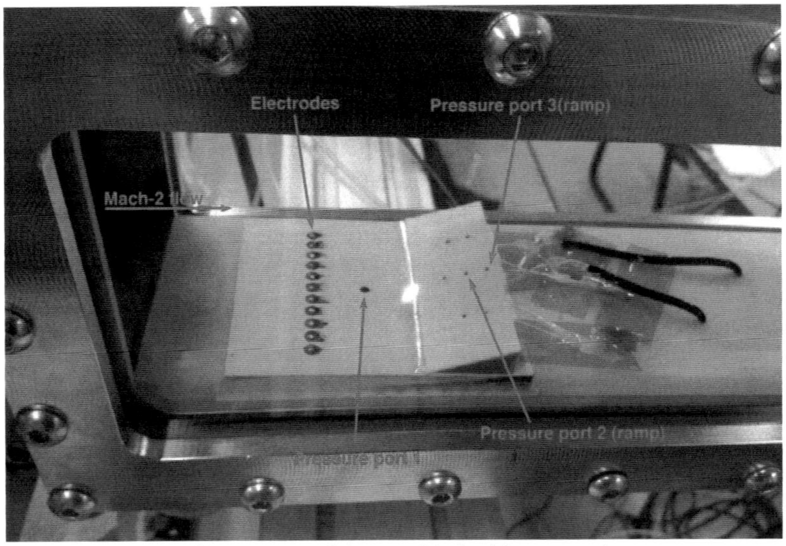

(b) Ramp model assembled with test section.
Figure 3. Ramp model.

2.3. Measurement
Flow visualization, pressure measurement, and measurement of plasma power were conducted in the wind tunnel tests for the purpose of clarifying changes to the flowfield.

2.3.1. Visualization
Flowfield and shock wave visualization were conducted by schlieren method in order to examine the change of flow structure induced by plasma. High-definition images were captured using a Basler camera with 2048×2048 pixel sensor at a rate of 133 fps. Although the imaging rate was low for supersonic flow, the exposure time for each image was <1 μs. This was achieved using a pulsed NIR laser diode. Using this method essentially "freezes" the flow allowing for a clear instantaneous view of turbulent structures within the flow.

2.3.2. Pressure measurement
Pressure measurements were made using a differential pressure sensor, TE NetScanner™ Model 9116. The reference pressure of this sensor was kept at the atmospheric pressure. Pressure data were recorded at the sampling rate of 400 Hz. In this work, there were 4 pressure ports which were especially important in evaluating the effect of plasma actuation: freestream pressure port, middle pressure port (port 1), and ramp pressure ports (port 2 & 3) as illustrated in Fig. 3(a). The freestream pressure port provides pressure at the nozzle exit, which is used to identify the stagnation pressure value by assuming the designed Mach number of the nozzle to be 2.0. The middle pressure port 1, located between electrode array and the ramp provides pressure value just at the downstream side of electrodes. Pressure ports 2 and 3 are located at the ramp surface. Pressure at these points is the direct indicator of plasma actuation effect over ramp geometry.

2.3.3. Plasma power measurement
The custom-made power supply [7] for plasma generation in the SBR-50 wind tunnel was used in the present flow-control experiments. Simplified schematics of the electrical circuit and typical waveforms of voltage traces are shown in Fig. 4. In this circuit, a high-voltage power supply and high-power capacitors are connected to 5 anodes through current-limiting resistors, R_a, as illustrated in Fig.

4(a). 6 cathodes are wired to the ground level through current-limiting resistors, R_c, which are also illustrated in Fig. 4(a). Total voltage supplied from the power supply circuit, denoted by V_{ps}, was recorded with an oscilloscope, Agilent Technologies DSO-X 4024A, through high-voltage probes at 800,000 samples per second. Anode and cathode voltage, V_a and V_c, were also measured together with V_{ps} and typical voltage traces are shown in Fig. 4(b). These voltage values were used to identify averaged discharge power. In this work, averaged plasma power P_{pl} was calculated as follows: (1) calculate total current I based on anode current I_a as $I = N_a\, I_a = N_a\,(V_{ps} - V_a) / R_a$ where $N_a = 5$ is the number of anodes and $R_a = 1$ kΩ is the resistance of each anode resistor. It is also possible to find the total current I based on cathode current, but it is less accurate because the wind tunnel facility itself is at the ground level voltage and a small portion of the current is lost by going through the metal walls of the wind tunnel. Therefore, in this work, total current I was calculated based on the anode current. (2) Then, plasma power at each moment W_{pl} is calculated by $W_{pl} = (V_a - V_c)\, I_a$. (3) By integrating W_{pl}, deposited plasma energy $E_{pl} = \int W_{pl}\, dt$ is obtained. Since the slope of E_{pl} during plasma actuation was found to increase almost linearly, the averaged plasma power P_{pl} was finally calculated by applying linear fitting to E_{pl} plot. In case of typical plot shown in Fig. 4(b), the averaged plasma power was found to be 6.3 kW.

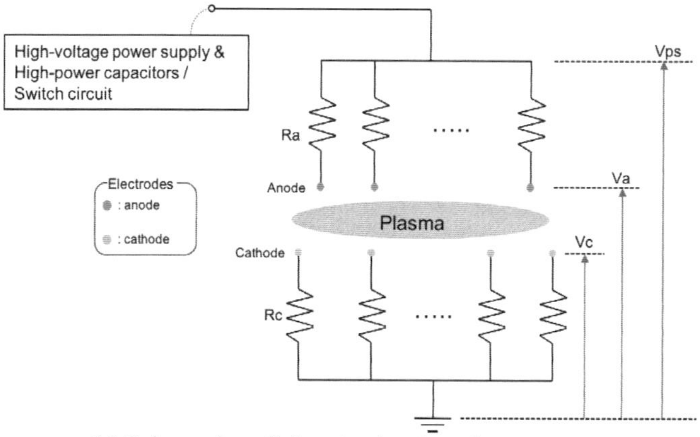

(a) Schematics of electric circuit and parameters.

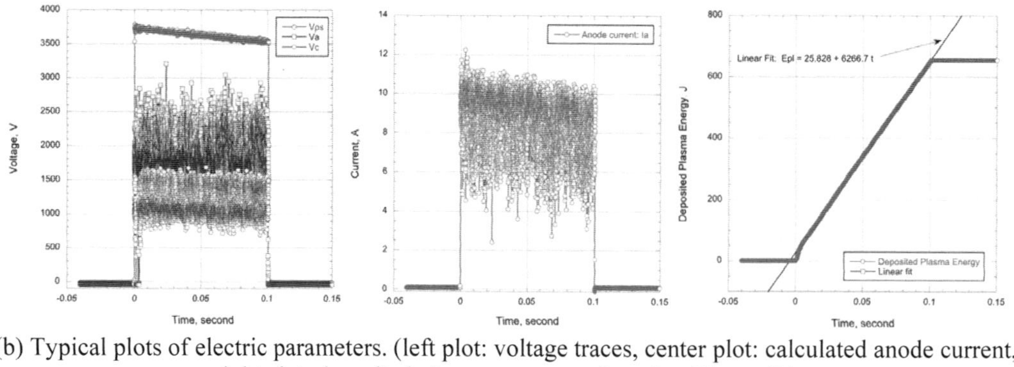

(b) Typical plots of electric parameters. (left plot: voltage traces, center plot: calculated anode current, right plot: deposited plasma energy and result of linear fit)

Figure 4. Schematics of circuit and typical traces of electric parameters.

3. Experimental result and discussion

The wind tunnel tests were conducted by varying stagnation pressure, stagnation temperature, and power supply voltage to attain various Reynolds number conditions. In the present study, stagnation temperature and pressure were varied as T_0=294-500K and P_0=1-3bar to attain Reynolds number ranging from 3.4×10^5 to 2.2×10^6. Quasi-steady plasma actuation results were observed in all test cases.

A pictorial view of a typical plasma-based flow control test is shown in Fig. 5. It is clearly seen that bright discharge zone was generated ahead of the ramp and the purple light emission from the vibrational mode of nitrogen molecules was observed during plasma actuation.

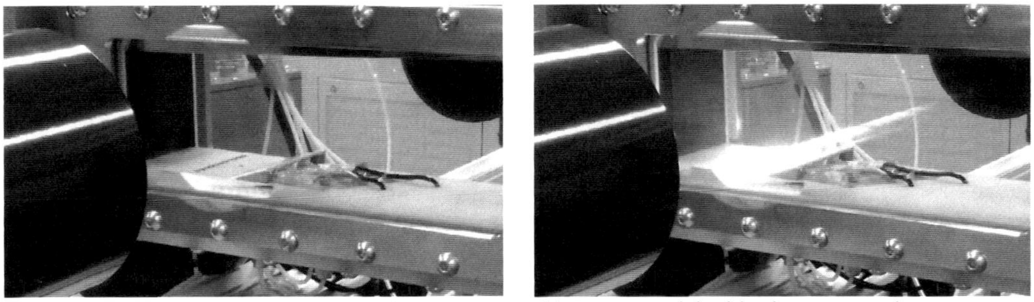

(a) Ramp model without plasma actuation. (b) Ramp model with plasma actuation.

Figure 5. Photographic view of typical plasma-assisted flow control test.

A snapshot image of a typical plasma filament, observed between electrodes and the ramp, is shown in Fig. 6. Periodical change in the transverse plasma filament location was confirmed in the images, but the change in streamwise plasma filament position was negligible. This can be because there was a separation zone formed ahead of the ramp which is on the downstream side of the electrodes. Therefore, the transverse filament was trapped inside the separation and exhibited periodical change.

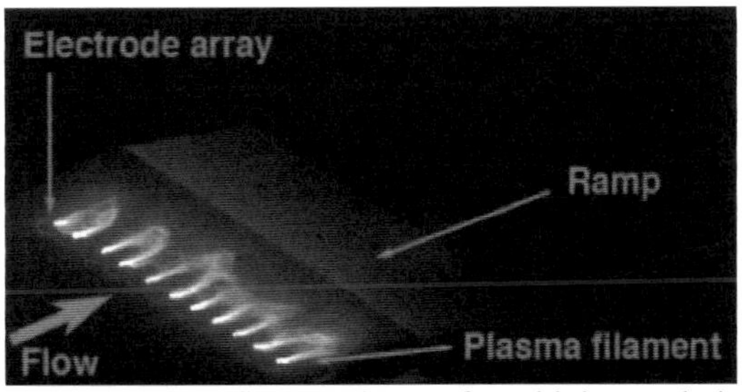

Figure 6. Typical plasma filament image from high-speed visualization.

From Fig. 6, it is suggested that significant flow heating took place because of the plasma filament and thereby it is expected that flowfield change was induced ahead of the ramp. Schlieren images are shown in Fig. 7, which exhibit significant change in shock positions. Similar change in the shock position was also observed in all the test cases. Prior to plasma actuation, the shock wave at the ramp was formed at almost 45 degrees to the flow. This shock wave was generated almost at the leading edge of the ramp but the shock front was a slightly ahead of the tip. This is because the ramp angle, 20 degrees, is very close to the shock detachment angle at Mach-2. While the plasma was turned on, the shock location shifted upstream and formed just ahead of the discharge. In Fig. 7(a), the shock wave at the ramp is almost 45 degrees to the flow, which is slightly smaller than the shock angle, 53 degrees, obtained from oblique shock relation. Similarly, in Fig. 7(b), the plasma-related shock wave formed at the electrodes is almost 34 degrees to the flow, which is also slightly smaller than the Mach angle: 30 degrees. The difference in actual shock angles from those obtained from theoretical relation can be caused by the flow acceleration due to 1-degree inclination of wind tunnel walls and also because of significant flow heating that took place at rear side of the shock waves.

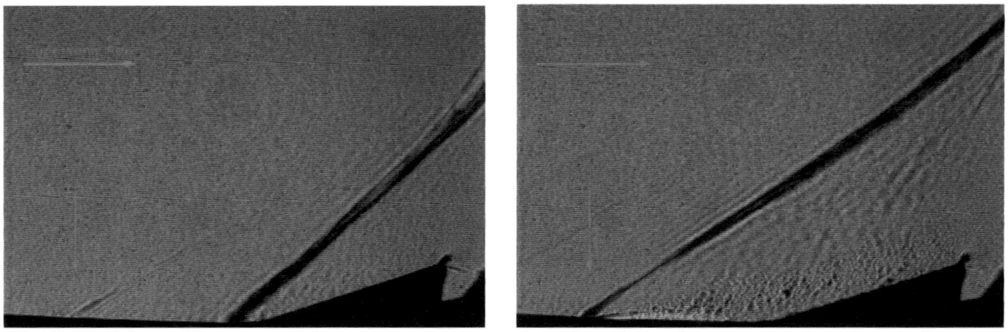

(a) Flow without plasma actuation. (b) Flow with plasma actuation ahead of ramp.

Figure 7. Snapshot images of typical flow actuation test.

Since there was significant change in shock locations, it is expected that there was a significant pressure change at the rear side of the electrodes and along the surface of 20-degree ramp. Typical pressure variation is shown in Fig. 8. A pressure rise was observed on the downstream side of the electrodes, which is a consistent result with previous research without the ramp [3]. However, a significant pressure drop was observed along the surface of the ramp at ports 2 and 3. This is evidence that the plasma actuation ahead of a ramp/flap can control pressure force and thereby cause a change in the pitching moment acting on the flaps. The pressure plot in Fig. 8(a) also suggests that the pressure curves of port 2 and 3 are almost identical and hence, separation ahead of the ramp was negligible when there is no plasma actuation. In the following discussion, the ramp pressure drop was evaluated based on the pressure at port 2 and quantified according to the definition of pressure drop illustrated in Fig. 8(b).

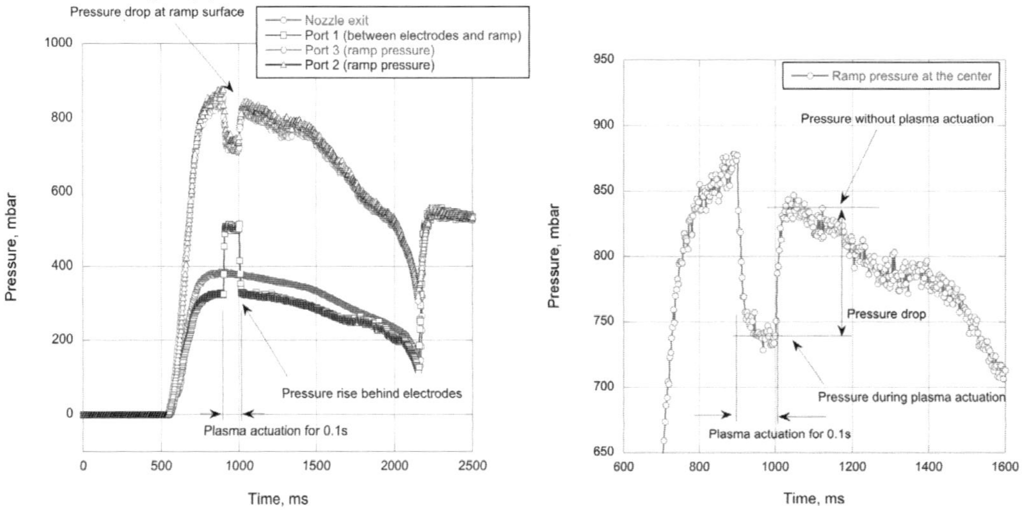

(a) Pressure curve during wind tunnel test. (b) Definition of pressure drop evaluation.

Figure 8. Pressure plot in typical test.

In this work, Reynolds number was varied by changing stagnation pressure and temperature independently. Averaged plasma power was also changed based on various flow conditions and it ranged from 4-12 kW. The averaged plasma power variation with Reynolds number is shown in Fig. 9. Since the tests were done at constant pressure condition and at constant temperature conditions, these results are plotted respectively in Fig. 9; red circle denotes P_0=1.7 bar constant pressure tests, blue rectangle for T_0=294 K tests, green triangle for T_0=400 K tests, and black rectangle for T_0=500 K tests. From Fig. 9, there is a trend that average plasma power increases with an increase in Reynolds number. This would be because, when the temperature T_0 is constant, the Reynolds number increase is attained by the increase in pressure P_0 and thereby increase in air density. Under a higher density environment, it is obvious that heat generation is realized in thicker air and it caused higher plasma power consumption. From the aerodynamic viewpoint, higher Reynolds number makes the boundary layer thickness smaller. This will cause the plasma filament to be kept close to the wall. Since the filament is formed inside the boundary layer and separation zones, smaller boundary layer volume would realize more dense plasma region to attain higher plasma power values.

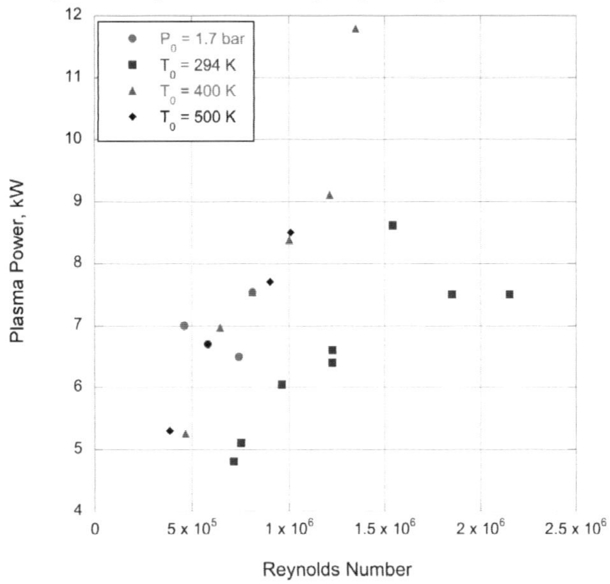

Figure 9. Averaged plasma power variation with Reynolds number.

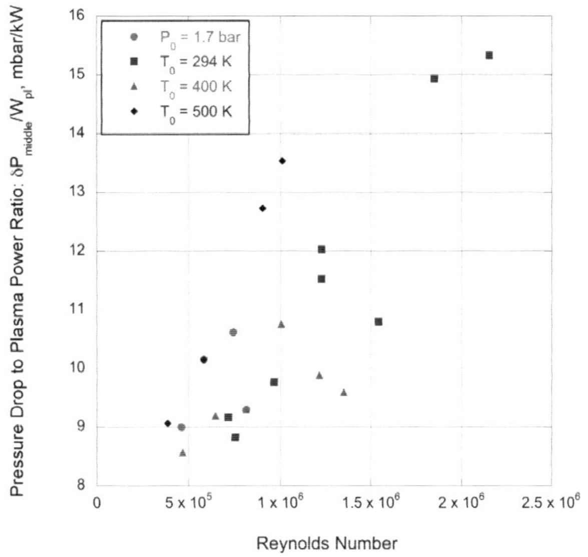

Figure 10. "Pressure drop per plasma power" variation with Reynolds number.

Although the ramp pressure drop and Reynolds number did not exhibit a distinct correlation, the quantity ramp pressure drop per averaged plasma power, which is a kind of pressure drop efficiency per power, showed linear dependency on Reynolds number as shown in Fig. 10. At constant temperature conditions, the increase in Reynolds number is realized by increasing the freestream pressure and density. Therefore, it is reasonable that, at higher Reynolds number, the pressure change will be greater than the flow with smaller freestream pressure. According to the experimental results, plasma power increases with increase in Reynolds number but the pressure drop increase seems to be much greater than those of plasma power. Since pressure drop per averaged plasma power is a kind of indicator for how efficiently plasma actuation is conducted, it is concluded that, in the present experimental conditions, the plasma actuation is more efficient at higher Reynolds number.

4. Conclusion

Plasma-assisted flow control ahead of 20-degree ramp geometry was examined with Mach-2 supersonic wind tunnel experiments and the effect of Reynolds number on flow control properties was also investigated. It was found that ramp pressure drop per averaged plasma power depends almost linearly on Reynolds number variation and, therefore it is concluded that plasma actuation is more efficient at higher Reynolds number flows.

Acknowledgements
This work was supported by Grant-in-Aid for Young Scientists (B) No. 16K18307. The first author appreciates the FlowPAC Institute, University of Notre Dame, for providing opportunity to use SBR-50 wind tunnel for this work.

References

[1] Shang, J. S., Surzhikov, S. T., Kimmel, R., Gaitonde, D., Menart, J. and Hayes, J.: Mechanisms of Plasma Actuators for Hypersonic Flow Control, Progress in Aerospace Sciences, 41, pp. 642–668, 2005.
[2] Fomin, V. M., Tretyakov, P. K. and Taran, J.-P.: Flow Control Using Various Plasma and Aerodynamic Approaches, Aerospace Science and Technology, 8, pp. 411-421, 2004.
[3] Leonov, S. B. and Yarantsev, D. A.: Near-Surface Electrical Discharge in Supersonic Airflow: Properties and Flow Control, Journal of Propulsion and Power, 24, pp. 1168-1181, 2008.
[4] Leonov, S. B., Napartovich, A. P. and Yarantsev, D. A.: Plasma-Assisted Combustion of Gaseous Fuel in Supersonic Duct, IEEE Transactions on Plasma Science, 34(6), 2006.
[5] Leonov, S. B., Firsov, A. A., Yarantsev, D. A., Falempin, F. and Miller, A.: Active Steering of Shock Waves in Compression Ramp by Nonuniform Plasma, AIAA Paper 2010-260, 2010.
[6] Leonov, S. B., Yarantsev, D. A. and Carter, C.: Experiments on Electrically Controlled Flameholding on a Plane Wall in Supersonic Airflow, Journal of Propulsion and Power, 25(2), 2009.
[7] Leonov, S. B., Houpt, A. and Hedlund, B.: Experimental Demonstration of Plasma-Based Flameholder in a Model Scramjet, AIAA Paper 2017-2249, 2017.
[8] Watanabe, Y. and Suzuki, K.: Investigation of Arc Plasma Discharge in Hypersonic Flow over Compression and Expansion Corner, AIAA Paper 2013-3130, 2013.
[9] Watanabe, Y. and Suzuki, K.: Aerodynamic Control Effect of Surface DC Plasma Discharge at Mach-7 Hypersonic Flow, AIAA Paper 2015-2342, 2015.

IOP Conf. Series: Materials Science and Engineering **249** (2017) 012007 doi:10.1088/1757-899X/249/1/012007

Supersonic Cavity-Based Flow Control Using a Quasi-DC Discharge

A Houpt[1], S Leonov[1], B Hedlund[1], T Ombrello[2] and C Carter[2]

University of Notre Dame, Notre Dame, IN 46556, USA[1]; U.S. Air Force Research Laboratory, Wright-Patterson Air Force Base, OH 45433, USA[2]

Alec.W.Houpt.3@nd.edu

Abstract. The Quasi-DC (Q-DC) discharge is studied as an active flow control authority on a rear-facing cavity in a supersonic duct by creating an oblique shockwave that impinges the cavity. This geometry simulates the geometry of a typical scramjet flameholding scheme. The tests were performed at the University of Notre Dame in the SBR-50 supersonic blowdown rig with dried air at M=2. Schlieren imaging is used to view the flow field with and without the Q-DC discharge in operation. A significant change in the flow field structure is observed. Pressure sensors detect a pressure increase throughout the entire rear-facing cavity while the Q-DC discharge is operating. This reveals that the cavity redistributes the pressure increase from the shockwave as a result of the flow within the cavity being subsonic. As a result of this pressure absorption and redistribution, the impinging shockwave created by the Q-DC is almost completely absorbed. This absorption is confirmed by the schlieren images. The data reveal that the discharge power is the dominating influence, as compared to electrode/discharge geometry, on the pressure increase produced in the cavity. There is a nearly linear correlation between the power of the discharge and the pressure increase produced directly behind the discharge, in the cavity, and on the ramp of the cavity (to varying magnitudes). It is suggested that the 11 electrode system may be slightly more effective than the 7 electrode system.

1. Introduction

Plasma-based flow control techniques have been studied extensively in recent decades. Electric discharges provide high-frequency capabilities and a robust, surface-flush installation for active flow control such that it can add momentum/energy into the boundary layer without obstructing the flow while it is not in operation. The large majority of plasma-based flow control studies focus on the use of dielectric barrier discharges (DBDs), but this type of discharge is not powerful enough to provide flow control authority in a highly compressible flow. There have been non-DBD plasma-based flow control studies most of which focus on shockwave generation and control of shockwave/boundary layer interactions [1, 2, 3, 4, 5, 6, 7, 8, 9, 10]. The quasi-DC (Q-DC) discharge has proven its potential as a flow control technique in high-density supersonic duct driven airflows, such as the environment typical of a scramjet [7, 8, 11, 12, 13, 14]. The velocity of the core flow in a supersonic duct flow is an order of magnitude faster than the combustion propagation speed of a typical gaseous hydrocarbon fuel (e.g. ethylene) making it difficult to maintain combustion. The prevailing method/geometry used for flameholding in scramjet engines is a rear-facing cavity. The rear-facing cavity creates a subsonic region of flow which provides an environment in which the flame will not be blown downstream. Supersonic duct flow is not favorable for fast mixing of the fuel and oxidizer for a non-premixed flow.

The interaction between the core flow and the cavity flow is limited to the free shear layer which does not provide sufficient mixing between the two if fuel from the core flow is wanted to enter the cavity. In order to overcome this, fuel can be injected simultaneously into the core flow upstream of the cavity and into the cavity. This overcomes the lack of mixing between the two flow regions, and the heat released from combustion can be sufficient to ignite and maintain combustion in the core flow. The Q-DC discharge provides a method for enhancing multiple aspects of off-design flow conditions in a supersonic combustion environment. The strong spanwise non-uniformity of the Q-DC discharge can provide enhanced mixing and the discharge can potentially be used for fast/cold re-ignition. The Q-DC discharge can also potentially be employed as a method to enhance the interaction between the core flow and the cavity flow.

The Q-DC discharge as it is applied to a supersonic duct driven airflow has been studied extensively [7, 8, 13, 14]. The discharge produces a rapid and local heating of the gas leading to a pressure increase which forms an oblique shockwave. This shockwave impinges the rear-facing cavity on the opposite wall leading to a pressure increase in the cavity and a significant change in the flow structure of the cavity [15]. In addition, the strong spanwise non-uniformity of the discharge may provide streamwise vorticity which would enhance the mixing of the fuel and oxidizer. The work performed for this paper focuses on the characterization of the effect of plasma power and electrode geometry on the pressure increase produced in the cavity.

2. Experimental Facility

The experiments performed for this work were carried out using the SBR-50 supersonic blowdown tunnel. The tunnel was equipped with a Mach 2 Laval nozzle arranged inline with the rectangular test section. The cross-section at the nozzle exit is Y×Z=76.2 mm (width) × 76.2 mm (height) with the top and bottom walls expanding at a 1° half angle over a length of X=610 mm as measured to the beginning of the diffuser. The test section is a modular design which allows for pieces to be interchanged/designed for each experiment and for the side walls to be quartz to allow for excellent optical access. The top and bottom walls are inserts, made of stainless steel or nylon, and are equipped with 48 pressure taps (24 on the top and 24 on the bottom) which are connected to a 16 channel PSI 9116 pressure scanner. This provides a time series of static pressure measurements along the length of the test section sampled at 400 Hz. Schlieren imaging and high-speed imaging with a Photron Fastcam are used for viewing the flow field and Q-DC discharge, respectively. The basic configuration of the SBR-50 test section with an electrode array and planar walls is shown in Fig. 1.

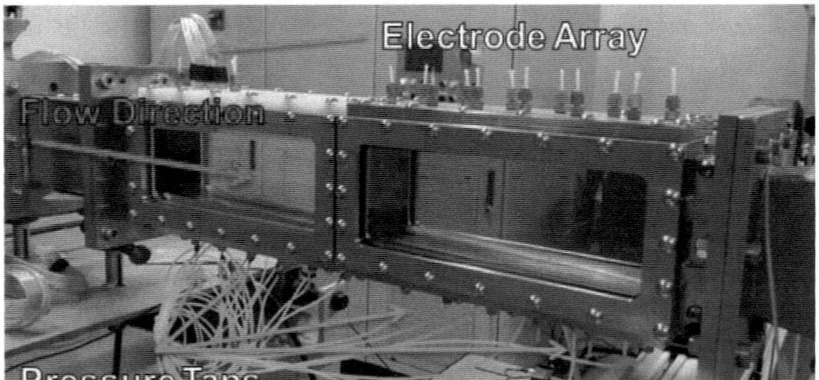

Fig. 1. SBR-50 test section at the University of Notre Dame equipped with planar walls.

An array of 7 or 11 brass electrodes are installed on the top wall with alternating anode and cathode in the spanwise direction (eg. ACACACA). This is used as a controllable shockwave generator and has been studied extensively in this facility [13, 14]. The wall insert is made of nylon, as shown in Fig.

1, and has a ceramic insert with the electrodes flush mounted on the test section surface. The wall insert with the electrode configuration is better shown in Fig. 2 along with the electrical diagram.

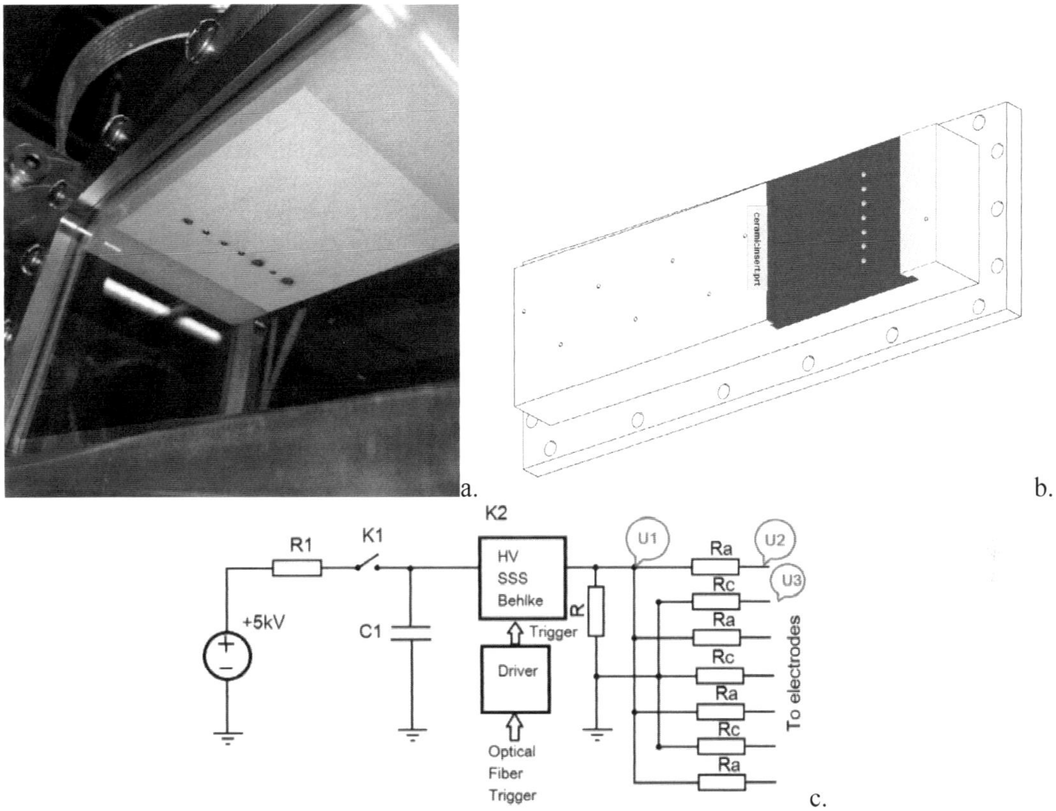

Fig. 2. 7 electrode array used for controllable shockwave generation a) installed in the test section b) 3D rendering showing the ceramic insert c) electrical diagram.

A typical test duration is approximately 1 second, with the Q-DC discharge operated for up to a few hundred milliseconds. This work concerns the application of the controllable shockwave generator used as a flow control authority in a test configuration with a rear-facing cavity on the opposite wall. This geometry is meant to simulate a typical flameholding scheme used for supersonic combustion. The generated shockwave impinges the cavity thus increasing the pressure and changing the structure of the flow field. This method is illustrated in Fig. 3.

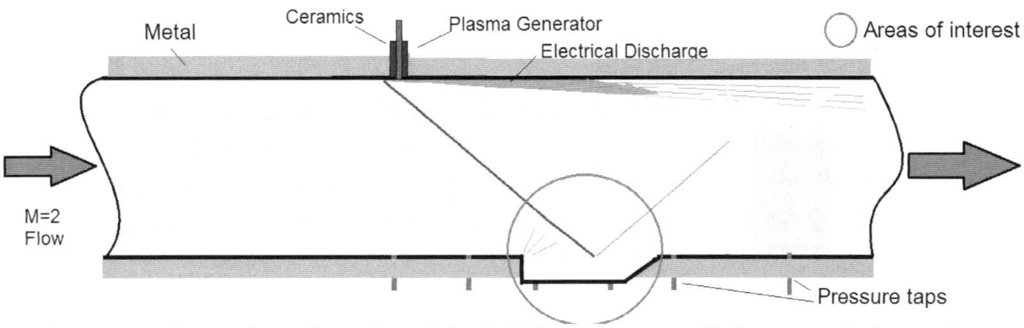

Fig. 3. Experimental configuration of the Q-DC discharge applied to a rear-facing cavity.

A custom made power supply is used for the operation of the high-voltage, high-power Q-DC discharge. It operates with a steep falling voltage-current characteristic with individual control of each output channel. Varying the output voltage of the power supply, U1, controls the output power of the plasma. The power supply operates in a nearly current-stabilized mode with a fixed ballast resistance on each channel. High-voltage probes and current probes measure the output voltage and corresponding current of the power supply U1, voltage across an anode U2, and voltage across a cathode U3. Typical electrical measurements are shown in Fig. 4. The instantaneous power deposition of the plasma fluctuates in the range of 8-20 kHz (depending on flow conditions) and is a reflection of the unsteady behavior of the plasma filaments. This oscillation is due to the strong coupling of the discharge to the flow. The process of the discharge oscillations is as follows: breakdown occurs across a gap between an anode and cathode, the flow velocity carries the ions downstream which increases the voltage across the gap, a critical length of the plasma filaments is reached and a new breakdown occurs closer to the electrodes [13, 14]. A sample image of the Q-DC discharge showcasing the structure of the plasma filaments is shown in Fig. 5.

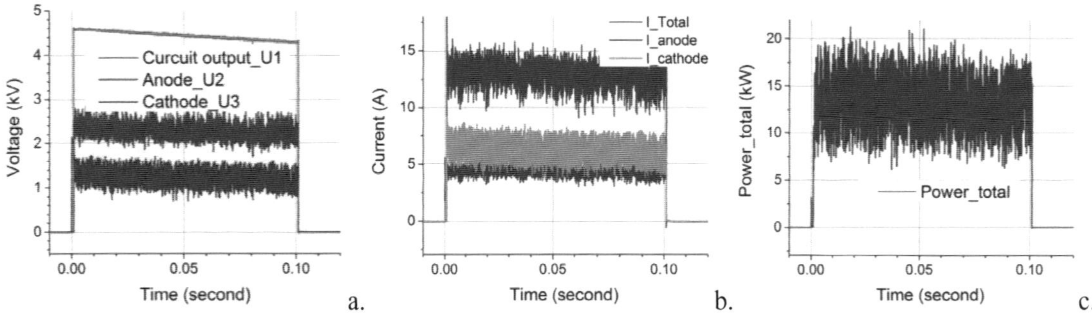

Fig. 4. Representative oscillograms of: a) voltage b) current and c) instantaneous power

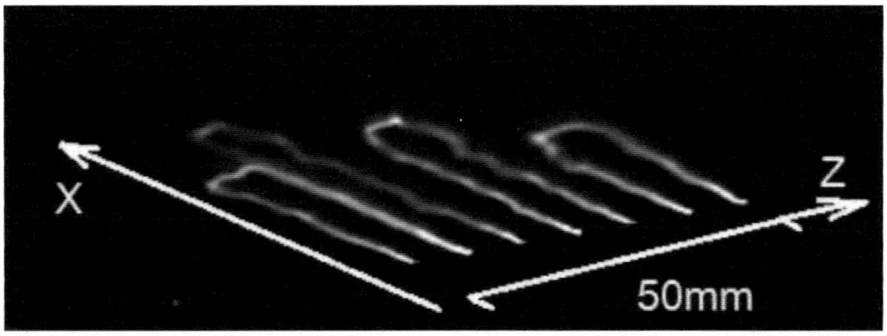

Fig. 5. Standard image of the Q-DC discharge with 4 μs exposure.

3. Experimental Results

Schlieren imaging provides a qualitative insight of how the Q-DC discharge affects the flow field. Figures 6 and 7 show the flow field at the front edge of the cavity and at the ramp on the downstream side of the cavity, respectively, with and without the Q-DC discharge in operation. The schlieren images have an exposure time of 50 μs, which essentially renders the schlieren images as time averages due to the speed of the flow. The bottom portion of the cavity is inaccessible for schlieren imaging due to the steel frames which hold the windows in place, as are some other portions due to the aperture of the schlieren system. Red lines denote the edges of the schlieren system and frame in Fig. 6. Figure 6b shows the plasma related shockwave impinging the cavity and interacting with the cavity/free shear layer resulting in a much weaker reflected shockwave. A new shockwave forms at the front edge of the cavity as a result of the Q-DC generated shockwave increasing the pressure in the cavity. As a part of the change in the flow field, the free shear layer shifts up toward the core flow slightly. The angle of the compression shock from the ramp, with respect to the ramp surface, increases as is shown in Fig. 7.

Fig. 6. Schlieren images of the top portion of the rear-facing cavity with plasma a) on b) off.

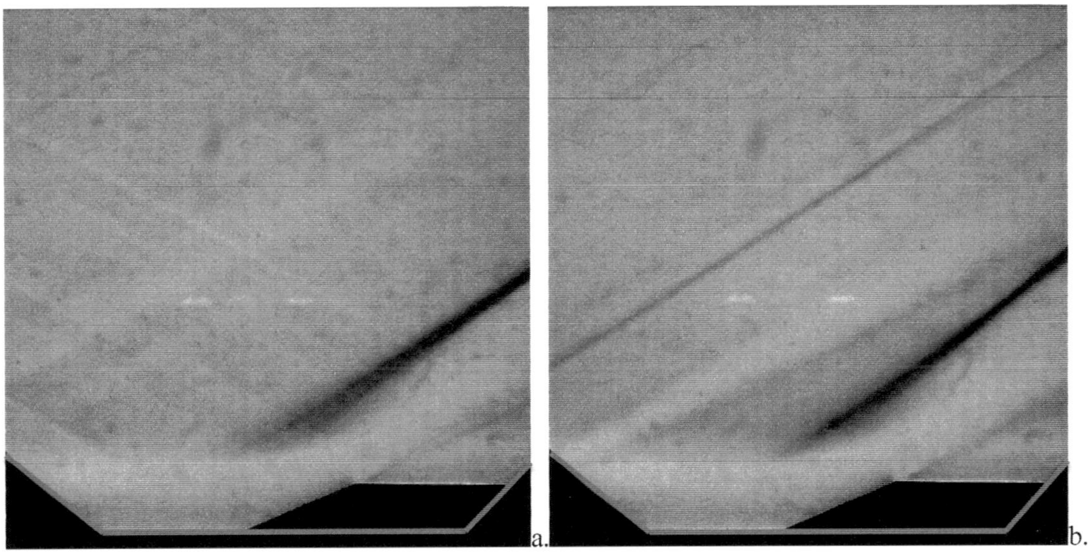

Fig. 7. The ramp at the downstream side of the cavity with plasma a) off b) on

The 16 channel pressure scanner provides static pressure distributions in the test section. Figure 8 shows the pressure distribution along the upper and bottom walls with 5 anodes and 6 cathodes in operation for the Q-DC discharge. Datasets are shown before the plasma is turned on, during plasma operation, and after plasma operation. The static pressure throughout the cavity increases during plasma operation in Fig. 8a. The cavity flow is a subsonic region which absorbs the pressure increase on the downstream side of the shockwave and redistributes it throughout the subsonic region. There is a sharp increase in pressure immediately downstream ($x<20$ mm) of the electrodes on the top wall where the plasma is influencing the flow the most. This is followed by a region along the top wall (50 mm$<x<$200 mm) where the pressure decreases due to separation and heating of the flow. The compression shock due to the ramp in the cavity produces a shockwave for which a pressure increase is seen without plasma operation approximately 290 mm downstream of the electrodes. When the plasma is turned on the pressure at this location decreases while the pressure increases at the neighboring upstream region. This indicates the compression shock due to the ramp changes angle, as shown in Fig. 7, most likely due to a lower Mach number and stronger pressure in the cavity. These same measurements were made for only 2 anodes in operation for the discharge and are shown in Fig. 9. The plasma shockwave generator with 2 anodes operating has the same effect but a lesser magnitude as when 5 anodes are operating. Since the cavity is a subsonic region, the power of the discharge rather than the geometry should be the dominating factor. This is confirmed in Fig. 10, where the pressure is measured directly behind the plasma generator, in the cavity, and on the ramp for 7 and 11 electrode arrays with varying power. The pressure increase due to the Q-DC shockwave generator is nearly the same linear trend for the 7 and 11 electrode arrays, with the 11 electrode array producing slightly more of a pressure increase. This may be due to the longer filaments of the 7 electrode array releasing its power over a larger area and some of this pressure increase may reach the opposite wall too far downstream to interact with the cavity. The 11 electrode system is slightly more effective than the 11 electrode system. As it is considered, the 11 electrode system has a smaller area of influence and can produce a larger pressure increase with equal power. It should be noted that the data points labeled as A1 in Fig. 10 are anomalous. The discharge did not connect across the gap which is in line with the pressure tap at that location.

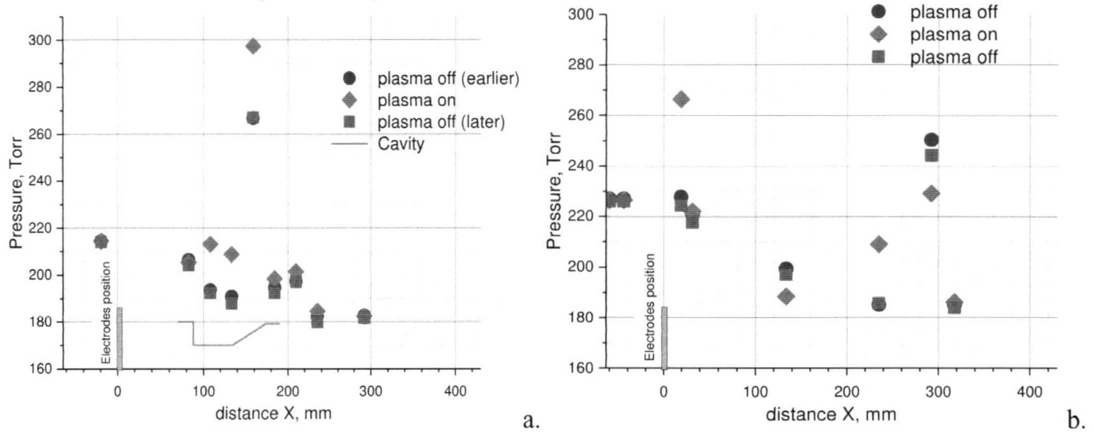

Fig. 8. Static pressure distribution along a) bottom wall with cavity b) top wall with plasma generator with 5 anodes on.

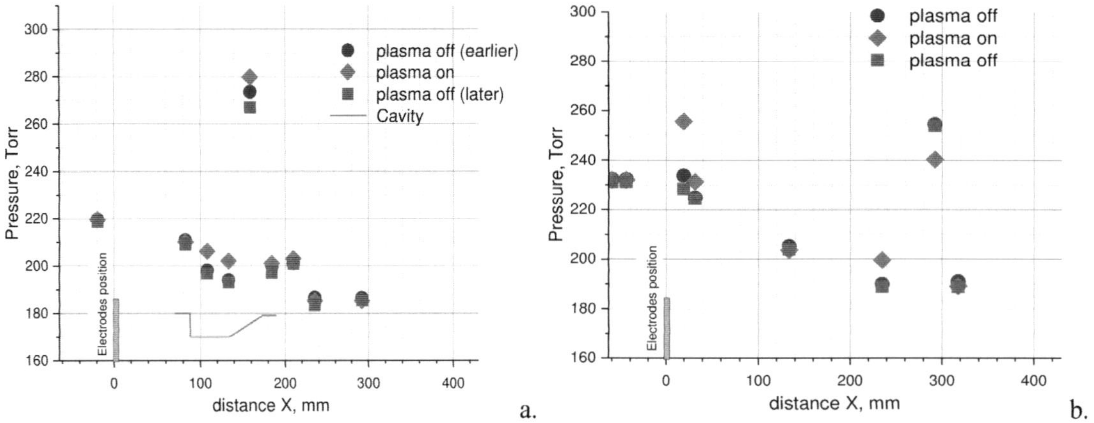

Fig. 9. Static pressure distribution along a) bottom wall with cavity b) top wall with plasma generator with 2 anodes on.

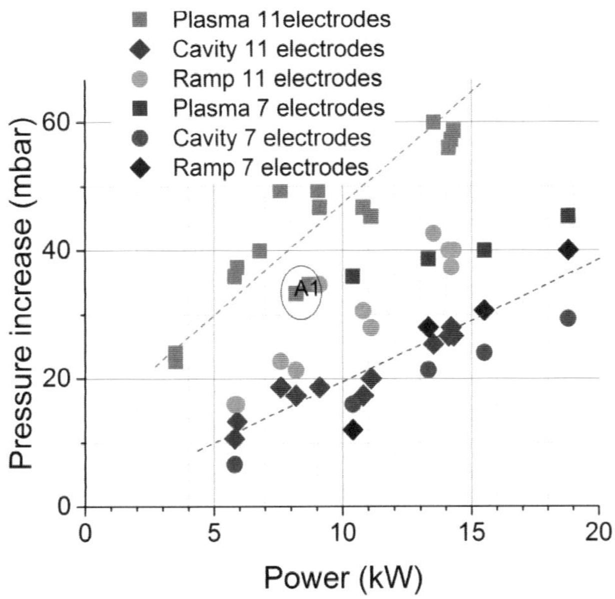

Fig. 10. Pressure increase as a function of average power for two different electrode arrays.

4. Summary

The quasi-DC (Q-DC) discharge was studied as a control authority to a supersonic flow with a rear-facing cavity, a geometry that imitates a typical flameholding scheme for a scramjet engine. The Q-DC discharge was used to generate an oblique shockwave which impinged a rear-facing cavity on a planar wall opposite the electrode array. Schlieren imaging shows the generated shockwave interaction with the cavity and the free shear layer which developed between the cavity and core flows. The cavity is a subsonic region that absorbs and redistributes the pressure increase of the shockwave. This causes the static pressure in the cavity to increase and the cavity to slightly enlarge which creates a new shockwave at the front edge of the cavity. This new flow field structure slightly moves the free shear layer up towards the core flow. Different electrode geometries and powers were tested, revealing that the plasma power is the dominating factor for increasing the pressure in the cavity rather than electrode geometry. The cavity absorbs any pressure increase and redistributes it throughout the entire recirculation flow because it is a subsonic flow region. Thus having a wider area of influence on the cavity is not as important as the power of the electrical discharge. There is a nearly linear relation between power of the Q-DC discharge and pressure increase experienced by the cavity. It is suggested that the 11 electrode system is slightly more effective than the 7 electrode system. This is due to formers ability to produce a larger pressure increase with equal power along with its ability to be better targeted due to its smaller region of influence produced from shorter plasma filaments.

The Q-DC discharge can be used in this arrangement to positively influence the combustion environment of a scramjet engine at off-design conditions. Increasing the pressure in the cavity increases the amount of oxidizer present in the cavity. This allows for more fuel to be in the cavity, if the same stoichiometric ratio is to be used, and thus can increase the heat release in this region. The upward shift of the free shear layer into the core flow can increase the entrainment of fuel for non-premixed fuel injection along the wall in which the cavity is installed. More studies need to be done in order to fully characterize the capabilities and flow control authority of thc Q-DC discharge with this flow geometry. Future studies will include varying flow temperature, implementing the Q-DC discharge into a metal wall (as opposed to the nylon and ceramic wall used thus far), moving the impingement location of the shockwave, and placing the discharge on the same wall as the cavity.

Acknowledgements
The authors would like to acknowledge the support of this work from the Air Force Research Laboratory and Innovative Scientific Solutions, Inc. (Subcontract # SB20181).

References

[1] B. R. Greene, N. T. Clemens, P. Magari and D. Micka, "Control of mean separation in shock boundary layer interaction using pulsed plasma jets," *Shock Waves,* vol. 25, no. 5, pp. 495-505, 2015.

[2] F. Falempin, A. Firsov, D. Yarantsev, M. Goldfeld, K. Timofeev and S. Leonov, "Plasma control of shock wave configuration in off-design mode of M=2 inlet," *Experiments in Fluids,* vol. 56, no. 54, 2015.

[3] S. Merriman, E. Plonjes and P. Palm, "Shock wave control by nonequilibrium plasmas in cold supersonic gas flows," *AIAA Journal,* vol. 39, pp. 1547-1552, 2001.

[4] N. Webb, C. Clifford and M. Samimy, "Control of oblique shock wave/boundary layer interactions using plasma actuators," *Experiments in Fluids,* vol. 54, no. 6, p. 1545, 2013.

[5] J. Wang, Y. H. Li, B. Q. Cheng, C. B. Su, H. M. Song and Y. Wu, "Effects of plasma aerodynamic actuation on oblique shock wave in a cold supersonic flow," *J. Phys. D: Appl. Phys,* vol. 42, no. 165503, 2009.

[6] V. Narayanaswamy, L. L. Raja and N. T. Clemens, "Control of a Shock/Boundary-Layer Interaction by Using a Pulsed-Plasma Jet Actuator," *AIAA Journal,* vol. 50, no. 1, pp. 246-249, 2012.

[7] S. B. Leonov and D. A. Yarantsev, "Near Surface Electrical Discharge in Supersonic Airflow: Properties and Flow Control," *Journal of Propulsion and Power,* vol. 24, no. 6, pp. 1168-1181, 2008.

[8] S. B. Leonov, I. V. Adamovich and V. R. Soloviev, "Dynamics of near-surface electric discharges and mechanisms of their interaction with the airflow," *Plasma Sources Sciences and Technology,* vol. 25, 2016.

[9] M. Kotsonis, "Diagnostics for characterisation of plasma actuators," *Measurement Science and Technology,* vol. 26, 2015.

[10] K.-S. Choi, T. Jukes and R. Whalley, "Turbulent boundary-layer control with plasma actuators," *Philosophical Transactions of the Royal Society,* vol. 369, pp. 1443-1458, 2011.

[11] S. Leonov, D. Yarantsev, A. Kuriachy and V. Soloviev, "Supersonic/Transonic Flow Control By Electro-Discharge Plasma Technique," in *Proceedings of the 25th international Congress of the Aeronautical Sciences (ICAS)*, Hamburg, Germany, 2006.

[12] A. Firsov, K. Savelkin, D. Yarantsev and S. Leonov, "Plasma-enhanced mixing and flameholding in supersonic flow," *Philosophical Transactions A,* vol. 373, no. 2048, 2015.

[13] A. Houpt, B. Hedlund, T. Ombrello and S. Leonov, "Controllable shock wave generation by near-surface electrical discharge," in *47th AIAA Plasmadynamics and Lasers Conference*, Washington, D.C., 2016.

[14] A. Houpt, B. Hedlund, S. Elliot, S. Leonov, T. Ombrello and C. Campbell, "Characterization of transient operation of a quasi-DC electrical discharge in supersonic flow," in *AIAA Aviation*, Denver, 2017.

[15] T. Ombrello, S. Peltier and C. Carter, "Effects of inlet distortion on cavity ignition in supersonic flow," in *53rd AIAA Aerospace Sciences Meeting, AIAA Scitech Forum*, Kissimmee, Florida, 2015.

IOP Conf. Series: Materials Science and Engineering **249** (2017) 012008 doi:10.1088/1757-899X/249/1/012008

Dynamic and Heat Effects in Sinusoidal Surface DBD

V.A. Bityurin and A.N. Bocharov

Joint Institute for High Temperatures of Russian Academy of Sciences, Izhorskaya str. 13, Bld. 2, Moscow, Russia

valentin.bityurin@gmail.com

Abstract. Numerical simulation of sinusoidal Surface Dielectric Barrier Discharge (SDBD) is carried out. The configuration and parameters of the discharge are those considered experimentally in paper [1]. In this paper, dielectric disc was surrounded by high-voltage metallic electrode. 34 kHz 5.5 kV applied to HV electrode at atmospheric pressure. The objective of the current paper is to study mechanisms inducing flow motion over the disc. Numerical simulations are made with the two-dimensional model developed earlier in paper [2]. The model incorporates the transport of charged particles, Poisson equation for the electric field strength, and Navier-Stokes equations for the fluid. In addition to the bulk processes (mainly, ionization, recombination, drift and diffusion) the deposition of the charge on the dielectric surface is considered.

1. Computational model for pulse-periodic discharge

The problem setup simulates, in general, the discharge studied in paper [1]. The pulse-periodic discharge over the disc dielectric plate surrounded with annular high-voltage (HV) electrode is considered. The diameter of dielectric disc was taken to be 14 mm. External diameter of HV electrode was taken to be 24 mm. The diameter of the grounded electrode is assumed to be the same as the dielectric one. Thickness of dielectric was taken as 0.3 mm; dielectric permeability is equal to 3. Sinusoidal 5.5kV voltage is applied to HV electrode with the frequency 33.3 kHz in air at atmospheric pressure. The discharge domain considered below is presented in Fig. 1.

The computational model is basically same as those described in paper [2,3]. We assume that the flow and electric discharge characteristics can be described by the set of equations reflecting conservation of mass, momentum and total energy for the whole fluid. Transport equations for charged particles, electrons and ions, as well as the Poisson equation for the electric field strength are solved assuming the drift-diffusion approach is valid. The set of governing equations is given below.

$$\frac{\partial \rho}{\partial t} + (\rho \mathbf{U}) = 0 \tag{1}$$

$$\frac{\partial \rho \mathbf{U}}{\partial t} + (\rho \mathbf{U}\mathbf{U}) + \nabla \boldsymbol{\tau} = \frac{\partial P}{\partial \mathbf{r}} \tag{2}$$

$$\frac{\partial \rho E}{\partial t} + (\rho \mathbf{U} H) + \nabla(\mathbf{U}\boldsymbol{\tau}) + \nabla \mathbf{q} = 0 \tag{3}$$

Here, ρ is the density, $\mathbf{U} = (U_x, U_y)$ is velocity, P is the thermodynamic pressure, E is specific total energy, H is specific total enthalpy.

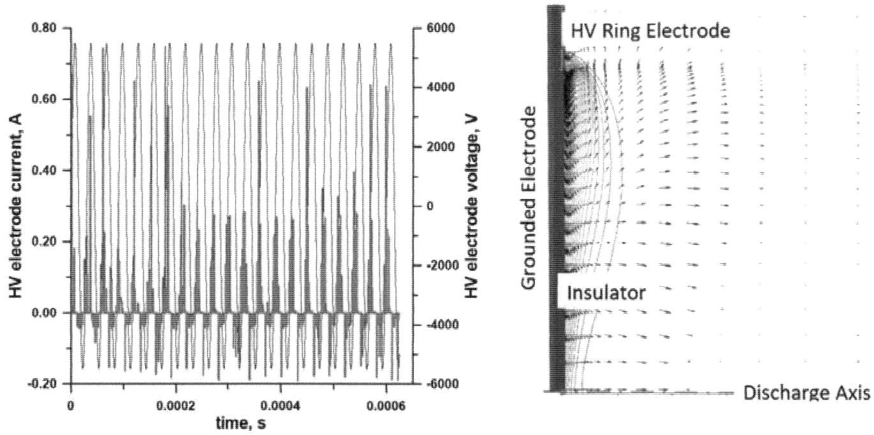

Fig.1 The time dependents of the HV electrode current (red line) and voltage (black line) at 34kHz 5.5kV SDBD (left) and temperature (colored lines) and velocity-vectors after 20 cycles. T_{max}=674K, V_{max}=0.88m/s (right).

$$H = E + \frac{P}{\rho}, E = e + \frac{U^2}{2}, (\gamma - 1)e = P/\rho, P = \rho RT \quad (4)$$

Here, γ= 1.4 is the ratio of specific heats, and T is the temperature. Viscous stress tensor components and heat fluxes are specified as usually:

$$\tau_{ij} = \frac{2}{3}\eta\delta_{ij}\nabla\mathbf{U} - \eta\left(\frac{\partial U_i}{\partial x_j} + \frac{\partial U_j}{\partial x_i}\right), \mathbf{q} = -\lambda\frac{\partial T}{\partial \mathbf{r}} \quad (5)$$

where δ_{ij} is Kronecker symbol, η is viscosity, λ is heat conductivity. Gradient and divergence operators are defined as follows:

$$\frac{\partial}{\partial \mathbf{r}} = \mathbf{e}_x\frac{\partial}{\partial x} + \mathbf{e}_y\frac{\partial}{\partial x}, \nabla = \frac{\partial}{\partial x} + \frac{1}{y^\xi}\frac{\partial}{\partial x}y^\xi, \quad (6)$$

Where $\mathbf{e}_x$ and $\mathbf{e}_y$ are unit base-vectors, ξ=0 for Cartesian coordinate system, and ξ=1 for axi-symmetric one.

The transport equations for concentrations and Poisson equation for electric field are read as follows:

$$\frac{\partial n_i}{\partial t} + \nabla\mathbf{\Gamma}_i = \dot{Q} \quad (7)$$

$$\frac{\partial n_e}{\partial t} + \nabla\mathbf{\Gamma}_e = \dot{Q} \quad (8)$$

$$\varepsilon\nabla\mathbf{E} = \rho_e, \rho_e = q(n_i - n_e) \quad (9)$$

In Eqs. (7) through (9) n_i is ion number density, n_e is electron number density, ρ_e is the electric charge density, $\mathbf{E}$ is electric field, $\mathbf{E} = -\frac{\partial\varphi}{\partial\mathbf{r}}$, φ is the electric potential, q is electron charge, ε is permittivity of vacuum, $\dot{Q}$ is the source term defined later. Strictly speaking the Poisson equation with zero right-hand-side should also be solved in the bulk of dielectric. In this paper, simplified approach is used. Namely, the dielectric surface is considered via special boundary conditions on the plasma-dielectric boundary.

Ion and electron fluxes, $\mathbf{\Gamma}_i$ and $\mathbf{\Gamma}_e$, are defined as follows

$$\mathbf{\Gamma}_i = n_i(\mathbf{U} + \mathbf{V}_{di}) - D_i\frac{\partial n_i}{\partial \boldsymbol{r}}$$

$$\boldsymbol{\Gamma}_e = n_e(\mathbf{U} + \mathbf{V}_{de}) - D_e \frac{\partial n_e}{\partial r} \tag{10}$$

Here, drift velocities $\boldsymbol{V}_{di}$ and $\boldsymbol{V}_{de}$ are defined as

$$V_{di} = \mu_i E, V_{de} = -\mu_e E \tag{11}$$

where μ_i and μ_e are ion and electron mobilities, respectively:

$$\mu_{i,e} = \frac{q/m_{i,e}}{\upsilon_{i,e}}, \upsilon_{i,e} = \frac{2\sqrt{2}}{3}\pi d_{i,e}^2 (\frac{8kT_{i,e}}{\pi m_{i,e}})^{1/2} \tag{12}$$

In (12) $d_{i,e}$ are mean collision cross-section diameters for ions and electrons, respectively. T_i and T_e are ion and electron temperature, respectively. m_i and m_i are ion and electron mass. $\nu_{i,e}$ are mean collision frequences for ion-neutral and electron-neutral collisions.
Diffusion coefficients, D_i and D_e in (10) are defined as

$$D_i = \mu_i k_B T_i/q, D_e = \mu_e k_B T_e/q \tag{13}$$

In (13) k_B is the Boltzman's constant. In paper [2] we used an approximation for electron temperature as function of the reduced electric field, $E_r = E/n$ (Td), where $n = P/k_B T$ is the total number density. Here, we simply take T_e=2ev, and T_i=T.

Electric current density is determined as

$$\boldsymbol{j} = \boldsymbol{j}_i + \boldsymbol{j}_e = q(\boldsymbol{\Gamma}_i - \boldsymbol{\Gamma}_e)$$

Source-terms in equations (2), (3), (7) and (8) are determined as follows

$$\dot{F} = \rho_e E,$$

$$\dot{W} = \eta_e j_e E + \eta_i j_i E \tag{14}$$

$$\dot{Q} = \alpha n_e - \beta n_e n_i \tag{15}$$

The following approximations for ionization coefficient, α, and recombination coefficient, β, obtained with using Boltzman's solver[3] are used in the paper [3]:

$\alpha = 10^{-6}\cdot(0.79K_1+0.21K_2)\cdot n$
$K_1 = 8.9\cdot10^{-9}\exp(-927.7/E_r)$, if $E_r < 300$, $K_1 = 4.27\cdot10^{-8}\exp(-927.7/E_r)$, if $300<E_r < 1423$,
$K_1 = 10^{-10}\cdot(-110.67 + 21.8\, E_r)$, otherwise.
$K_2 = 4.9\cdot10^{-9}\exp(-657.8/E_r)$, if $E_r < 260$, $K_2 = 2.88\cdot10^{-8}\exp(-1118/E_r)\cdot(1 + 4\cdot10^{-10}\cdot E_r^{\ 3})$,
if $260<E_r < 1000$,
$K_2 = 10^{-10}\cdot(-110.67 + 21.8\, E_r)$, otherwise.
$\beta = 2\cdot10^{-13}\cdot(300/T_e)^{0.7}$

First of terms (14) is the body force acting on the fluid. Second term is the heat from flowing electron and ion currents. It has been taken in this paper that, $\eta_e = 1$ and $\eta_i = 1$.

The boundary conditions for flow variables are specified as follows. No-slip adiabatic wall conditions are set on the surface of electrode and dielectric surface. Symmetry boundary conditions are set at the symmetry axis. Namely, **Un**=0, ∂/∂n=0 for other variables. Here, **n** – is the unit normal vector. Far-field boundary conditions are specified at other boundaries of the domain.

Boundary conditions for plasma variables are specified as follows. At symmetry axis zero normal gradients are set for n_e, n_i, and φ. At the outer boundary (except HV electrode and dielectric surfaces) zero normal gradients are also set for all plasma variables. At the electrode and dielectric surfaces, the type of boundary conditions depends on the operation mode of electrode. The latter is determined by the sign of the projection of electric field on to outer normal to the surface. If it is positive, anode-type boundary conditions are applied. Otherwise, cathode-type boundary conditions are used. Anode-type conditions are specified as follows. Zero normal gradient for the electron number density is set, as well as zero ion number density. Boundary condition for electric potential is specified depending on the metallic or dielectric surface is considered. At the electrode surface potential is specified as: $\varphi_{HV} = E(t) = E_0 \cdot \sin(2\pi\nu t)$, where $E_0 = 5.5$kV, $\nu = 33.3$kHz. Cathode-type boundary conditions are set as follows. Zero normal gradient is set for ion number density. Relation $(\Gamma_e\, \mathbf{n}) = -\gamma_e(\Gamma_i\, \mathbf{n})$ is used for electrons. In this relation, γ_e is coefficient of secondary electron emission. Calculations were carried out with $\gamma_e = 0.01$. Electric potential at the HV electrode is set as described above. At the dielectric surface, the jump condition for electric field strength projection is specified:

$$\varepsilon_P(\mathbf{E}_P\mathbf{n}) - \varepsilon_D(\mathbf{E}_D\mathbf{n}) + (1/q)\cdot\int \mathbf{jn}dt = 0 \tag{16}$$

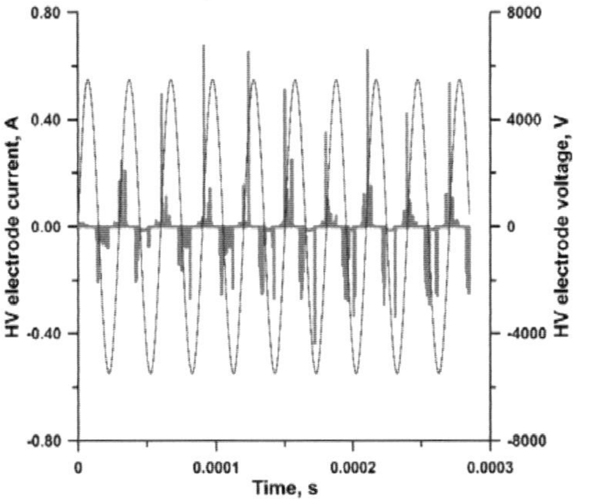

Fig.2. HV electrode current (red line) and voltage (blue line) vs time in interval (0 – 0.3) ms..

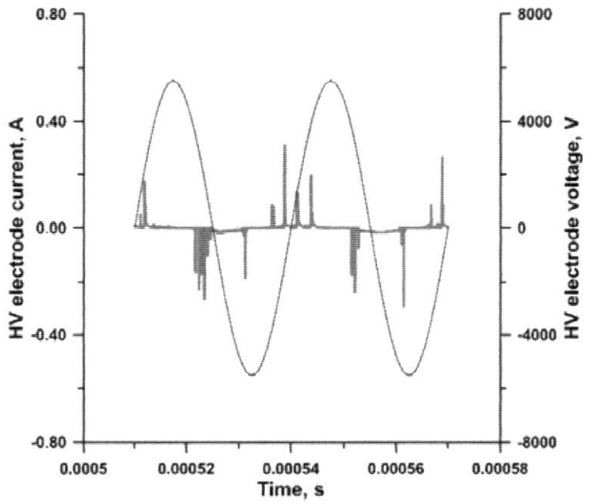

Fig.3. HV electrode current (red line) and voltage (blue line) vs time in interval (0.5 ms – 0.58) ms

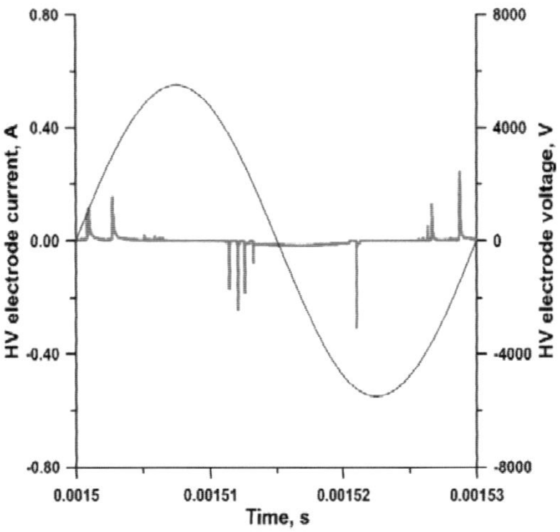

Fig.4. HV electrode current (red line) and voltage (blue line) vs time in interval 1.5 ms – 1.53) ms

Here, ε_P is electric permittivity of plasma, $\varepsilon_P = \varepsilon_0$. ε_D electric permittivity of dielectric material. In the paper $\varepsilon_D = 3\varepsilon_0$. $(\mathbf{E}_P\mathbf{n})$ is projection of electric field strength onto normal-to-surface from the side of plasma, and $(\mathbf{E}_D\mathbf{n})$ is projection of electric field from the side of dielectric. The latter term in (15) represents accumulation of the charge on the surface due to the flux of charged particles onto the surface. In general, equations (7) – (9) coupled with the equation for electric potential within the dielectric should be solved using conjugate condition (15) at the plasma-dielectric interface. In this paper simpler approach is used, in which the dielectric field strength, $\mathbf{E}_D$, is approximated as $(\mathbf{E}_D\mathbf{n}) = -(\varphi_D - \varphi_G)/\Delta$. Here, φ_D is the value of potential at the point of dielectric surface under consideration. φ_G is the potential of grounded metallic electrode built-in dielectric, $\varphi_G = 0$. Δ is characteristic length scale, which, in general, is a function of surface coordinate. With this, relation (15) is a 3rd kind boundary condition for plasma region only and can be used for rough estimations of charge accumulation effect.

Initial state is specified as follows. Quiet (no flow) atmospheric air is assumed over the computational domain. Plasma sheet as thick as 50 μm with n_e=n_i=10^{16} m^{-3} is set over the surface.

To estimate long-term action of pulse-periodic discharge on the flow field special technique developed in paper[3] is applied. Namely, the time-averaged source-terms (14) from equations (2) and (3) are used in time-integrating the Navier-Stokes equations. In turn, the time-averaged source-terms are computed while integrating the full set of equations during one or two cycles. Remind that the cycle time is 30 μs, which is much smaller than characteristic hydrodynamic time of order of 10 ms.

2. Results and Discussion

Consider the characteristics of the discharge during one cycle. The HV electrode current vs time is represented In Fig.2 – Fig.4. for different time instances during which the full set of equations is solved without time-averaging technique. Between these instances, time averaging technique is applied for Navier-Stokes equations only.

One could note several features of the discharge. The time evolution of current looks like almost chaotic one. However, one can distinguish several characteristic time zones. Current peaks corresponding to the growing part of HV voltage are due to "streamers" passing over the dielectric plate from HV electrode. Streamers, in general, are varied in length and life time. The magnitude of each current peak correlate well with the distance the streamer passes over the plate toward the symmetry axis. Examples of streamers evolution are presented in Fig.5. This behavior is typical for initial time period, when every next HV electrode operation cycle (30 μs) is not strongly influenced by changes in flow field. On later stages, heating of the air over the dielectric changes the air density, which, in general, increases the amplitude of one of the important quantity, E/n. In turn, increase in reduced electric field E/n results in faster ionization near the streamer head. As a result, more streamers pass full distance from the HV electrode to the axis of symmetry. It should be noted that positive-current streamer usually jumps over the portion of dielectric surface nearest to HV electrode

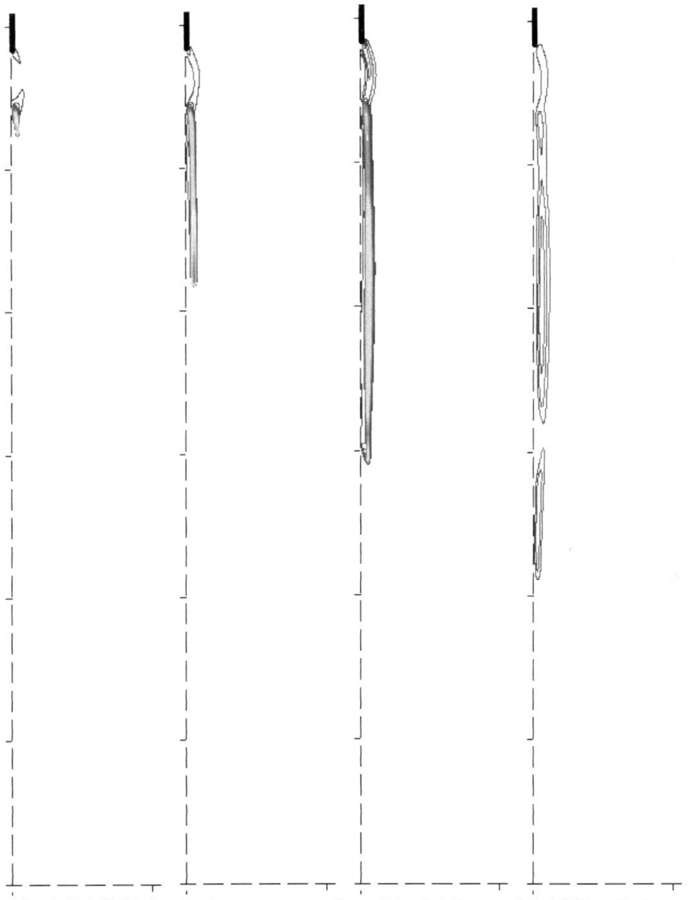

Fig.5. Ni field for the last streamer from Fig.4. Nimax=3.9·1020m-3. Part of HV electrode is designated by thick black line. Dashed vertical line is the dielectric surface. Dashed horizontal line is axis of symmetry.

edge. In general, the discharge behavior depends not only on the instant applied voltage, but also on the distribution of electric potential on the surface. Typically, both ionization rates in the head of streamer and the length the streamer passes are determined by the local values of near-surface electric field. The latter, in turn, is determined by the distribution of the surface charge. The distribution of electric charge is quite non-uniform over the surface because it depends on surface charge deposited earlier. Therefore, anode-type operation mode is observed for HV electrode even if the applied voltage is negative relative to its ground-value. Similarly, cathode-type operation mode is observed for the HV electrode even if the applied voltage is positive relative to the ground value.

Examples of streamer passing over the dielectric surface are given in Fig.5. These pictures correspond to the last current peak from Fig.4. More exact, they correspond mainly to the growing part of the peak, which takes about 90 ns. First two positive-current streamers from Fig.4 pass the complete distance between electrode and the axis. Negative-current streamers (cathode-type operation mode for HV electrode) produce typically smaller plasma concentrations over the dielectric surface in comparison with the near-edge values of ion or electron concentrations.

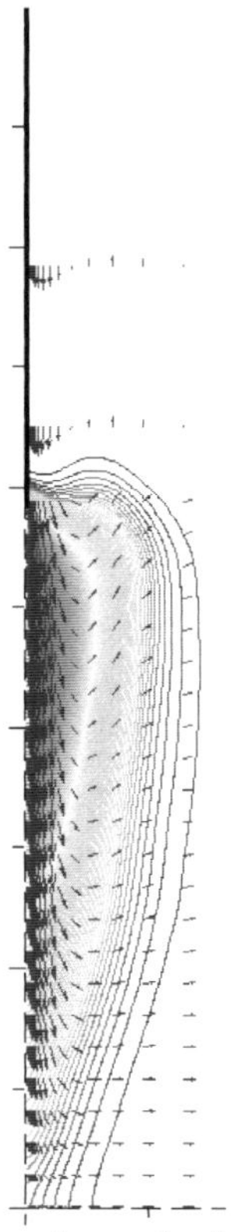

Fig.6. Temperature and velocity field at t=1.5ms calculated with time-averaging technique. Tmax=778K, Umax=1.12 m/s. HV electrode is shown by thick black line.

The main question of this work is how the pulse-periodic discharge effects on the flow field. As found from calculations, the main effect of the discharge comes from the region near the electrode edge. Local maxima of both heat release and force are located either just top or just below the leading edge of HV electrode depending on the operation mode. As far as near-edge heat release is concerned, it can be considered as pulse-periodic power source making the main contribution to the fluid disturbance. The near-edge dynamic source appearing in equation (2) due to the electric charge acts always toward the wall, which could lead to the change in the pressure field near the wall. The influence of negative-current streamers (the electrode edge is not considered) is small. The influence of positive-current streamers on the flow is remarkable though it is less than those near the electrode edge. When positive-current streamer passes over the dielectric surface, two places of disturbances can be mentioned. First, the head of streamer is always the source of both heat and force. The force directs from the electrode toward the axis. Second, the bulk of air between the quasi-neutral streamer body and the dielectric surface. This is cathode layer in which the force is directed mainly toward the wall. Combined effect of both heat release and electrostatic force is the vortex production over the plate. Example of long-term flow field obtained with time-averaged source-terms is shown in Fig.6. Note that instant flow field can be quite different from the time-averaged one.

3. Concluding remarks and future work

Direct fully self-consistent numerical simulation of the heat and dynamics effects of the sinusoidal surface dielectric barrier discharge is used to predict the flow field evolution under conditions closed to experimental ones described in the paper [1]. The preliminary results have shown that induced air flow is formed both by heated gas expansion (mostly isotropic factor) and the (vectorized) electrostatic body force. The instant distributions of the gas velocity and temperature are highly non-uniform due to the strong non-uniformities of joule heat release, pressure gradient and body force.

It is important to outline that for conditions considered (as well as for very wide range of SDBD conditions) the joule heating power is much greater than the body force work power [4]. The detailed mechanism of the well-recognized streaming flow is the goal of our next steps of the study. The combined technique of the numerical simulation of time multi-scale evolution in plasma aerodynamics systems has been developed and successfully tested for the realistic configuration.

The qualitatively correlation with the experimental data presented in paper [1] has been demonstrated. The quantitative correlation will be the task of our next step efforts.

Acknowledgement
This work was partially supported by the Russian Science Foundation, grant No. 14-50-00124. The authors are grateful to Dr. I. Moralev and to Dr. V. Soloviev for fruitful discussions.

References
[1] M Taglioli, A Shaw, A Wright, B FitzPatrick, G Neretti, P Seri, C A Borghi and F Iza, “EHD-driven mass transport enhancement in surface dielectric barrier discharges”, Plasma Sources Sci. Technol. 25 (2016) 06LT01 (5pp), doi:10.1088/0963-0252/25/6/06LT01e
[2] Bityurin V., Bocharov A., and Popov N. “Numerical Simulation of the Discharge in Supersonic Flow Around a Sphere”, Paper 2007-0223, Proc. 45th AIAA Aerospace Sciences Meeting & Exhibit, 5-8 January 2007, Reno, NV.
[3] V. A. Bityurin, A. N. Bocharov, and N. A. Popov, “A High_Frequency Discharge on a Dielectric Surface”, High Temperature, 2011, Vol. 49, No. 5, pp. 758–761
[4] V. A. Bityurin, Dynamics and Heat Release Effects in Magneto-Plasma Aerodynamics, In: WSMPA 16, Moscow, JIHT, 2016

Computational Aeroacoustics Simulation of an Air Micro-Injector with Dimples Shape on the Surface of Divergent Nozzle

Jun Liu[1*], Toshiharu Kagawa[2]

[1] *Keihin-Corp., Hoshakuji, 2021-8, Takanezawa-machi, Shioya-gun, Tochigi, 329-1233, Japan*
[2] *Tokyo Institute of Technology, 4259 Nagatsuta-cho, Midori-ku, Yokohama 226-8503, Japan*
[*] liujun3@hotmail.com (corresponding author e-mail)

Abstract. As for jet noise reduction of an air micro-injector with divergent nozzle, dimples shape in the inner surface of divergent nozzle is proposed, in the paper, the jet noise reduction effect of dimples shape is numerically confirmed at delivery pressure 1.33 atm. and 2.0 atm. (fully expanded jet Mach number 0.65 and 1.05), using Ffows Williams and Hawkings (FW-H) method base on time-accurate solutions of Large-Eddy Simulation (LES). In addition, the influence of dimples nozzle on the jet flow field is numerically investigated.

1. Introduction

Recently, jet noise is an interesting and important topic in fluid mechanics, and has received much attention from scientists and engineers. On the one hand, for scientists, the ultimate goal is elucidation of the fluid dynamic mechanisms for noise generation and construction of jet noise prediction methodology. On the other hand, engineers are significant concern in jet noise reduction. However, an understanding of noise generation mechanisms may be helpful in guiding noise reduction studies.

The early investigations on jet noise were started by Lighthill [1], where the total acoustic power of the jet is obtained with the eight power of jet velocity, and the eight power law matches surprisingly well with many jet experiments, e.g. Viswanathan [2], and thereafter, the most general form of the Lighthill acoustic analogy were developed by several studies, Ffows Williams [3] accounted the effect that the turbulent eddies that give rise to the Reynolds stresses are convected by jet velocity and their motion alters the radiated sound, the refraction effect of the mean jet velocity on sound is investigated by Mani [4] and Colonius [5], and a more accurate acoustic formulation of linear propagation effects is discussed by Goldstein [6], the nonlinear problems of sound generation is challenged by Colonius [7].

For high-speed jets, the different of jet-noise properties at small and large observation angles to the flow are studied in numerous experimental and numerical investigations, e.g. Tam [8,9,10], Morris [11,12] and Viswanathan [13], they found that the fitting functions for high observer angles are referred to as the fin-scale spectrum (FSS) and the large-scale spectrum (LSS) for the small observer angles, and the semi-empirical two-source FSS-LSS model is suggested.

Nowadays, because of the rapid development of computers and in particular the growing computational power, Computational fluid dynamics (CFD) / Computational aeroacoustics (CAA) has been very successful in solving fluid and acoustic problem for industrial applications. Direct numerical simulation (DNS) has been used to predict radiated noise from turbulent flows at low Reynolds numbers by Freund [14,15], and large eddy simulation (LES) is attractive for higher Reynolds number applications, e.g. chevron nozzle by Ali Uzun [16], beveled nozzle by Viswanathan [17], dual nozzle by Michael [18], cold and hot nozzle by Bodony [19], rectangular jet by Nichols [20], convergent–divergent nozzle by Liu [21], etc.

In the paper, attention is focused on jet noise reduction of an air micro-injector that in use for blowing away the metal powder on the machined surface (e.g. hole drilling in Aluminum) in small town factory, in many cases, jet noise of air micro-injector is a large component of ambient noise levels that in pair worker comfort and productivity, and so for reduction of jet noise level, a dimples shape in the inner surface of divergent nozzle is considered, a computational aeroacoustics simulations of this air injector is presented to confirm the jet noise reduction effect of dimples nozzle, using FW-H acoustic analogy method base on transient jet flow solutions of LES, and furthermore, the influence of dimples nozzle on the jet flow field is numerically investigated.

2. Simulation Method

2.1. Physical Model

An air micro-injector used in the study was a long injection tube with divergence nozzle and a short delivery large diameter tube, for jet noise reduction, dimples shape is engraved on the inner surface of divergence nozzle, Fig.1 shows the parameters of injector configuration in detail, injection tube length is 21mm (including divergence nozzle length 5mm), and delivery tube length is 4mm, diverging angle of nozzle is 8 deg., and the dimples shape is consist of 15 dents that its radius is 0.4 mm, the inner diameter of injection tube D_j and delivery tube D_t, are 2.5mm and 5.0mm, respectively.

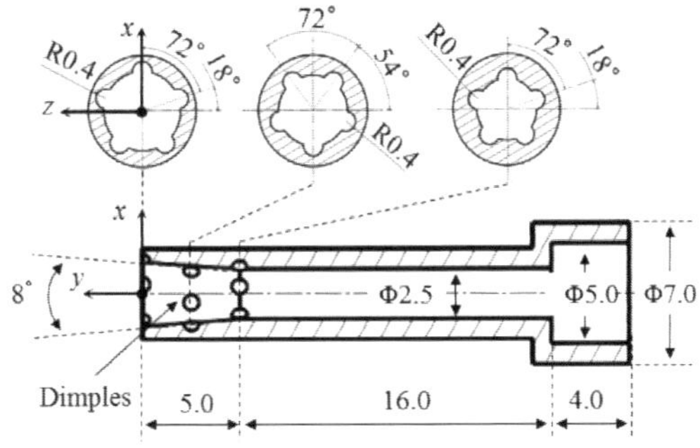

Fig.1 An air micro-injector with dimples shape on the inner surface of divergence nozzle

Here, two cases of delivery pressure are considered such as 1.33 atm. and 2.0 atm., and the fully expanded jet Mach number corresponding to delivery pressure are obtained as 0.65 and 1.05, using follows formula

$$M_j = \sqrt{\frac{2}{\kappa-1}\left\{\left(\frac{p_1}{p_a}\right)^{\frac{\kappa-1}{\kappa}} - 1\right\}} \quad , \tag{1}$$

where, κ which is specific heat ratio, equals 1.4 in case of air, p_1 is delivery pressure and p_a is atmospheric pressure.

2.2. Numerical Method

Due to dimples shape on the inner surface of divergent nozzle, it is obviously non-axisymmetric jet flow, three-dimensional computational domain is dealt with a prolate spheroid, air injector is fixed a pole of spheroid, the major axis which is arranged along air jet direction is $110D_j$ (275 mm), and the minor axis is $80D_j$ (200 mm).

Jet noise is generated from jet flow, which is governed by well-known mass conservation and Navier-Stokes momentum equations as follows

$$\frac{\partial \rho}{\partial t}+\frac{\partial \rho u_i}{\partial x_i}=0 \quad , \tag{2}$$

$$\frac{\partial \rho u_i}{\partial t}+\frac{\partial \rho u_i u_j}{\partial x_i}=-\frac{\partial p}{\partial x_i}+\frac{\partial \sigma_{ij}}{\partial x_i} \quad , \tag{3}$$

where, ρ is density, u_i and u_j are the velocity components, p is the statistic pressure, and σ_{ij} is the viscous stress tensor as

$$\sigma_{ij}=\mu\left[\frac{\partial u_i}{\partial x_j}+\frac{\partial u_j}{\partial x_i}-\frac{2}{3}\left(\frac{\partial u_k}{\partial x_k}\right)\delta_{ij}\right] \quad , \tag{4}$$

where, δ_{ij} is Kronecker delta.

For the turbulence simulation, the air was set up as a compressible ideal-gas with temperature 20 degrees Celsius so that boundary conditions for air outflow from computational domain were defined as atmospheric pressure P_a, and the injection pressure on the jet inlet is given as delivery pressure P_1. The first step, a steady-state solution of jet flow was obtained by realizable *k-ε* turbulent model, and the second step, LES of unsteady air jet flow was carried out on time step (1×10^{-5} sec.) to total time step 2.0×10^{-1} sec. with which subgrid-scal model is chosen dynamics Smagorinsky-Lilly model, and the Werner-Wengle wall function [22] is used on the wall surface of the air injector.

For the acoustic computation, the mid- and far-field formulation of the permeable FW-H surface integral method is used base on transient jet flow solutions of LES, time-accurate solutions of the jet flow field variables, such as pressure, velocity components and density on the selected interior (permeable) surface, are required to evaluate the permeable surface integrals (monopole & dipoles sound source) and partially volume integrals (quadrupoles sound source within the region enclosed by the permeable surface), and the sound pressure in the acoustic observation point is given as

$$\begin{aligned} p'(\mathbf{x},t)=&\frac{1}{4\pi a_0^4}\int_V \frac{(x_i-y_i)(x_j-y_j)}{R^3}\frac{\partial^2 T_{ij}(\mathbf{y},t')}{\partial t'^2}dV(\mathbf{y})+\frac{1}{4\pi a_0}\int_S \frac{(x_i-y_i)l_i}{R^2}\frac{\partial p(\mathbf{y},t')}{\partial t'}dS(\mathbf{y}) \\ &+\frac{1}{4\pi}\int_S \frac{(x_i-y_i)l_i}{R^3}p(\mathbf{y},t')dS(\mathbf{y}) \end{aligned} \quad , \tag{5}$$

where, T_{ij} is the Lighthill stress tensor as $T_{ij}=\rho u_i u_j-\sigma_{ij}+p_0\delta_{ij}$, a_0 is speed of sound, $\mathbf{x}(x_i, x_j)$ is the acoustic observation point where acoustic quantities are measured, $\mathbf{y}(y_i, y_j)$ is the point in the flow field where sound is generated, $R=|\mathbf{x}-\mathbf{y}|$ is therefore the distance between the acoustic observation point and the point in the flow field where sound is generated, l_j is the unit direction vector of the solid boundary, pointing toward the fluid, t is the current observation time measured at $\mathbf{x}$, respectively.

Since the permeable FW-H surface integral method is only enable to account for the contributions from the quadrupoles source (volume integral) enclosed by permeable surface, and does not include external quadrupoles source in the region outside the permeable surface, in the case of jet noise, it is necessary to select the permeable surface through the jet far downstream of the jet exit, which encloses all quadrupoles source from fine grain turbulence and large scale turbulence of turbulence jet flow (e.g. see the discussion in Michael et al. [18]). Therefore, in the paper, two integral surfaces with conical divergence of S_1=20 deg. and S_2=50 deg. which around jet flow are considered as shown in Fig.2a, and the influence of different integral surface on acoustic result is investigated by comparing with each other.

Here, 15 observation points were assumed set-up on an arc at a distance of 150 mm from the nozzle exit, and they were placed 10 deg. apart from 10 deg to 170 deg. as shown in Fig.2b.

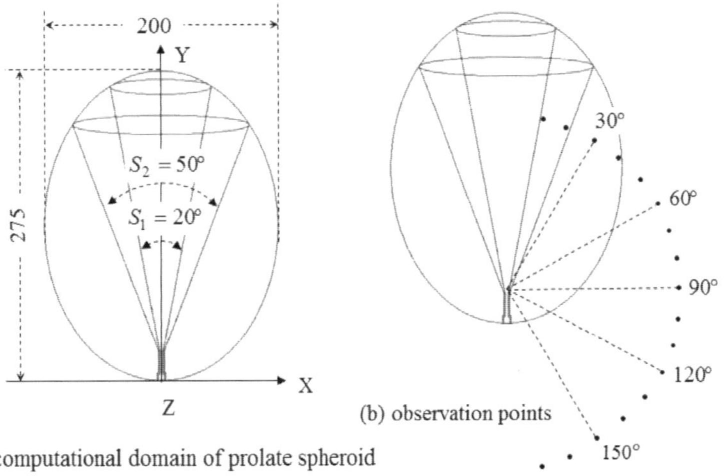

(a) computational domain of prolate spheroid (b) observation points

Fig.2 Computational domain (a) and acoustic observation points (b)

In case of using the permeable surface, the mesh resolution needs to be fine enough to resolve the transient flow structures inside the volume enclosed by permeable surface, the computing mesh grid of turbulent simulation was generated, and it also needs a fine distribution mesh at interior of dimples nozzle, due to which the numerical result is sufficiently accurate for estimate the influence of dimples geometric configuration on the turbulent jet flow, see in Fig.3, there are 2.56 million meshes for injector with dimples nozzle and 1.98 million meshes for base injector without dimples, respectively.

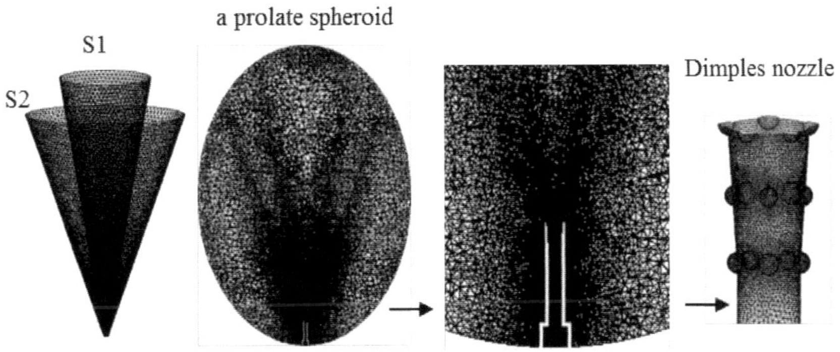

Fig.3 A fine distribution mesh at interior of dimples nozzle and permeable surface

3. Results and Discussion

In the section, first, the influence of dimples nozzle on the jet flow field is numerically investigated. Second, followed by discussion of acoustic result in the different FW-H surface, the jet-noise reduction efficiency of dimples nozzle is numerically confirmed with variation of observer points.

3.1. Numerical Results of Flow Filed

LES of unsteady air jet flow simulation was carried out by the commercial software FLUENT version 14.5, numerical solution convergence was determined by the mass flow rate and velocity components, the numerical solution was considered converged when these values changed less than 0.01% at every time step.

Figures 4a-4b display the contours of static pressure in the jet-center plane (xy-plane at z=0 and zy-plane at x=0), and the oscillation of static pressure with along the centerline of air micro-injector is plotted as shown in Fig.5, it shows that oscillation of static pressure is greatly reduced due to dimples effect in case of M_j=0.65, but in the case of M_j=1.05, oscillation of static pressure is increased at jet exit in reverse.

Additionally, the variations of axial velocity and logarithmic vorticity with along centerlines axial of jet flow are plotted as shown in Fig.6-7, it is clear that the vorticity magnitude is great reduced due to drag reduction effect of dimples configuration in the inner region of nozzle (y/Dj = -5 ~ 0), and in the region of nozzle exit, the axial velocity and vorticity of base nozzle are decreased more quickly than that of dimples nozzle in the case of M_j=0.65. Figures 8a-8b display the contours of vorticity near nozzle exit in the xy-plane at z=0 and zy-plane at x=0, where FW-H surface S_1 and S_2 are expressed as the white dashed lines, it can be found that dimples shape reduces turbulence in case of subsonic jet-flow (M_j=0.65), but the turbulence near nozzle exit is enhanced by dimples nozzle in case of supersonic jet-flow (M_j=1.05).

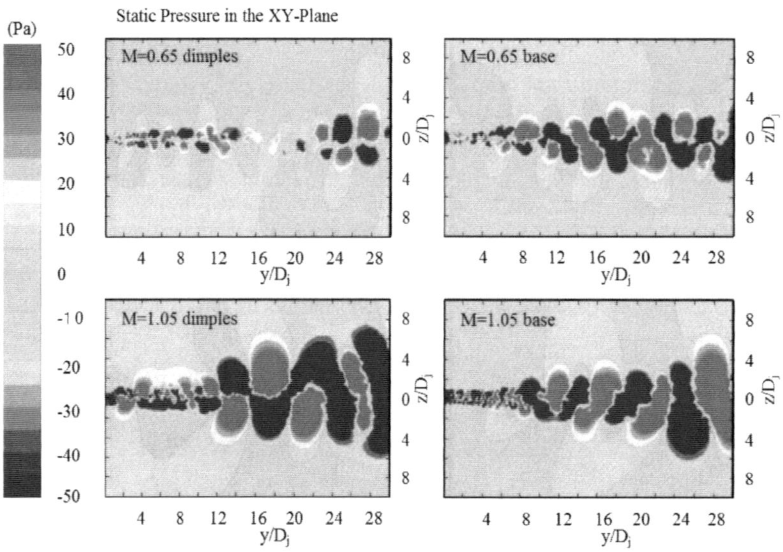

Fig.4a. Distribution of static pressure in the XY-Plane

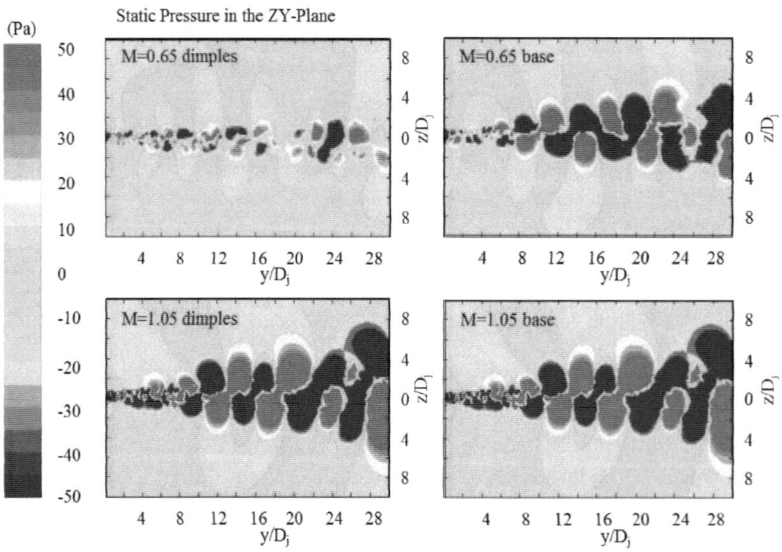

Fig.4b. Distribution of static pressure in the ZY-Plane

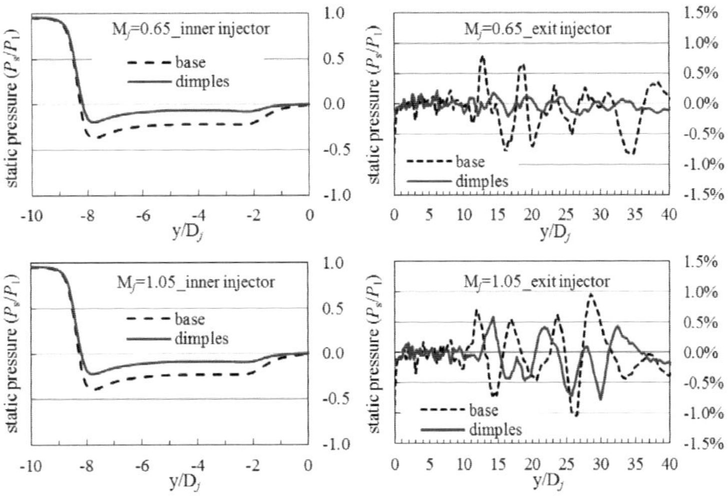

Fig.5. Oscillation of static pressure along jet centreline

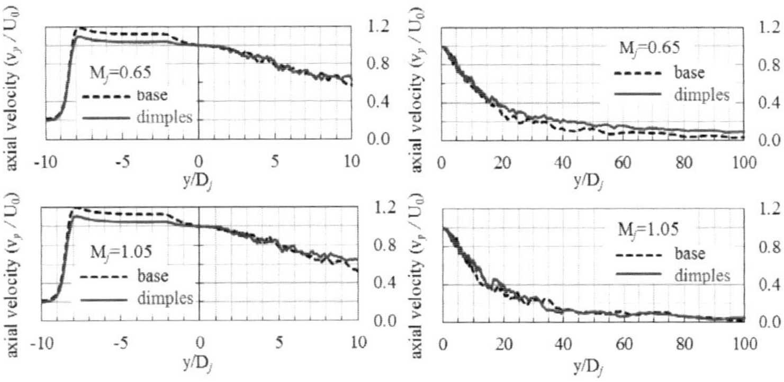

Fig.6. Reduction of axial velocity along jet centreline

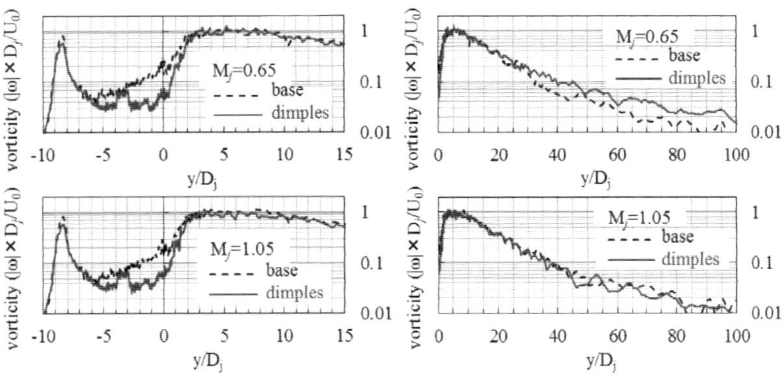

Fig.7. Variation of logarithmic vorticity magnitude along jet centerline

IOP Conf. Series: Materials Science and Engineering **249** (2017) 012009 doi:10.1088/1757-899X/249/1/012009

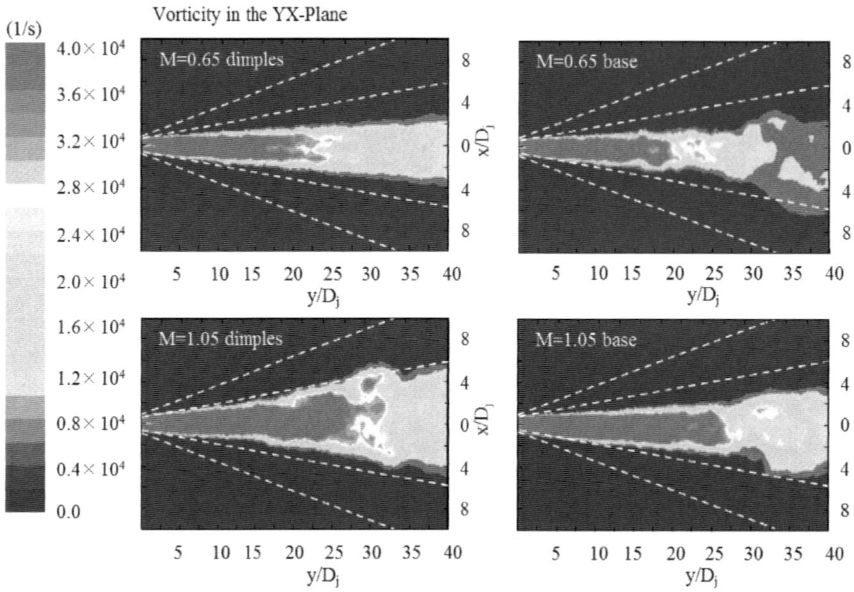

Fig.8a. Contours of vorticity near nozzle exit in the xy-Plane at z=0

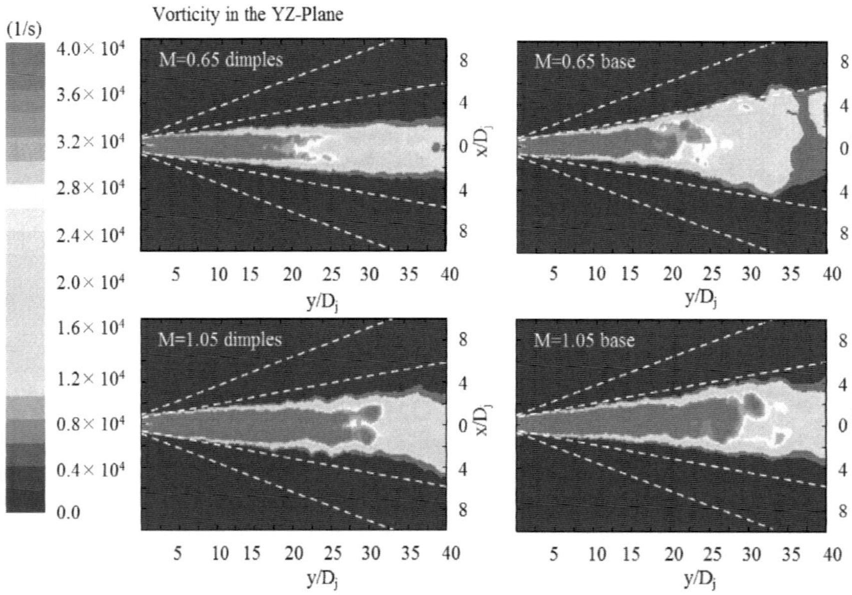

Fig.8b. Contours of vorticity near nozzle exit in the zy-plane at x=0

Moreover, as shown in the influence of dimples configuration nozzle on the structure of turbulent jet flow, four distinct zones in the region of nozzle exit have been divided with related to axial velocity decay from many studies (e.g. Yue, Z., [23])., in the paper, the variation of centerline vorticity is described in each zone of jet flow as shown in Fig.9. Immediately downstream of the nozzle is called is convergent zone (potential core) of the jet where the axial velocity is equal to the nozzle outlet, jet turbulence is confined to the annular shear layer which surrounds the potential core, and the centerline vorticity does not change also in the zone, as a simulation result, the length of the generalized potential core (the region with uniform centerline velocity ~ 99% U_0, where U_0 is axial velocity at nozzle exit of the jet-flow) is about $0.74D_j$ for base nozzle and $1.25D_j$ for dimples nozzle in case of M_j=0.65, 0.75 D_j

for base nozzle and 1.34D_j for dimples nozzle in case of M_j=1.05, respectively, it shows that the length of potential core was extended by drag reduction effect of dimples configuration. After the potential core of jet ends the axial velocity falls downstream of this point, which velocity decay can be approximated as proportional to $y^{-0.5}$, where y is the axial distance from nozzle exit), and an intermediate zone (so called transition region) appears, where the large-scale eddies which are generated with the rolling-up of the annular shear layer states to mix, and the centerline vorticity starts to increase rapidly, then fully mixed at maximum valve of centerline vorticity, from the simulation results, fully mixed (which is defined as logarithmic of $|\omega|\times D_j / U_0 \sim 0.9$) at $y/D_j = 3.98$ for base nozzle and $y/D_j = 5.24$ for dimples nozzle in case of subsonic jet (M_j=0.65), $y/D_j = 7.66$ for base nozzle and $y/D_j = 3.40$ for dimples nozzle in case of supersonic jet (M_j=1.05), respectively, it indicates that the dimples configuration enhances mixing rate of turbulent eddies at centerline in case of supersonic jet (M_j=1.05), and delays mixing speed of turbulent eddies in case of subsonic jet (M_j=0.65). however, the intermediate region of jet is often regarded as the region of the dominant noise generation.

Beyond the fully mixed point, the turbulent flow towards the self-similar zone jet, the centerline vorticity starts to reduce, the axial velocity decay is approximated as proportional to $y^{-1.0}$. And the last zone of turbulent jet flow is termination zone, the rates of evolution of the flow field and turbulent fluctuations are quite different, the actual mechanisms in this zone are not understood properly.

By Yue.Z [1999]

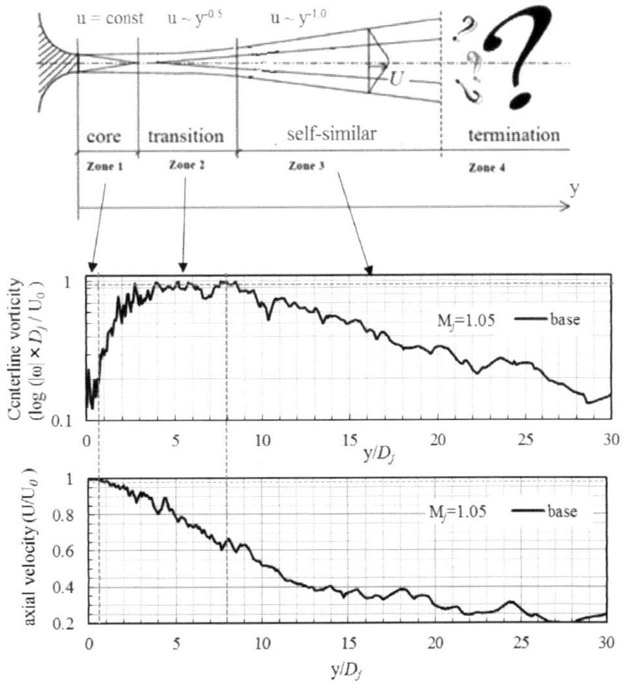

Fig.9 Four distinct zones region of nozzle exit

3.2. *Results of Acoustic Field*

In FLUENT version 14.5, results of acoustic field are obtained by the permeable FW-H surface integral method, using acoustic signals files with stocked time histories of jet flow field variables on the selected surface, the acoustic signals file was stocked at every 10 time steps of LES unsteady air jet flow simulation, and total 2000 files are used for acoustic computation.

3.2.1. Effect of FW-H surface on acoustic results

Figures 10 give the comparisons of spectrum of sound pressure level (SPL) at several observation points (observer angle = 30, 60, 90, 120, 150) in the different FW-H, it shows that the acoustic results is difference in particular lower frequency of both base nozzle and dimples nozzle at all observer angle, it is clear that the acoustic computation should select larger integral surfaces with conical divergence of S_2=50 deg., which encloses all quadrupoles source from large scale turbulence of turbulence jet flow as shown in Fig.8 again.

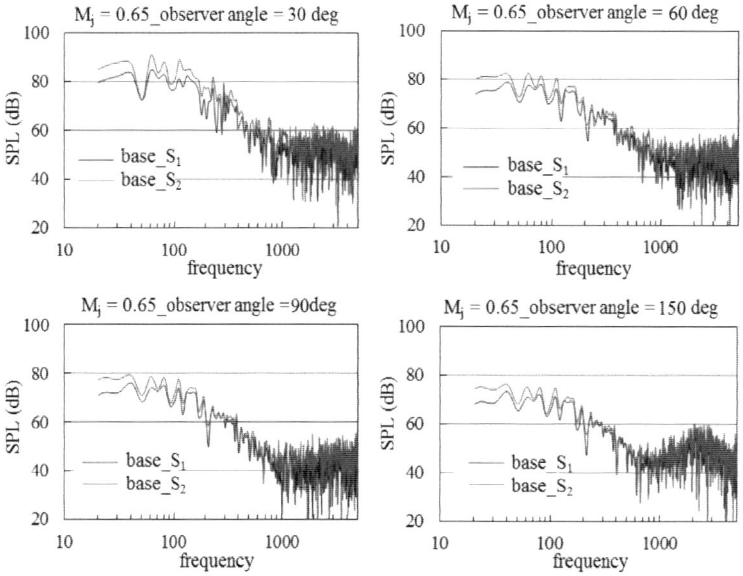

Fig.10a Effect of FW-H surface on acoustic results in case of M_j=0.65

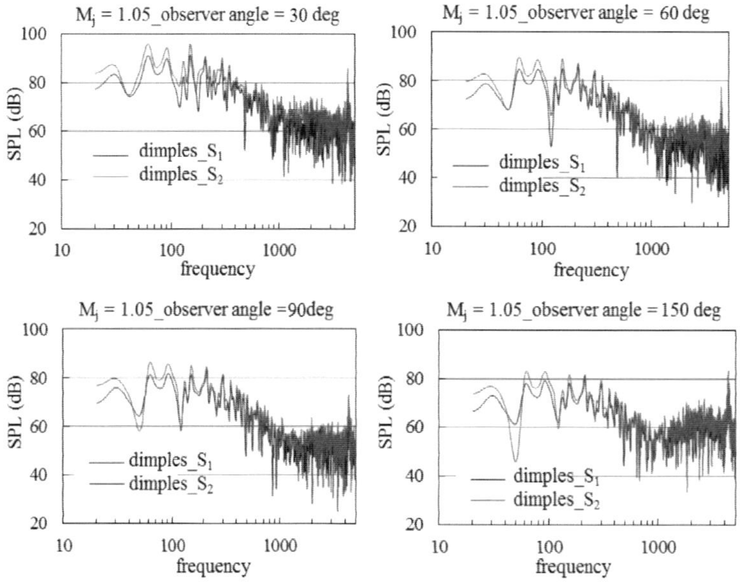

Fig.10b Effect of FW-H surface on acoustic results in case of M_j=1.05

3.2.2. Acoustic results with variation of observer angles

Jet noise reduction effect of dimples nozzle in the all observer points (from 10 to 170 deg.) are plotted as shown in Fig.11, it can be found that noise reduction effect of dimples nozzle is only valid in case of subsonic jet (M_j=0.65), especially in the zone of high frequency noise. Conversely, the jet noise is increased behind of nozzle exit (at observer angles 120 and 130 deg.) in case of dimples nozzle for supersonic jet (M_j=1.05), due to the notch lip of dimples nozzles exit, in which the notch itself caused additional noise by increasing the sound pressure level at the nozzle exit.

In addition, Fig.12 compares the spectrum of sound pressure level with base nozzle and dimples nozzle at observer angles 30, 60, 90, 120, 150deg., it also shows the differences in noise spectrum with variation zone of observer angles, it means that distinct physical mechanisms of jet noise should exist, noise from subsonic jets is mainly due to turbulent mixing, and two jet noise source are fine-scale turbulence structures at larger observer angle and larger-scale turbulent structures at small observer angle, which the high frequency noise increases at lager observer angle and is un-variant at small observer angle. Supersonic jet noise consists of three main components, the first source is Mach wave emission (turbulent mixing noise) which dominates in the downstream of jet flow, the second source is screech tones for observer angles behind jet exit, the third source is broadband shock noise at observer angles near jet exit, the latter two noise components are generated only when the jet is imperfectly expanded and a shock cell structure is formed in the jet plume (see image graphic Fig.13 by B. Henderson [24]), moreover, in case of C-D (Convergent-Divergent) nozzle, transonic tones which is generated by internal shock cell structure of nozzle effects all observer angles, and the frequency of that is differs from the frequency of screech tones (Zaman[25]), however, in the present study, the high pick value which is considered as transonic tones is clearly visible at the high frequency (~ 4318 Hz) in case of dimples nozzle for supersonic jet (M_j=1.05), due to the internal shock cell structure in tube inlet of injector as shown in Fig.14, and despite the fact that the internal shock cell structure is also occurred in case of base nozzle, the transonic tones is not found within ranges of less than 5000 Hz, the reason for this can be considered that the dimples shape on the nozzle surface reduces the inflection from internal wall of nozzle, which is advantageous for transonic tones propagation in the nozzle.

Furthermore, Fig.15 compares the results of base nozzle with predictions from the Michel model [26] and the experimental data by Tam [10] for a Mach number M ($\equiv U_0/a_0$) = 0.9 jet, for the variation tendency of SPL with observer angles changes, the aeroacoustics characteristics with observation angle are in qualitative agreement with those published data.

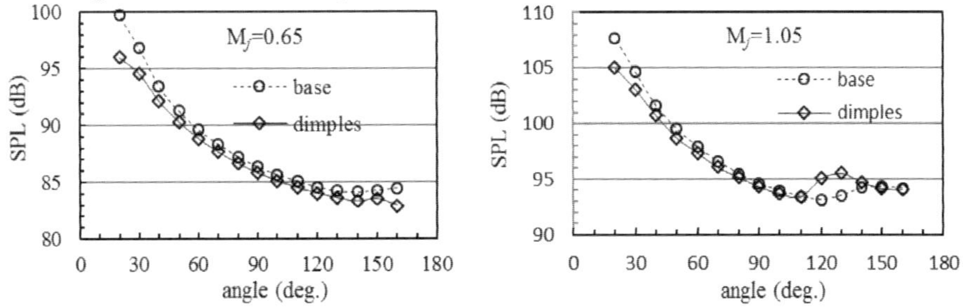

Fig.11 Jet noise reduction effect of dimples nozzle in the all observer points

IOP Conf. Series: Materials Science and Engineering **249** (2017) 012009 doi:10.1088/1757-899X/249/1/012009

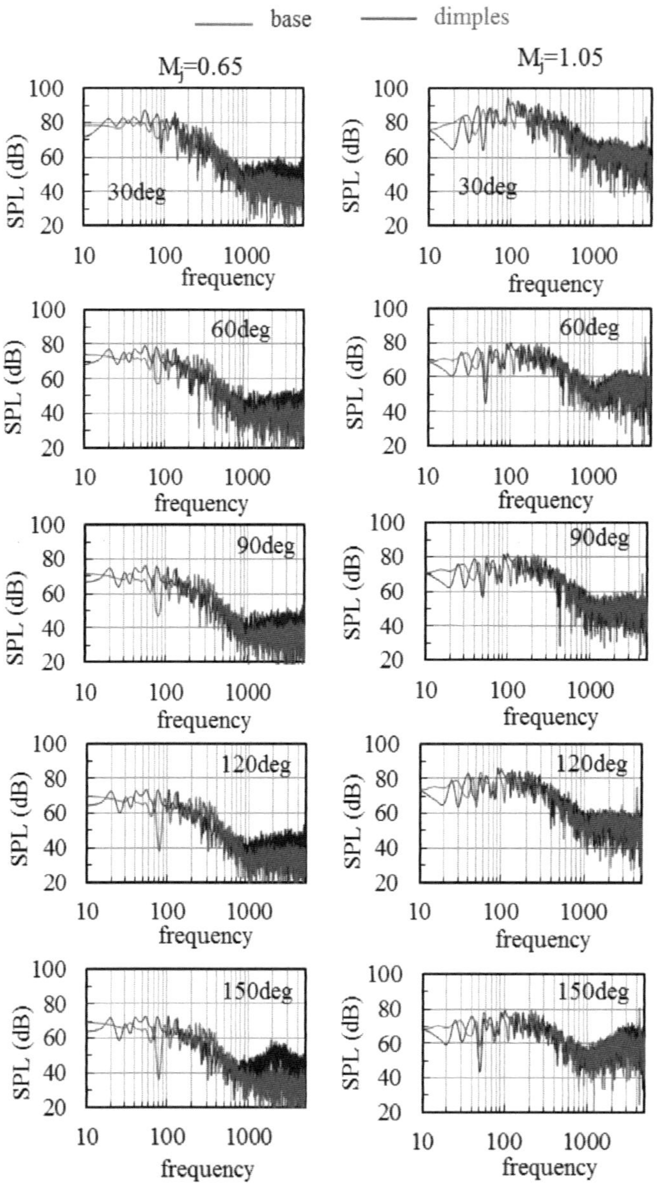

Fig.12 Compares spectrum of sound pressure level with base nozzle and dimples nozzle

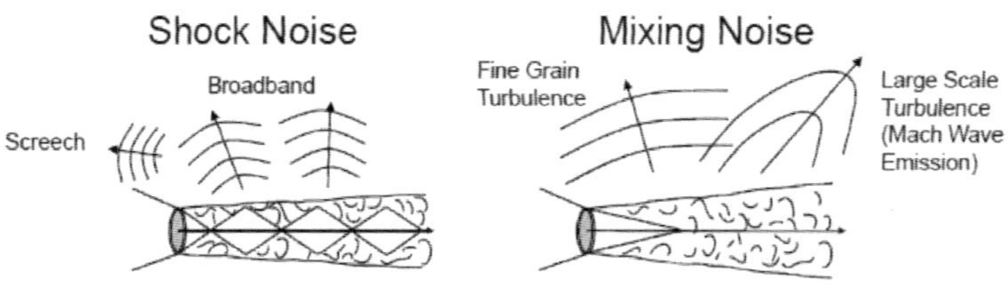

Fig.13 Jet noise sources by B. Henderson [24]

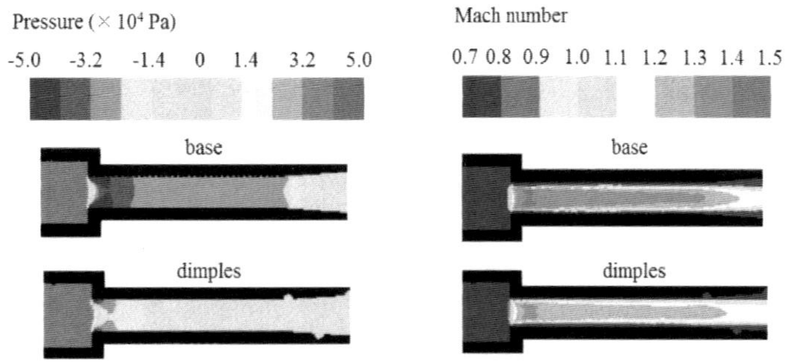

Fig.14 Internal shock cell structure in tube inlet of injector in case of M_j=1.05

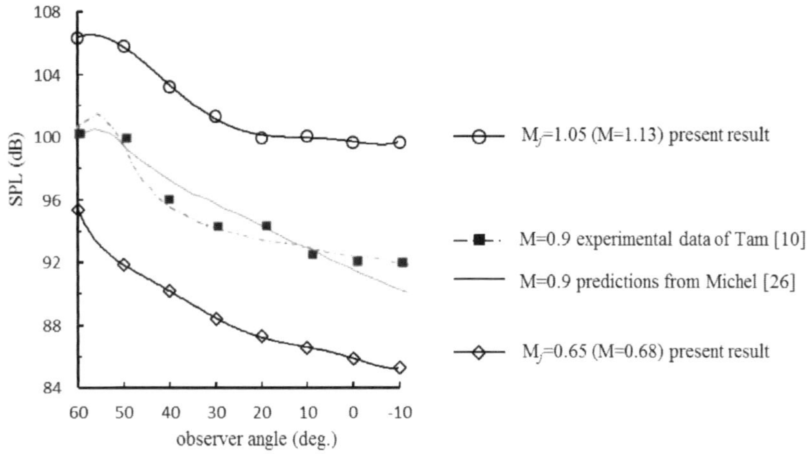

Fig.15 Comparison with Michel model and experimental data for a Mach number M =0.9 jet.

4. Conclusion

In the paper, jet noise computation of an air micro-injector with dimples shape on the inner surface of divergent nozzle is carried out in case of subsonic and supersonic jet flow, using LES/FW-H simulation method, and the influence of dimples nozzle effect on the turbulent jet flow and the jet noise reduction effect of dimples shape are numerically investigated. As for results, the noise reduction effect of dimples nozzle is only confirmed in case of subsonic jet, in which the fine-scale turbulent structure near jet exit becomes weak due to drag force reduction effect of dimples nozzle, and conversely, the jet noise level of dimples nozzle is increased in case of supersonic jet, because the dimples shape disorders the jet flow and enhances turbulent structures near jet exit.

5. References

[1] Lighthill, M.J., "On sound generated aerodynamically: I. General theory," *Proc. Royal Soc. of London A*, Vol.222, 1952, pp.564-587.
[2] Viswanathan, K., "Mechanisms of jet-noise generation: classical theories and recent developments," *Int. J. Aeroacoustics*, Vol.8, 2009, pp.355–407.
[3] Ffowcs Williams, J.E. and Hawkins, D.L., "Sound generation by turbulence and surface in arbitrary motion," *Proc. Royal Soc. of London A*, vol.264, 1969, pp.321-32.
[4] Mani, R., "The influence of jet flow on jet-noise, parts 1 and 2," *J. Fluid Mech.*, Vol.73, 1976, pp.753–793,
[5] Colonius, T., Lele S. K., Moin, P., "Sound generation in a mixing layer," *J. Fluid Mech.*, Vol.330, 1997, pp.375–409.
[6] Goldstein, M. E., "A generalized acoustic analogy," *J. Fluid Mech.*, Vol.488, 2003, pp.315–333.

[7] Colonius, T., and Lele, S. K., "Computational Aeroacoustics: Progress on Nonlinear Problems of Sound Generation," *Prog. Aerospa. Sci.*, Vol.40, 2004, pp.345-416.
[8] Tam, C. K. W., Golebiowski, M. & Seiner, J. M., "Two components of turbulent mixing noise from supersonic jets," *In 2nd AIAA/CEAS Aeroacoustic Conf.*, State College, PA, 6–8 May 1996. AIAA-96-1716.
[9] Tam, C. K. W., Auriault, L., "Jet mixing noise from fine scale turbulence," *AIAA Journal*, Vol.22, 1999, pp.145–153.
[10] Tam, C. K. W., Viswanathan, K., Ahuja, K. K., and Panda, J., "The Sources of Jet Noise: Experimental Evidence," *J. Fluid Mech.*, Vol. 615, 2008, pp. 253–292.
[11] Morris P. J., Farassat F., "Acoustic analogy and alternative theories for jet-noise prediction," *AIAA Journal*, Vol.40, 2002, pp.71–680
[12] Morris P. J., "A note on noise generation by large scale turbulent structures in subsonic and supersonic jets," *Int. J. Aeroacoustics*, Vol.8, 2009, pp.301–316.
[13] Viswanathan, K., "Aeroacoustics of hot jets," *J. Fluid Mech.*, Vol.516, 2004, pp.39-82.
[14] Freund, J. B., Lele, S.K. and Moin, P., "Numerical Simulation of a Mach 1.92 Turbulent Jet and Its Sound Field", *AIAA Journal*, Vol.38, No.11, 2000, pp.2023-2031
[15] Freund J. B., "Noise sources in a low Reynolds number turbulent jet at mach 0.9," *J. Fluid Mech.*, Vol.438, 2001, pp.277–305
[16] Ali Uzun and M. Yousuff Hussaini, "Simulation of Noise Generation in the Near-Nozzle Region of a Chevron Nozzle Jet," *AIAA Journal*, Vol.47, No.8, 2009, pp. 1793-1810.
[17] Viswanathan, K., Shur M.L., Spalart P.S., Strelets M.Kh., "Numerical prediction of noise from round and beveled nozzles," CD-ROM, *Proc.of Euromech Colloquium*, Vol.467: "Turbulent Flow and Noise Generation", Marseille, France, July 18-20, 2005.
[18] Michael L. Shur , Philippe Spalart , Michael Strelets, "Noise Prediction for Increasingly Complex Jets. Part I, Methods and tests; Part II, Applications," *Int.J. Aeroacoustics*, Vol.4, No.4, 2005, pp.247-266.
[19] Bodony, D. J. and Lele, S. K., "Jet Noise Prediction of Cold and Hot Subnoic Jets Using Large-eddy Simulation," AIAA Paper No. 2004-3022, May 2004
[20] Nichols, J. W., Ham, F. E. & Lele, S. K., "High-fidelity large-eddy simulation for supersonic rectangular jet noise prediction," AIAA Paper 2011-2919, 2011.
[21] Junhui Liu, K. Kailasanath, Ravi Ramamurti, David Munday, Ephraim Gutmark, and Rainald Lohner, "Large-Eddy Simulations of a Supersonic Jet and Its Near-Field Acoustic Properties," *AIAA Journal*, Vol. 47, No. 8 (2009), pp. 1849-1865.
[22] Werner, H. and Wengle, H., "Large-eddy simulation of turbulent flow over and around a cube in a plate channel," *Proc. 8th Symp. On Turbulent Shear Flows*, Munich, Germany, pp.19.4.1-19.4.5, 1991.
[23] Yue, Z., "Air jet velocity decay in ventilation applications," Installationsteknik Bullettin, 1999, n. 48, ISSN 0248-141X.
[24] Henderson, B., Norum, T., "Impact of Air Injection on Jet Noise," NASA-REPORT, No.20080002268, Fall Acoustics Technical Working Group, December 4 –5, 2007, Cleveland, OH.
[25] Zaman, K.B.M.Q., Dahl, M.D., Bencic, T.J., and Loh, C.Y., "Investigation of a Transonic Resonance with ConvergentDivergent Nozzles," Journal of Fluid Mechanics, Vol. 263, 2002, pp. 313-343
[26] Michel, U., "The role of source interference in jet noise," AIAA Paper 2009-3377, *In 15th AIAA/CEAS Aeroacoustics Conf.*, 11–13 May 2009, Miami, FL.

IOP Conf. Series: Materials Science and Engineering **249** (2017) 012010 doi:10.1088/1757-899X/249/1/012010

Liquid Crystals, PIV and IR-Photography in Selected Technical and Biomedical Applications

Jan Stasiek, Marcin Jewartowski

Gdansk University of Technology, ul. Narutowicza 11/12, 80-233 Gdansk, Poland

jstasiek@pg.edu.pl

Abstract. Thermochromic liquid crystals (TLC), Particle Image Velocimetry (PIV), Infrared Imaging Themography (IR) and True-Colour Digital Image Processing (TDIP) have been successfully used in non-intrusive technical, industrial and biomedical studies and applications. These four tools (based on the desktop computers) have come together during the past two decades to produce a powerful advanced experimental technique as a judgment of quality of information that cannot be obtained from any other imaging procedure. The brief summary of the history of this technique is reviewed, principal methods and tools are described and some examples are presented. With this objective, a new experimental technique have been developed and applied to the study of heat and mass transfer and for biomedical diagnosis. Automated evaluation allows determining the heat and flow visualisation and locate the area of suspicious tissue of human body.

1. Introduction

In order to demonstrate the feasibility of thermochromic liquid crystals techniques, true-colour image processing and particle image velocimetry in practical heat transfer and flow visualization contexts several experiments has been performed. The first set was carried out to investigate temperature and then heat transfer coefficient distributions (in Nusselt number form) on film-heated models in channels with ribs/turbulators. Two experimental techniques were employed in the research for comparison: firstly, the steady-state and secondly transient technique. The flow pattern produced by transverse vortex generators (ribs) was visualized using planar beam of double-impulse laser tailored by a cylindrical lens and oil particles. Sequential images of the particles in a cross sectional plane taken with CCD video camera from the downstream side of the flow were stored on a personal computer to obtain distributions of velocity vectors by means of the PIV method.

Also in experimental investigation of the film cooling effectiveness for model turbine blades using TLC technique is presented. The film cooling and visualization of the few different holes configuration in steady-state and transient performance were analyzed experimentally.

In biomedical situations, e.g. skin diseases, breast cancer, blood circulation and other medical application, TLC, IR and image processing are successfully used as an additional non-invasive diagnostic method especially useful for screening large groups of potential patients. Biomedical data presented in the paper is based on results taken from literature and from medical researchers collaborating with authors.

Using the automated evaluation allows determining the heat transfer coefficient without arbitrary influence of human interpretation. Most of described studies and obtained data (liquid crystals experiments, pressure gradient, flow visualization and PIV) provide useful information for better understanding of thermal performances of many industrial installations and medical screening.

2. Liquid Crystals

Liquid crystals constitute a class of matter unique in exhibiting mechanical properties of liquids (fluidity and surface tension) and optical properties of solids (anisotropy to light, birefringence). Although they were first observed more than 120 years ago by an Austrian botanist, Friedrich Reinitzer, studies of liquid crystals were confined to material laboratories until the 60's when applications to science and engineering began [3]

The original use of the term 'liquid crystal' is credited to Lehman, a German physicist. In many publications of the time, Lehman use terms like Fliesende Krystalle (flowing crystals) and Flussige Krystalle (fluid crystals) to describe the unusual phase that existed between the two melting points of not only cholesteryl benzoate, but also a large number of other compounds he studied.

Liquid crystals are temperature indicators that modify incident white light and display colour whose wavelength is proportional to temperature. They can be painted on a surface or suspended in the fluid and used to make visible the distribution of temperature. Normally clear, or slightly milky in appearance, liquid crystals change in appearance over a narrow range of temperature called the 'colour-play interval' (the temperature interval between first red and last blue reflection), centered around the nominal 'event temperature'. The displayed colour is red at the low temperature margin in the colour-paly interval and blue at the high end. Within the colour-play interval, the colours change smoothly from red to blue as a function of temperature [4]. The effect is reversible which, in practical therms, means that a single application of liquid crystals to a model can be used repeatedly (usually until coating is oblated or otherwise damaged). Both the colour play interval and the event temperature range of a liquid crystal can be selected by adjusting its composition, and materials are available with event temperatures from -30°C to 120°C and with colour play bands from 0.5°C to 20°C [4-5] although not all combinations exist of event temperature and colour play band widths. Widths of 1°C or less will be called narrow band materials, while those whose band width exceeds 5°C will be called wide band. The types of material to be specified for a given task depend on the type of image interpretation technique to be used [1].

Liquid crystals are classified into three categories according to their molecular structures: smectic, nematic and cholesteric. However, from a structural viewpoint, it can be argued that there are only two basic types of liquid crystals, chiral-nematic and smectic, with cholesteric being regarded as a special kind of nematic. Their structures are shown schematically in figure 1 [5].

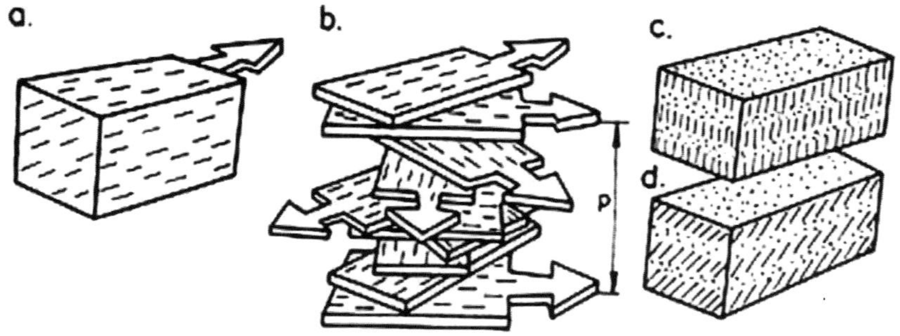

Figure 1. A different view of the mesophase structure. The orientation of the molecular director is represented by the arrows: a – nematic, b – cholesteric, c – smectic A, d – smectic B [5]

For flow analysis the suspension of thermochromic liquid crystals can be used to make visible the temperature and velocity fields in liquids. By dispersing the liquid crystal material into the liquid they become not only classical tracers used for flow visualisation but simultaneously small thermometers monitoring local fluid temperature [6]. A collimated source of white light must be used to illuminate a selected cross-section of the flow (light sheet technique) and colour images are acquired in a perpendicular direction. In the following examples the un-encapsulated TLC's tracers have been applied to measure both temperature and velocity flow fields. We found that light scattered by the capsule shell inevitably diminishes saturation of the observed colours. Because in a slow motion the stress effects are negligible we prefer application of fine dispersed raw material in flow measurements.

The liquid crystals can be also used as temperature sensors in the case of investigations in strong magnetic field [7-8]. Magnetic field was the source of additional force, which was able to enhance heat transfer of paramagnetic fluids connected directly with temperature changes indicated by the liquid crystals. To get the qualitative information about temperature field an image analysis is necessary. Some publications described the suitability of color space transformations for such optical temperature measurements [9].

Many image processing operations developed for processing grey-level (intensity) images can be readily applied to HSI colour images. These include image transformations, enhancements, analysis, compression and restoration. A schematic view of image processing system (developed by one of the authors) is shown in figure 2.

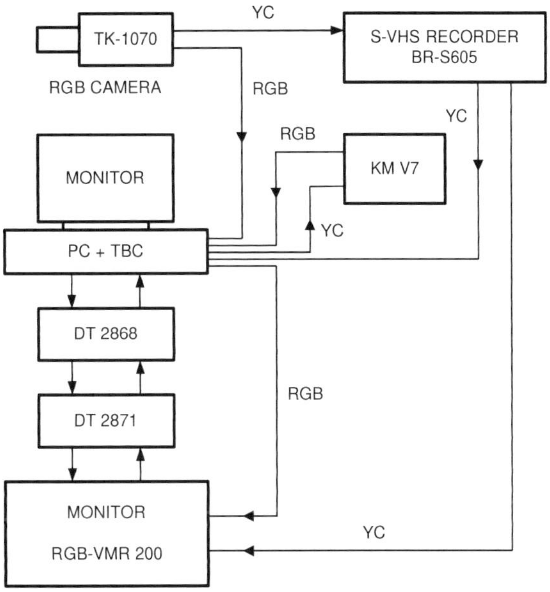

Figure 2. The information flow chart for a true-colour image interpreter and computer-controlled video system

Before the execution of a thermal or flow visualization experiment, we should recognize the characteristics of the overall combination of TLC, the light source, the optical camera system (ordinary, CCD black and white or RGB camera), and make a rational plan for the total measurement system. The relationship between the temperature of the crystal and the measured Hue of the reflected lights defines the calibration curve for the liquid crystal. The result is a curve relating the Hue of the reflected light to the surface temperature. A known temperature distribution exists on a 'calibration plate' (brass plate) to which is attached the light crystal layer. In order to maintain a linear temperature distribution with desired temperature gradients, one end of a brass plate was cooled by stabilized water and the other end controlled electrically to give a constant temperature [10-11]. The brass plate with the liquid crystal layer is calibrated in place in the wind tunnel with the same lighting level and viewing angle used during the data acquisition phase of the experiment [11]. The distribution of the colour component pattern on the liquid crystal layer was measured by RGB colour camera and series of images at a different temperatures defines the calibration. A representative calibration curve including Hue and temperature distribution along the plate is shown in figure 3.

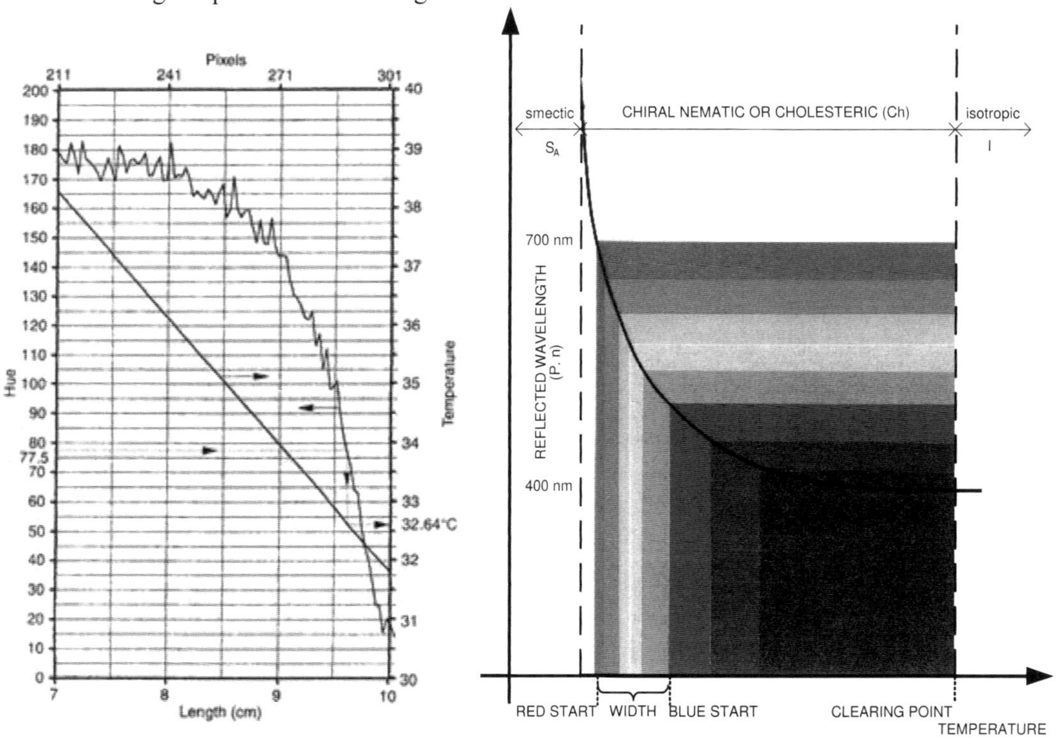

Figure 3. Representative calibration curve – Hue and temperature distribution along the test plate (left; for Hue = 77.5 temperature is 32.64°C) and typical reflected wavelength (colour) temperature response of a TLC mixture (right)

Two main methods of surface temperature measurement are performed involving steady state and transient techniques.

2.1. Steady State Analyses - Constant Flux Method

The steady state techniques employ a heated model and the TLC is used to monitor the surface temperature. Usually a surface electric heater is employed such that the local flux, q is known and this, together with the local surface temperature, T_w, (found from the TLC), gives the local heat transfer coefficient, h,

$$q = I^2 r \tag{1}$$

and

$$h = \frac{q}{T_a - T_w} \tag{2}$$

where T_a is a convenient driving gas temperature, I is the current and r is the electrical resistance per square of the heater.

2.2. Steady State Analyses - Uniform Temperature Method

The TLC-coated test specimen forms one side of a constant temperature water bath and is exposed to a cool/hot air flow. The resulting thermograph is recorded on film or video and further measurement positions are obtained by adjusting the water bath temperature. This method is more time consuming

IOP Conf. Series: Materials Science and Engineering **249** (2017) 012010 doi:10.1088/1757-899X/249/1/012010

due to the large volume of water that needs to be heated. In this case, the heat transfer coefficient is determined by equating convection to the conduction at the surface,

$$h(T_a - T_w) = \frac{k}{x}(T_w - T_b) \tag{3}$$

where, T_b is a water-side temperature of the wall, x the wall thickness and k the thermal conductivity.

2.3. *Transient Method*

This technique requires measurement of the elapsed time to increase the surface temperature of the TLC-coated test specimen from a known initial temperature predetermined value. The rate of heating is recorded by monitoring the colour change patterns of the TLC with respect to time. If the specimen is made from a material of low thermal diffusivity and chosen to be sufficiently thick, then the heat transfer process can be considered to be one-dimensional (1-D) in a semi-infinite block. Numerical and analytical techniques can be used to solve the 1-D transient conduction equation. The relationship between wall surface temperature, T, and heat transfer coefficient, h, for the semi-infinite case is,

$$\frac{T-T_i}{T_a-T_i} = 1 - e^{\beta^2} erfc(\beta) \tag{4}$$

$$\beta = h\left(\frac{t}{\rho c k}\right)^{0,5} \tag{5}$$

where, ρ, c and k are the model density, specific heat and thermal conductivity. T_i and T_a are the initial wall and gas temperatures and t is time from initiation of the flow. More recent formula for evaluation of heat transfer coefficient h is in the following form:

$$h = -\frac{\delta \rho c}{2t} ln\left[\frac{T_a-T}{T_a-T_i}\right] \tag{6}$$

where, δ is a plate wall thickness and the transient local surface temperature T is detected after a time interval t [1,5,12-13].

3. Particle Image Velocimetry (PIV)

Particle Image Velocimetry technique is another well-established experimental method in fluid mechanics, that allows quantitative measurement of two-dimensional flow structure.

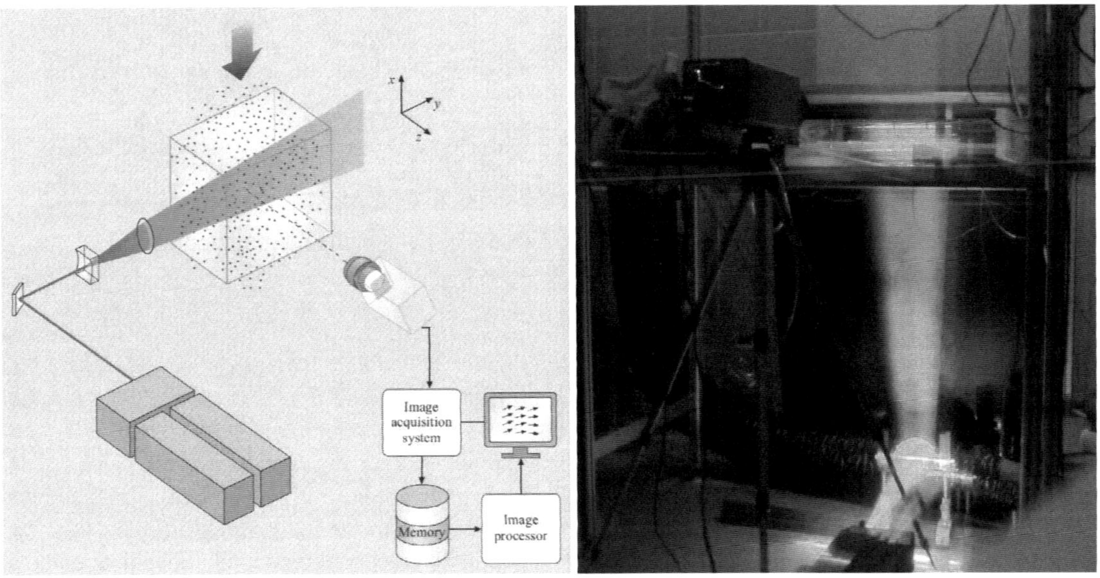

Figure 4. Schematic of a typical PIV measurements (left) and "laser light sheet" (right)

It enables measurements of the instantaneous in-plane velocity vector field within a planar section of the flow field and allows to calculate spatial gradients, dissipation of turbulent energy, spatial correlations, and the like. In PIV technique selected cross-section of the investigated seeded flow is illuminated by laser light formed in thin "light sheet" (figure 4). Images of the flow are recorded by CCD camera and correlated to calculate instantaneous velocity fields.

4. IR-Photography

Infrared photography is a non-contact measurement method, which enables visualisation and registration of temperature distribution on the surface. Every object with temperature above 0 K emits radiation as an electromagnetic wave. Figure 5 presents blackbody radiation as a function of wavelength and temperature given by Planck. Radiation in the range of 0.8 μm up to 1000 μm is non visible by human eye and is called infrared radiation (IR). Its intensity is a function of temperature and emissivity of the object surface. IR-photography measures from the distance intensity of infrared radiation emitted by the body and converts it into digital signal. That signal is enhanced and transformed, which leads to obtaining a temperature field of measured object as a thermogram (map of temperature field). It can be later analysed using computer software.

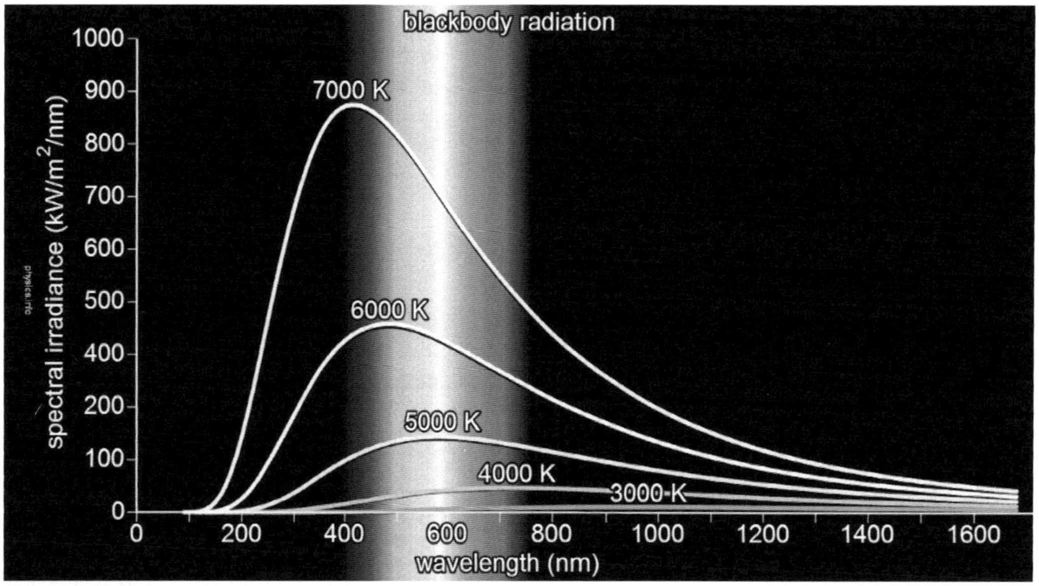

Figure 5. Spectral irradiance graphs as a function of wavelength and temperature [14]

This technique has wide applications in industrial [15], scientific [16] and medical examination [17].

5. Experimental facility for heat and flow visualisation

The experimental study was carried out using an open low-speed wind tunnel consisting of entrance section with fan and heaters, large settling chambers with diffusing screen and honeycomb, and then working sections (figure 6). Air is drawn through the tunnel using a fan able to attain the Reynolds numbers in the range from 500 to 40 000. The working air temperature in the rig ranges from 15°C to 65°C produced by the heater or cooler positioned just downstream of the inlet. The major construction material of the wind tunnel is perspex. Local and mean velocity are measured using conventional Pitot tubes and DISA hot-wire velocity probe.

The alternative effects of constant wall temperature and constant heat flux boundary conditions are obtained using plate electric heater. Photographs are taken using RGB video-camera and a true-color image processing technique. The liquid crystals used here, manufactured in sheet form by Hallcrest

[18], had an event temperature range from 30.0 to 35.0°C. In this particular experiment uncertainty for temperature measurement was estimated at about ± 0,05°C by considering only the section of the surface used in the experiment, span-wise non-uniformities in Hue value are minimized.

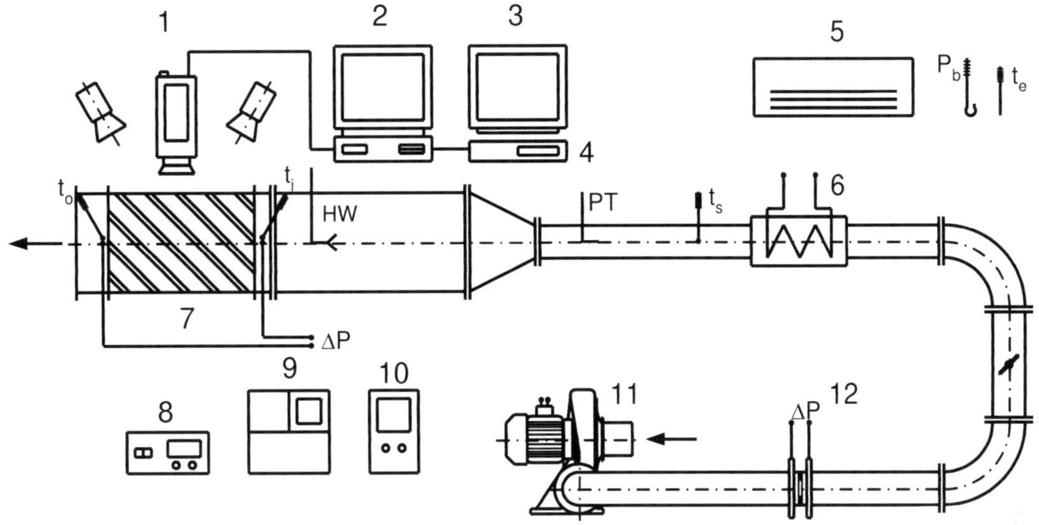

Figure. 6. Open low speed wind tunnel, 1 – RGB camera (TK-1070), 2 – PC, 3 – monitor (RGB-VMR 200), 4 – S-VHS recorder, 5 – air conditioning system, 6 – heater, 7 – LC mapping section, 8 – digital micro manometer FCO 12, 9 – DISA hot wire system, 10 – variac, 11 – fan, 12 – orifice

In many cases, as mentioned above, remarkable enhancement of local and spatially averaged surface heat transfer rates are possible with rib turbulators, in spite of the lower local Nusselt number at certain locations along the ribbed surfaces. The test surface that is analyzed contains a collection of rib turbulators that are perpendicular and angled with respect to the flow stream (figure 7) [19].

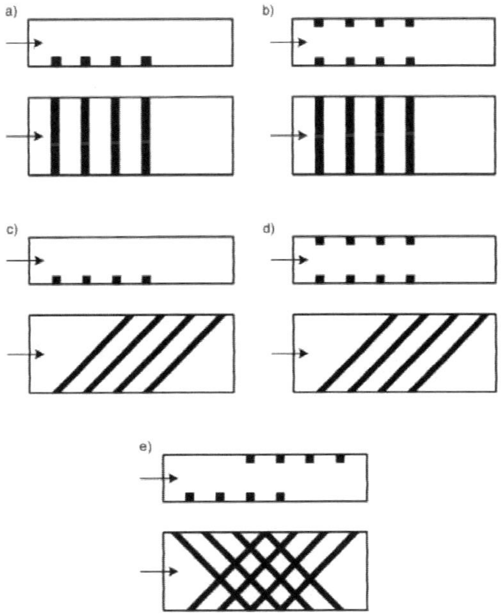

Figure 7. Schematic view of five types of transverse vortex generators (rectangular ribs)

To determine the surface heat flux (used to calculate heat transfer coefficients and Nusselt numbers), the convective power levels provided by the thermofoil heaters are divided by flat test surface area. Spatially resolved temperature distributions along the bottom rib turbulator test surface are determined using liquid crystals thermography (LCT) and true-color image processing system commercially available from Hallcrest [18].

The main part of PIV set-up as a part of wind tunnel consists of a transparent model of ribbed channel, laser light source (30 mJ double pulse Nd:YAG laser SoloPIV, New Wave Research, Inc.) and high resolution 12 bit digital CCD camera (1280×1024 pixels, PCO SensiCam). This system permits acquisition of two images at the minimum time interval of 200ns, exposition time of 5ns, and about 3.75 Hz repetition rate. The PIV recording system installed on 3 GHz Pentium 4 computer with 3GB RAM capable for acquisition over 200 pairs of images during a single experimental run. The PIV measurements were performed for pure air seeded with small droplets (few micrometers in diameter) of synthetic oil DEHS (Di-Ethyl-Heksyl-Sebacat). The oil drops volumetric concentration was very low (< 0.001), hence they did not affect the flow structure.

Experimental facility is also equipped with Infrared Camera FLIR T440bx. T-Series camera offer superior sensitivity, high resolutions, excellent infrared image quality and reliable temperature measurement accuracy at an affordable price, which is ideal for scientific applications. FLIR T440bx has been specifically designed to have 8 times continuous digital zoom with a temperature range from -20°C to 1200°C, MSX Thermal Image Enhancement and many other features.

6. Selected results

In the following examples the TLC's plastic sheet and capsulated TLC's tracers have been applied to measure both temperature and velocity flow fields. For human body screening or biomedical situations few images using TLC and IR-Photography techniques are also presented.

6.1. Full field surface temperature measurements

Liquid crystals can be used to determine the distribution of the surface temperature, and if the surface heat flux can be found, this allows evaluation of the heat transfer coefficient or the Nusselt number. The alternative effects of constant wall temperature and constant heat flux boundary conditions are obtained using a water bath. Photo-graphs are taken using a RGB video-camera and a true-colour image processing technique. Usually several isotherms (each corresponding to a different heat flux) are taken by RGB video-camera to record the local Nusselt contours under an oblique Reynolds number. The locations of each isotherm and colour (adjusted to each Nusselt number) are digitised following a projection of the false colour image on a digitising image respectively (this particular method can be called 'Image Combination Technique' ICT). Figure 8 shows photographs of the colour distribution of the liquid crystal layer around a square section column, image of the computer display after segmentation processing (Hue: 45-55) and false colour image processing (ICT) respectively on the right.

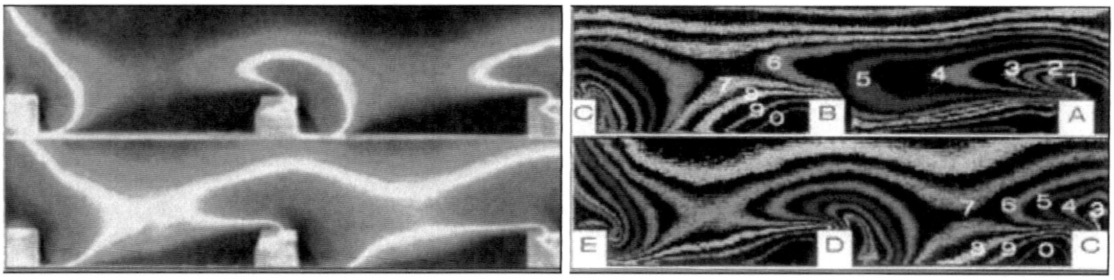

Figure 8. True-colour images from liquid crystal thermography for an endwall surface with in-line square ribs (up); Pattern of 10 Nusselt numbers Nu reconstructed by false colour images of the heat transfer surface for Re = 20000 (0: Nu = 79; 9: 99; 8: 113; 7: 123; 6: 136; 5: 147; 4: 160; 3: 175; 2: 185; 1: 209) (bottom)

Improvements in flow and thermal characteristics of heat exchanger do not require any demonstrations and would substantially reduce fuel and production costs. The measuring technique comprising the use of LC flexible sheets and true-colour processing may also be used for a great variety of applications and should be of considerable use in improving the design of all types of compact heat exchanger. Experimental procedures cover full field flow patterns in classic heat exchanger elements (flat plate with fine-tubes in-line, staggered and with vortex generators) describing local heat transfer coefficient and Nusselt number on the surfaces. An examples of such results are presented in figure 9.

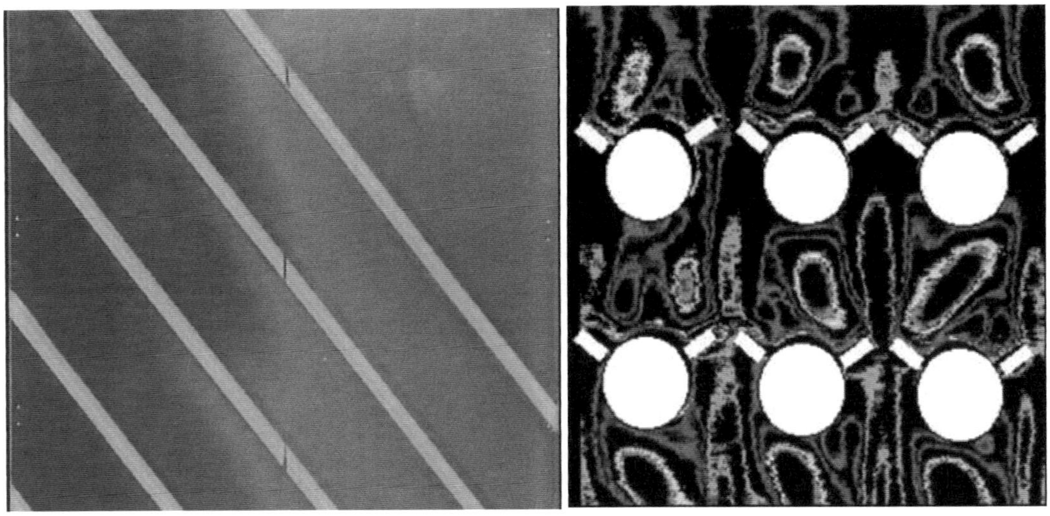

Figure 9. Visualization of model heat exchanger with vortex generators: ribbed surface - true-colur image (left) and winglets – false image (right)

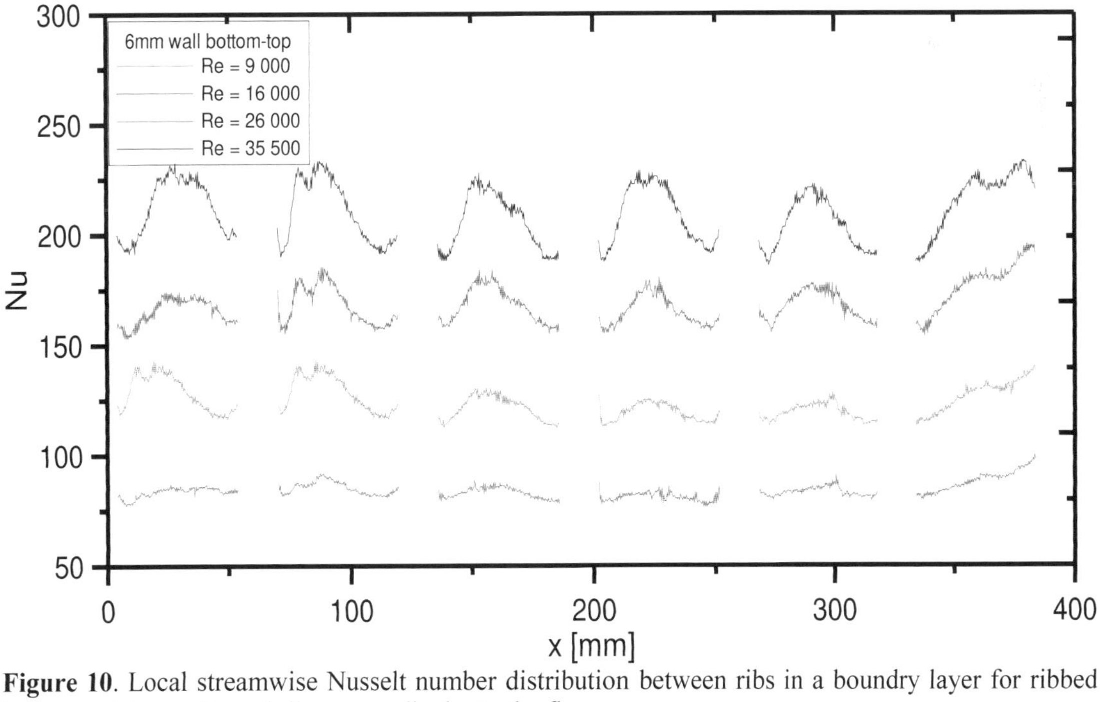

Figure 10. Local streamwise Nusselt number distribution between ribs in a boundry layer for ribbed bottom and top walls and ribs perpendicular to the flow

Nusselt number distribution for channel with vortex generators (rib turbulators) for one geometry with ribbed bottom and top walls (geometry b in figure 7) for different Reynolds numbers (9,000; 16,000; 26,000; 35,500) in the boundary layer is shown in figure 10. The analysis of the boundary layer was carried out to assess the impact of the channel wall on heat transfer. More information can be found in [19]

6.2. Visualization of film cooling of model turbine blades

Internal cooling technique for gas turbine blades has been studied for many years. The designers need detailed data for heat transfer and temperature distributions along with the flow and streamwise pressure gradient to understand the flow physics and to improve the current internal cooling designs. Examples of modified gas turbine blades are shown in figure 11 [20]. Within this objective, TLC technique also has been applied to study of aerodynamic and aerothermal blade turbine design with effusive cooling concept. View of test section of low-speed wind tunnel is presented in figure 11. Examples of steady-state and transient LCT measurements and visualization are shown in figure 12.

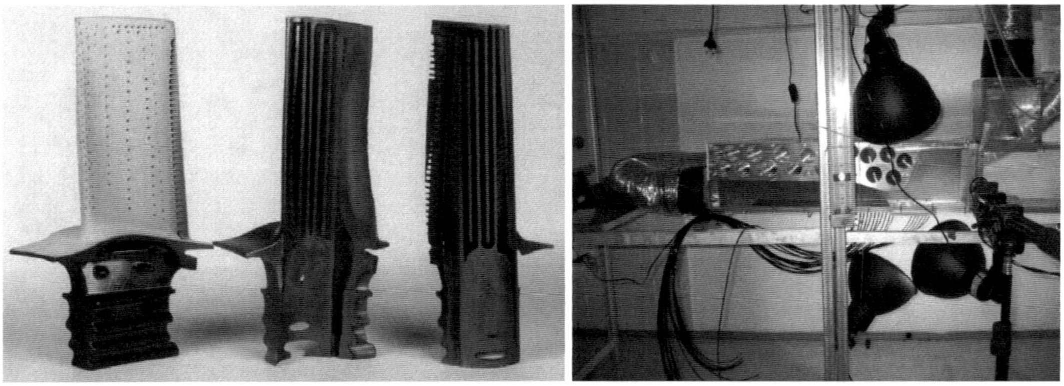

Figure 11. gas turbine blades with cooling holes (left) and view of test section for model turbine blade cooling (right)

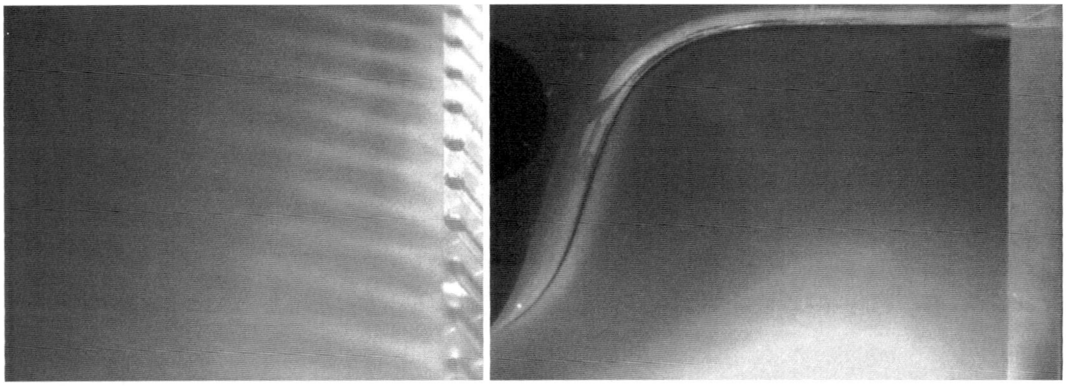

Figure 12. Steady-state LTC visualization of air flow on a blade surface with air jets (left) and end-wall temperature distribution by transient TLC method (right)

6.3. Natural convection in a closed cavity

Vertical temperature gradients are mainly responsible for the atmospheric or oceanographic fluid motion. Development of the nocturnal stable stratification over flat areas, inversion of temperature gradient and break-up of the convective boundary layer developed at the ground levels can be simulated using small-scale laboratory experiments. One of the typical features characterizing instabilities

IOP Conf. Series: Materials Science and Engineering **249** (2017) 012010 doi:10.1088/1757-899X/249/1/012010

generated by vertical temperature gradient are plumes or ejections appearing, when the thermal boundary layer breaks up. Figure 13 shows temperature and velocity visualization in glycerol-filled cavity under free convection in horizontal and vertical position.

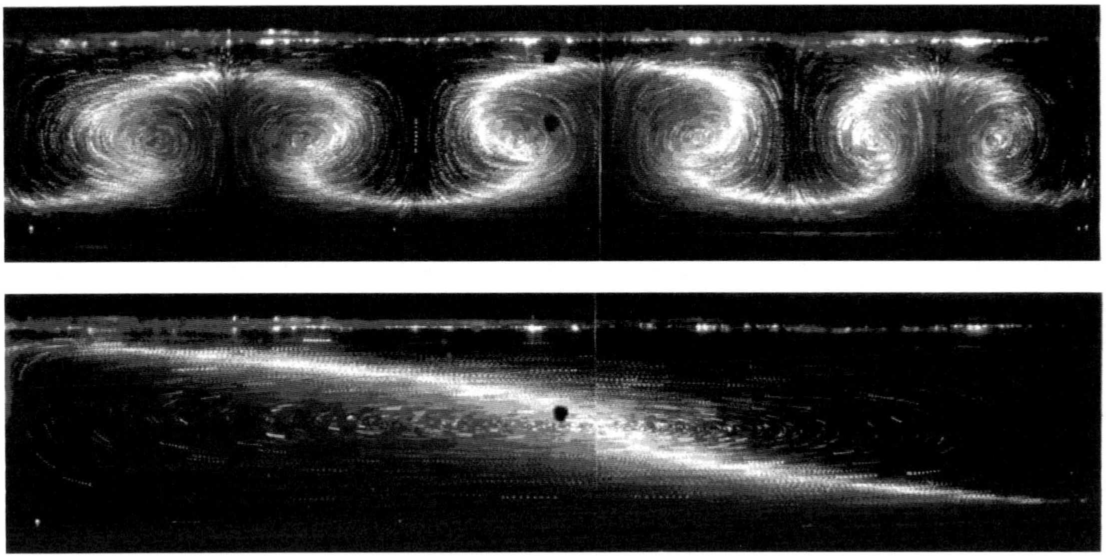

Figure 13. Visualisation of Benard cells in the horizontal (top) and vertical (bottom) cavity using TLC

6.4. *PIV measurements*

Velocity fields in case of ribbed channel were obtained from the PIV measurements for different ribs geometries (figure 7) and different Reynolds numbers (Re = 9000, 16,000 and 26,000). The area scanned by the PIV method was in all cases located in the mid-vertical-plane between side walls (one position of the laser light sheet). Figure 14 shows an average velocity field (averaged over 100 instantaneous velocity fields) for a geometry with ribbed bottom wall only (ribs height is 6 mm) and flow perpendicular to the ribs, while figure 15 show geometry with ribbed top and bottom walls. Reynolds number is 16,000. Maximum velocity for this cases is about 5.5 m/s and is located in the top part of the channel. Figures 14-15 show also a turbulence intensity computed from 100 instantaneous velocity fields.

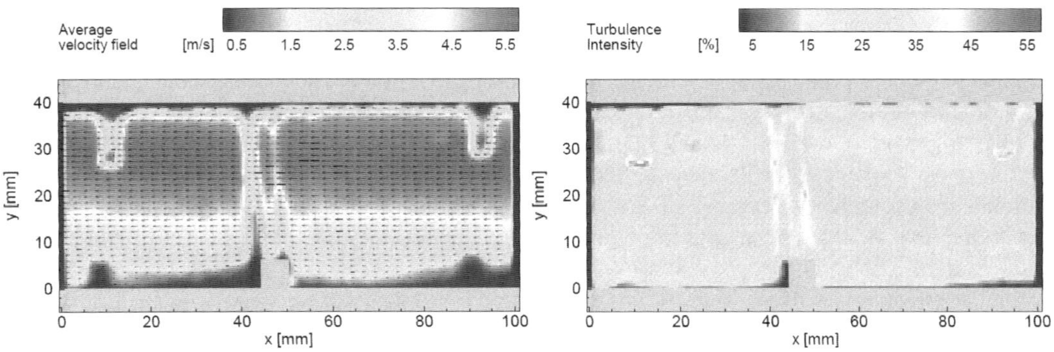

Figure 14. PIV measurements for ribbed channel: averaged velocity field (left) and turbulence intensity (right) for geometry with ribbed bottom wall

IOP Conf. Series: Materials Science and Engineering **249** (2017) 012010 doi:10.1088/1757-899X/249/1/012010

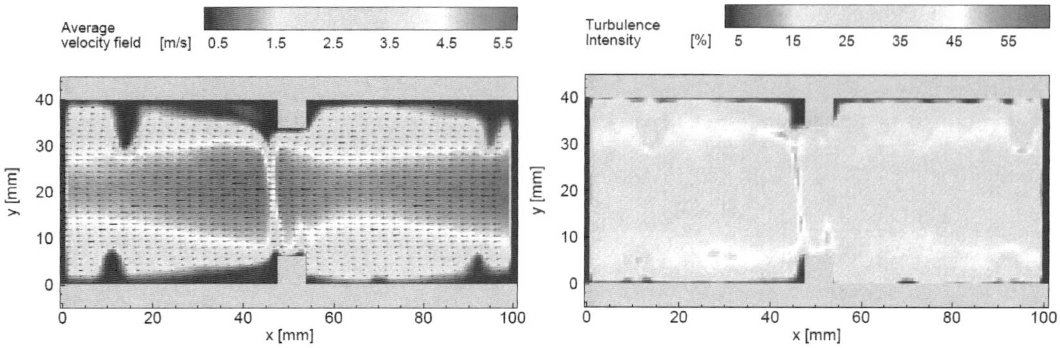

Figure 15. PIV measurements for ribbed channel: averaged velocity field (left) and turbulence intensity (right) for geometry with ribbed bottom and top walls

6.5. Medical applications

Both, liquid crystal thermography and infrared photography, as non-invasive diagnostic techniques, can be used for skin surface temperature measurements. Obtained images can give information about pathological phenomena in the patient organism.

Liquid crystals in the mode of thermographic examination, can be also used in the case of diagnostics and monitoring of different diseases e.g.: diagnosis of cancers, tumours and inflammatory states (liver, lung, etc.), evolution of skin tests (allergy, tuberculin), traumatology and forensic medicine. Those methods are especially useful for children because they are non-invasive, fast, reliable and visually attractive. Because the temperature of human skin depends upon the metabolic level of neighbouring tissues, heat transfer can occur via muscular and adipose tissues, blood circulation, muscular tensions, body position or effect of drugs. This kind of medical diagnosis should be confirmed by other independent methods in the case of diseases suspicion [13].

Requirements for thermographic examination are well-known [21]. The minimum area of a testing room has to be 9 square metres. All devices emitting heat should be removed, if possible. Air movements have to be minimised. The preferred temperature of the testing room should be 25 to 27 °C, while humidity should be 50 to 70 %. Both these parameters should be carefully controlled during measurement. The effect of sunlight should be minimised. Light sources have to secure scattered light of low intensity to avoid light reflections. Patients should abstain from physical activity during 20 minutes before measurement. The skin area studied should be exposed for at least 15 minutes before a test to ensure thermal equilibrium with the environment [13].

The patient's position should be similar to their physiological (normal) one to avoid temperature changes due to enforced changes of blood circulation. Body areas studied should not contact other body areas. Medicines, drugs (alcohol, nicotine and so on) and treatments affecting blood circulation should not be taken, if possible, for 24 hours before measurement. Also patients should not have cold and hot drinks and not eat large meals before measurement. Cosmetics can also affect thermal properties of skin. All subsequent measurements have to be done at the same hour of the day. The thermological report should contain all conditions of measurement, including the states of the patient and environment. The patient should be allowed to see the thermographic images because their colours have a positive psychological effect especially in the case of children. Various cases have been studied and typical results of LCT are shown in figure 16 [21], while figure 17 shows examples of infrared images of a human body taken by FLIT T440bx camera.

IOP Conf. Series: Materials Science and Engineering **249** (2017) 012010 doi:10.1088/1757-899X/249/1/012010

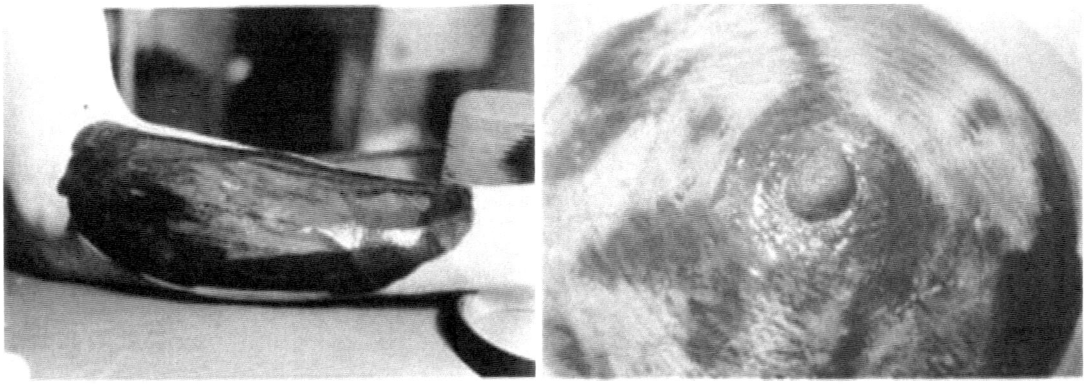

Figure 16. LC thermogram of forearm (left) and breast cancer (dark region) (right)

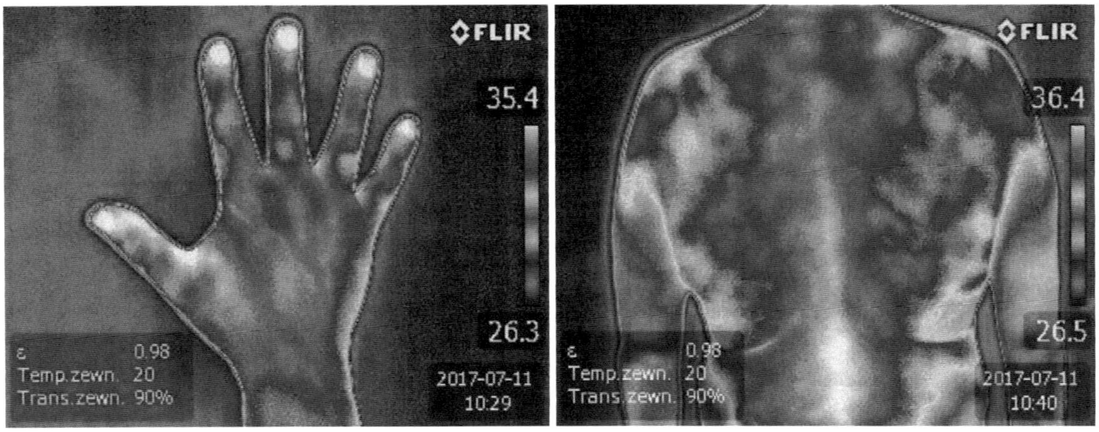

Figure 17. IR thermogram of human hand and human back

7. Conclusions

Advanced experimental techniques, in this case true-colur image processing of liquid crystal patterns, particle image velocimetry and infrared imaging, allow new approaches to old problems and open up new areas of research and applications. Image processing data make available quantitative, full field information about the distribution of temperature, heat transfer and adjunctive medical pathogenesis and diagnostics. It is evidence from our research that liquid crystal and infrared imaging technique have many applications, not only in testing, but also in industry, medicine and the home.

The presented overview provides an introduction to thermochromic liquid crystals, digital infrared imaging and particle image velocimetry, how this measurements techniques and methods work and how they are used.

References

[1] Stasiek J 1997 *Heat and Mass Transfer* **33** pp 27-39

[2] Fergason J L 1969 Applied Optics **7** pp 1729-1737

[3] Reinitzer R 1899 *Monatschr, Chem, Wien* **9** pp 421-441

[4] Moffat R J 1991 *Proc. 9th Intl. Heat Transfer Conference* Jerusalem Vol. 1 pp 308-310

[5] Jones T V, Wang Z, Ireland P T 1992 Proc. Of the First I.Mech.E. *Seminar on optical methods and Data Processing in Heat and Fluid Flow* City University London pp 51-65

[6] Stasiek J A and Kowalewski T A 2002 *Optoelectronics Review* **10** No.1 pp 1-10

[7] Fornalik E, Filar P, Tagawa T, Ozoe H and Szmyd J S 2006 *Intl Journal of Heat and*

Mass Transfer 49 pp. 2642-2651
[8] Fornalik E 2007 *Journal of Theoretical and Applied Mechanics* 45 pp 557-568
[9] Ziemba A and Fornalik E 2016 *Journal of Physics Conference Series* 745 032108
[10] Akino N, Kunugi T, Ichimiya K, Mitsuchiro K and Ueda M 1989 *ASME J.Heat Transfer* **111** pp 558-565
[11] Stasiek J, Collins M W 1996 *Atlas of Visualization* Vol 2 CRC Press Inc USA
[12] Stasiek J, Jewartowski M and Kowalewski T A 2014 *Journal of Crystalization Process and Technology* **4** pp 46-59
[13] Stasiek J, Stasiek A, Jewartowski M and Collins M W 2006 *Optics & Laser Technology* 38 pp 243-256
[14] http://260h.pbworks.com/w/page/69071363/Beginnings%20of%20Quantum%20Mechanics
[15] Bagavathiappan S, Lahiri B B, Saravanan T, Philip J and Jayakumar T 2013 *Infrared Physics & Technology* **60** pp 35-55
[16] Mikielewicz D, Wajs J, Glinski M and Zrooga A-B R S 2013 *Experimental Thermal and Fluid Science* 44 pp 556-564
[17] Polidori G, Renard Y, Lorimier S, Pron H, Derrau S and Taiar R 2017 *Int. Journal of Surgery Case Reports* **34** pp 56-59
[18] HALLCREST 2014 *Handbook of Thermochromic Liquid Crystal Technology* LCR Hallcrest
[19] Mikielewicz D, Stasiek A, Jewartowski M and Stasiek J 2012 Applied Thermal Engineering 49 pp 61-72
[20] Giampaolo A 2006 *Gas turbine handbook: principles and practices* The Fairmont Press, Inc.
[21] Zmija J, Klosowicz S and Borys S 1989 *Cholesteric liquid crystals in a detection of radiation* (Warsaw: Poland/WNT in Polish)

IOP Conf. Series: Materials Science and Engineering **249** (2017) 012011 doi:10.1088/1757-899X/249/1/012011

Ultrasonic Measurement of Velocity Profile on Bubbly Flow Using Fast Fourier Transform (FFT) Technique

W Wongsaroj[1], A Hamdani[1], N Thong-un[2], H Takahashi[1], and H Kikura[1]

[1] Laboratory for Advanced Nuclear Energy, Institute of Innovative Research, Tokyo Institute of Technology, 2-12-1-N1-7, Ookayama, Meguro-ku, Tokyo, Japan
[2] Department of Instrumentation and Electronics Engineering, Faculty of Engineering, King Mongkut's University of Technology North Bangkok, 1518 Pracharat 1 Road, Wongsawang, Bangsue, Bangkok, Thailand

*wongsaroj.w.aa@m.titech.ac.jp

Abstract. In two-phase bubbly flow, measurement of liquid and bubble velocity is a necessity to understand fluid characteristic. The conventional ultrasonic velocity profiler (UVP), which has been known as a nonintrusive measurement technique, can measure velocity profile of liquid and bubble simultaneously by applying a separation technique for both phases (liquid and bubble) and transparent test section is unnecessary. The aim of this study was to develop a new technique for separating liquid and bubble velocity data in UVP method to measure liquid and bubble velocity profiles separately. The technique employs only single resonant frequency transducer and a simple UVP system. An extra equipment is not required. Fast Fourier Transform (FFT) based frequency estimator paralleled with other signal processing techniques, which is called as proposed technique, was proposed to measure liquid and bubble velocity separately. The experimental facility of two-phase bubbly flow in the vertical pipe was constructed. Firstly, the Doppler frequency estimation by using the FFT technique was evaluated in single-phase liquid flow. Results showed that FFT technique showed a good agreement with autocorrelation and maximum likelihood estimator. Then, separation of liquid and bubble velocity was demonstrated experimentally in the two-phase bubbly flow. The proposed technique confirmed that liquid and bubble velocity could be measured efficiently.

1. Introduction

Bubbly flow has been more attractive to many researchers to investigate its property. Some parameters of the phenomena have complicated to understand such as a liquid flow rate and the bubble motion. To understand the characteristic of two-phase bubbly flow, the important parameters in two-phase bubbly flow i.e. velocity are necessary to be known accurately. The velocity profile of liquid and bubble in bubbly flow is an important parameter to signify fluid characteristic, relate with liquid flow rate, and bubble motion. Hence, classification of liquid and bubble velocity is a necessity. Several intrusive techniques were utilized in two-phase flow such as hot-film anemometry [1]. This technique disturbs the flow, and intrusive effects on flow structures reduce its lifetime. To eliminate these effects, many nonintrusive measurement techniques have been developed, such as laser Doppler anemometry (LDA) [2] and Particle Image Velocimetry (PIV) [3]. Nonintrusive measurement methods have usually used for the measurements of the rising velocity of a single or few bubbles. Moreover, these measurement techniques for velocity measurements require transparent test section, and it obviously fails if working fluid is opaque.

The ultrasonic measurement technique is a powerful tool to obtain the flow characteristic. It is a nonintrusive method and capable of measuring velocity profile on the non-transparent test section. Ultrasonic Velocity Profile (UVP) measurement uses a pulsed echography of an ultrasonic wave reflected from reflector such as small particle dispersed in a liquid to measure an instantaneous velocity profile along its measurement line. Takeda *et al.* [4] originally developed the UVP method to measure single-phase liquid flow. The advantage of this technique is that it can perform spatial-temporal measurements of the flow. Aritomi *et al.* [5] developed a system, which combined the UVP system and video data processing unit to measure the liquid velocity distributions in bubbly flows, the average bubble diameter, and the void fraction. They proposed the detection of bubble positions if bubbles cross the ultrasonic measuring line by applying statistical methods to the UVP. When the UVP method was applied to measure two-phase bubbly flows, the echo signal reflected by both the dispersed small particles in the liquid phase and the bubble. Hence, the velocity data measured by the UVP method included velocity information of both phases. Suzuki *et al.* [6] developed a phase-separation technique using the UVP method. The method applied to pattern recognition to distinguish a phase of the velocity profile. Yamanaka *et al.* [7] tried to develop a separation technique, which based on the differences of ultrasonic intensity of signals reflected from tracer particles and bubble. Ultrasonic intensity reflected from a bubble's surface is stronger than a small particle dispersed in a liquid. It is due to the differences in acoustic impedances and diameters between them. Later on, Murakawa *et al.* [8] employed multi-wave TDX transducer and time domain cross-correlation method (UTDC) to separate liquid and bubble velocity profile. This technique could synchronise the instantaneous velocities of liquid and bubbles. Some measurement technique for two-phase bubbly flow by UVP was developed to measure velocity profile of bubbly flow, and their system requires several types of equipment. For instance, the measurement system combines the UVP system and video data processing unit [5], and two basic frequencies were required [8]. As a consequence, the number of pulser-receiver and accessories has been doubled or more. Therefore, the measurement system will be complex, and furthermore, the cost will also be higher.

In this paper, we study the effect of bubbly flow that influences to the Doppler signal in Doppler pulse repetition method. The size of the bubble and small particle in bubbly flow are significantly different (bubble is bigger than small particle 10^3 times), and its effects on Doppler signal amplitude size. Also, Doppler shift frequency, which obtains from the bubble and small particle, are different due to their velocities difference. As a result, multi-frequency and different amplitude of Doppler signal are observed. Therefore, our purpose is to develop a powerful signal processing techniques such as Fast Fourier Transform (FFT), Windowing and Signal Enveloping, to separate instantaneous velocity profile of both bubble and liquid. The proposed technique analyses multi-frequency and different amplitude phenomena of Doppler signal using single transducer. Also, it provides not only velocity data but also bubble diameter and number of bubble data. In this study, the classification of liquid and bubble velocity is demonstrated experimentally.

2. Methodology

2.1. Ultrasonic Velocity Profile (UVP) principle

The UVP method can obtain instantaneous velocity profiles of fluid, and flow rate can be calculated by integrating the measured velocity profile over a pipe or channel diameter. An ultrasound pulse is emitted from transducer along the measurement line, and the echo reflected from the surface of the reflector such as small particle is derived in the same transducer. The position of a particle can be calculated from the speed of sound c and traveling time of ultrasound τ from the start of the pulse burst to its reception.

$$x = \frac{c\tau}{2} \tag{1}$$

The particles in working fluid follow the main flow, and its density is nearly equal to water (0.98 g/cm^3). Therefore, the velocity of a particle is assumed to be same as the velocity of fluid at that position The echo signal received from moving particles contains Doppler shift frequency f_D (x) and the velocity of particle V (x) is described as

$$V(x) = \frac{cf_D(x)}{2f_0 \sin\theta} \tag{2}$$

where f_0 is the basic frequency of ultrasound and θ is the incident angle.

2.2. Doppler pulse repetition method and Doppler signal

Over a decade, Doppler pulse repetition method has been used to extract Doppler signal from echo signal reflected by the reflector under a certain number of repetitions (N_{rep}). This Doppler signal is used to estimate Doppler frequency respectively. Figure 1 shows a block diagram of UVP system based on Doppler pulse repetition technique. An ultrasonic pulser/receiver, operated in a pulse-echo mode, emits ultrasonic pulses through an ultrasonic transducer. Then, the echo signal received by the transducer are sent to pulser/receiver. In quadrature demodulation, the echo signals from the output of pulser-receiver are multiplied by the cosine and sine function separately to detect flow direction (forward and backward), a low-pass filter (LPF) is applied to eliminate the carrier wave components of basic frequency, and extract the Doppler signal. Frequency estimation section computes Doppler frequency, and finally, velocity can be calculated by equation (2)

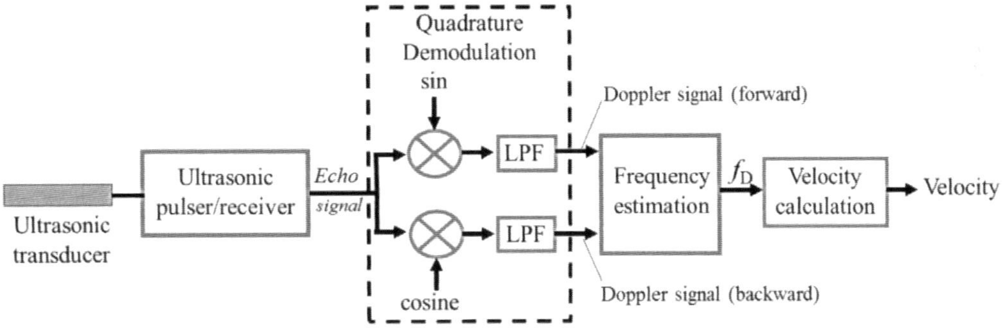

Figure 1. Block diagram of UVP system based on Doppler pulse repetition technique.

Doppler signal is obtained from echo signals reflected by the reflector in liquid; the reflector can be particle or bubble. Its frequency and amplitude inform the velocity and position of moving reflector respectively. In the case of bubbly flow, the size of the bubble and small particle are different, and its effects on the size of Doppler amplitude. Also, Doppler shift frequency which is obtained from the bubble and small particle are different due to their velocities difference. The bubble has a lower density than water and density of the particle is nearly equal to water (0.98 g/cm^3). Thus, bubbles move faster than particles. In some cases, the position of bubble and particle occurs in same measurement channels concurrently along the number of repetition (N_{rep}). In this case, it is difficult to distinguish their

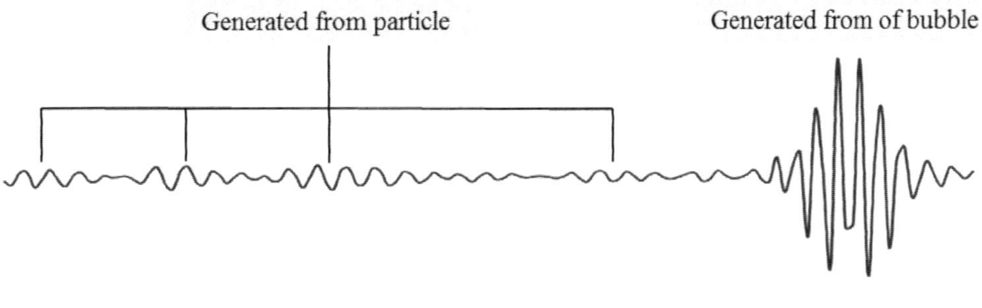

Figure 2. Multi-frequency and different amplitude effect of Doppler signal.

velocities. When this phenomenon occurs, multi-frequency and different amplitude of Doppler signal will be generated as shown in figure 2. Therefore, to separate velocity profile of both bubble and particle, powerful signal processing in post processing is required to analyze the effect of Doppler signal.

2.3. Doppler frequency estimation using Fast Fourier Transform (FFT)
The Fast Fourier Transform (FFT) was widely used in the conventional frequency measurement. It is selected as Doppler frequency estimator in this study and called as "FFT-based frequency estimator." The power spectrum of the frequency $P\ (f_k)$ is demonstrated as follows:

$$P_f = \left(\mathrm{Re}[X_L] - \mathrm{Im}[X_Q]\right)^2 + \left(\mathrm{Re}[X_Q] + \mathrm{Im}[X_L]\right)^2 \quad (3)$$

$$P_b = \left(\mathrm{Re}[X_L] + \mathrm{Im}[X_Q]\right)^2 + \left(\mathrm{Re}[X_Q] - \mathrm{Im}[X_L]\right)^2 \quad (4)$$

where Re and Im are a real and imaginary number, $X_L(f_k)$ and $X_Q(f_k)$ are the discrete Fourier transforms of the frequency demodulation, P_f and P_b are power spectra of a forward direction and backward direction respectively. The value of Doppler frequency (f_D) can be computed by averaging the spectrum as follows:

$$f_D = \frac{\sum_{k=0}^{N_{rep}} f_k\left(P_f(f_k) - P_b(f_k)\right)}{\sum_{k=0}^{N_{rep}} \left(P_f(f_k) + P_b(f_k)\right)} \quad (5)$$

where N_{rep} is the number of repetitions of signal emission.

2.4. Separation of particle and bubble velocity
Particle and bubble velocities are known separately by using proposed technique, which is called "Proposed FFT" as illustrated in figure 3. The technique based on the integration of FFT-based frequency estimator and several signal processing such as Windowing, Enveloping, and Threshold defining. The proposed technique is not only used to estimate the frequency of Doppler signal, which is extracted from quadrature demodulation section. Furthermore, the proposed technique is performed for analyzing both frequency and amplitude of the signal. As phenomena of Doppler signal in bubbly flow was explained in section 2.2, a separation of particle and bubble velocity is possible by replacing frequency estimation section in figure 1. Doppler signals on both forward and backward direction received from the quadrature demodulation section are divided into several windows (W_{n}). Doppler signals in each window are sent to FFT-based frequency estimator to calculate the Doppler frequencies and are sorted to be array form ($f_{\mathrm{D}i}$) as follows:

$$f_{Di} = \frac{\sum_{k=\frac{N_{rep}}{W_n}(i-1)}^{\frac{N_{rep}}{W_n}i-1} f_k\left(P_f(f_k) - P_b(f_k)\right)}{\sum_{k=\frac{N_{rep}}{W_n}(i-1)}^{\frac{N_{rep}}{W_n}i-1} \left(P_f(f_k) + P_b(f_k)\right)} \quad (6)$$

$$\mathrm{f_D} = \left[f_{D1}, f_{D2}, \ldots, f_{Dn}\right] \quad (7)$$

where $i = 1, 2, 3, \ldots, n$ is window index, and W_n is the number of windows.

On the other hand, the amplitude of Doppler signal in each window is known by making the envelope of the Doppler signal ($e_i\,(t)$) in the forward direction using Hilbert transform. The maximum of signal amplitudes in each window are arranged in array form (a_i) as equation (8) and (9); index number (i)

IOP Conf. Series: Materials Science and Engineering **249** (2017) 012011 doi:10.1088/1757-899X/249/1/012011

$$e_i(t) = \left(s_i^2(t) + \hat{s}_i^2(t)\right)^{1/2};\ \hat{s}_i(t) \overset{\text{def}}{=} \mathrm{H}\left[s_i(t)\right] \textit{ is the Hilbert Transform} \tag{8}$$

$$a_i = \max(e_i(t)) \tag{9}$$

$$f_{Di} \overset{\text{def}}{=} \begin{cases} f_{Db} \text{ if } a_i \geq \text{Threshold} \\ f_{Dp} \text{ if } a_i < \text{Threshold} \end{cases} \tag{10}$$

$$\mathrm{f}_{\mathrm{D}(bubble)} = \left\lfloor f_{Db}, f_{Db-1}, \ldots, f_{Db-(l-1)} \right\rfloor \tag{11}$$

$$\mathrm{f}_{\mathrm{D}(particle)} = \left\lfloor f_{Dp}, f_{Dp-1}, \ldots, f_{Dp-(m-1)} \right\rfloor \tag{12}$$

$$\bar{f}_{D(bubble)} = \frac{f_{Db} + f_{Db-1} +, \ldots, + f_{Db-(l-1)}}{l} \tag{13}$$

$$\bar{f}_{D(paticle)} = \frac{f_{Dp} + f_{Dp-1} +, \ldots, + f_{Dp-(m-1)}}{m} \tag{14}$$

where l is the number of bubble Doppler frequency and m is the number of particle Doppler frequency.

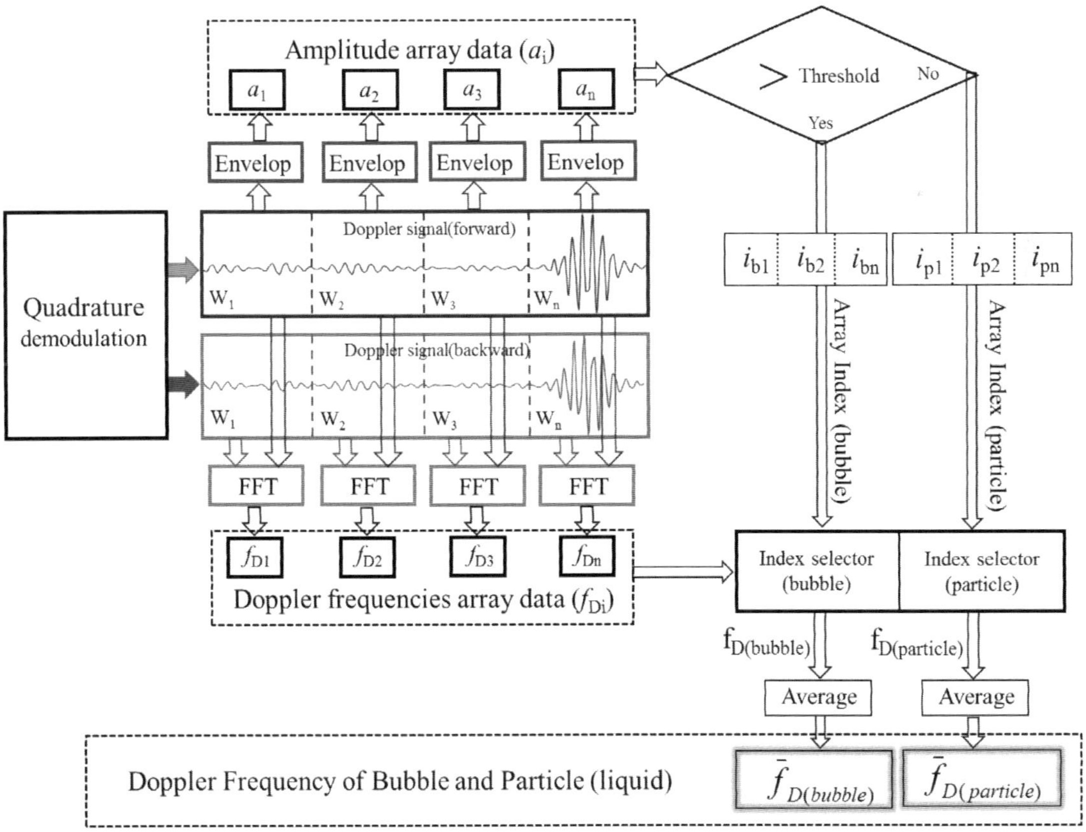

Figure 3. Block diagram of proposed algorithm to separate liquid and bubble velocity.

IOP Conf. Series: Materials Science and Engineering **249** (2017) 012011 doi:10.1088/1757-899X/249/1/012011

identifies window number. The data from each index is compared with the threshold value as equation (10); threshold value is defined to be higher than the maximum of Doppler amplitude of small particle, in which analyses of statistical data of 100,000 Doppler signals were only generated from small particles. Then, the index number of amplitude data will be separated into two categories. Firstly, if the amplitude value is higher than the threshold, its index will be defined as an array index of the bubble. Secondly, if the value is lower than the threshold, its index will be defined as an array index of the particle. These indexes are called "separation index." These indexes are sent to index selector. In this function, Doppler frequency in each window is classified to two group by pointing of the separation index; bubble and particle Doppler frequencies as shown in equation (11) and (12). Then, these Doppler frequencies are separately averaged in each group by using equation (13) and (14) for the bubble and the particle respectively. The Doppler frequency of bubble and particle in same measurement channel are known separately. Finally, velocity value for both liquid and bubble are separately calculated in the velocity calculation section.

3. Experimental setup

3.1. Experimental apparatus
Figure 4 shows a schematic diagram of the experimental apparatus in this study. The measurement systems consisted of ultrasonic pulser/receivers, 4 MHz ultrasonic transducers, digitizer, and computer. The detail of UVP system is shown in table 1 respectively. The pulser/receivers emitted and received ultrasonic pulses through an ultrasonic transducer. The echo signals received by pulser/receiver was converted to a digital signal by a digitizer, with a sampling rate of 100 MS/s. The digitizer was connected to a computer. The digitizer and the pulser/receivers were connected and synchronized to each other. The data from digitizer was sent to the computer and analyzed using LabVIEW software. The measurement test section was a vertical pipe made of acrylic with an inner diameter (D) of 20 mm. The transducer was installed on the outer surface with an incident angle (θ) 45°. The measurements were conducted in downstream at a distance of $50D$ from a bubble injector. Nylon tracer particles with the size of 80 μm were dispersed in the water. The water temperature was monitored using thermocouple, and it was controlled around 20 ± 2°C by the cooling coil. The air was injected from bubble injector using an air compressor. The measurement accuracy of liquid phase velocity was evaluated by making a comparison with an electromagnetic flow meter (EMF) (accuracy: ± 0.5% of reading). The comparison was done by using flow rate value. Also, bubble velocity measurement was compared with image processing method which calculates bubble velocity from bubble images were caught up by a high-speed camera (HSC).

Table 1. The equipment list of UVP system.

Parameter	Configuration
Transducer	4 MHz with beam diameter 5 mm, Imasonic, model: 0902 0119
Pulser-receiver	Met-Flow, Model: UVP-DUO
Digitizer	National Instrument, Model: NI USB 5133
Operating software	National Instrument, LabVIEW 2011
Computer	Laptop, VAIO

3.2. Experimental procedure
There were two experiment configurations in this study. Firstly, evaluation of the performance of FFT-based frequency estimator in single-phase liquid flow. The accuracy of the technique was compared with other techniques such as Autocorrelation [9] and Maximum Likelihood Estimation [10]. The experiment was conducted with single-phase liquid (water) at flow rate 3 L/min and 6 L/min. Secondly, the experiment was conducted to measure velocity profile of bubble and liquid in the two-phase bubbly flow. Proposed FFT was applied to distinguish bubble and liquid velocity, and to estimate bubble

diameter and a number of bubble data respectively. The bubbles diameter size in this experiment was about 0.5 ~ 4 mm. Experimental conditions and parameter of UVP system are shown in table 2 and table 3 respectively.

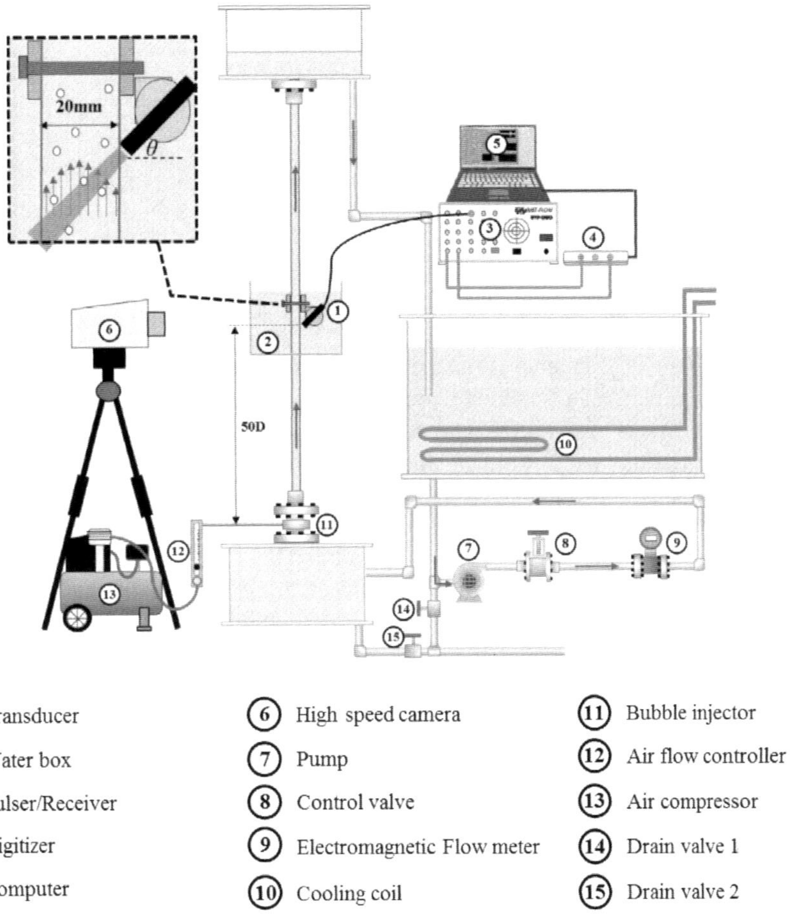

Figure 4. Schematic of the experimental apparatus.

Table 2. Experimental conditions.

Parameter	Condition
Fluid	Water
Fluid temperature	20°C ±2°C
Fluid flow rate	3 L/min and 6 L/min
Reynolds number	3,169 at 3 L/min 6,339 at 6 L/min
Particle type	Nylon particle
Particle size	80 μm
Bubble diameter	0.5 ~ 4 mm
Pipe size (inner diameter)	20 mm
Pipe material	Acrylic resin

Table 3. Parameter configuration of the UVP system.

Parameter	Configuration
Basic frequency	4 MHz
Pulse repetition frequency	8 kHz
Number of repetition	128
Channel width	0.74 mm
Measurement channel	38 channel
Sound velocity in water	1,480 m/s at 20°C
Number of windows (W_n)	4

4. Results and Discussion

4.1. Comparison of Doppler frequency estimation techniques

Figure 5 shows a comparison of velocity profile measurement with different Doppler frequency estimators. The graph shows the average data of 1,000 instantaneous velocity profiles. The experiment was conducted using single-phase liquid flow at flow rate 3 L/min and 6 L/min. The accuracy evaluation of each technique is shown in table 4. Good agreement is obtained between FFT-based frequency estimator and other techniques (Autocorrelation and Maximum Likelihood estimation (MLE)). Figure 6 illustrates standard deviation value of each technique. The standard deviation for both FFT-based frequency estimator and autocorrelation shows a similar trend. However, the standard deviation in MLE is scattered in particular position, especially at the near and far wall. It means that FFT-based frequency estimator and autocorrelation are better than MLE. Therefore, it can be concluded that performance of FFT-based frequency estimator, at least to be equal and is not worse than other well-known techniques.

Table 4. Measurement result and accuracy of average single-phase liquid velocity profile (1,000 profiles) at 3 L/min and 6 L/min in each Doppler frequency estimation techniques.

EMF	FFT		Autocorrelation		MLE	
Flow rate (L/min)	Flow rate (L/min)	Accuracy (% of reading)	Flow rate (L/min)	Accuracy (% of reading)	Flow rate (L/min)	Accuracy (% of reading)
3 L/min	3.009	0.30	2.775	7.50	2.458	18.06
6 L/min	5.738	4.36	5.484	8.60	4.721	21.31

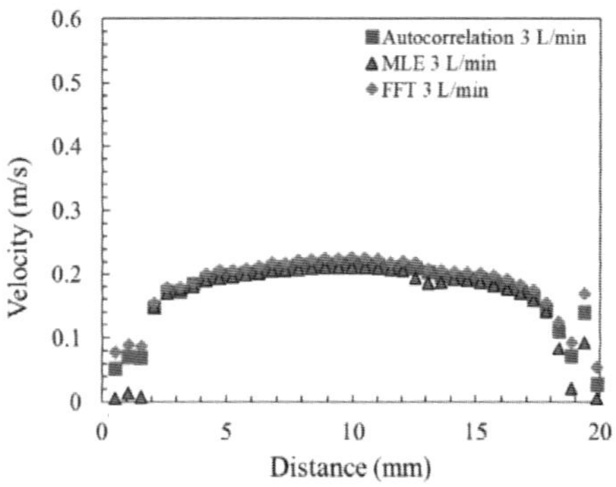

(a)

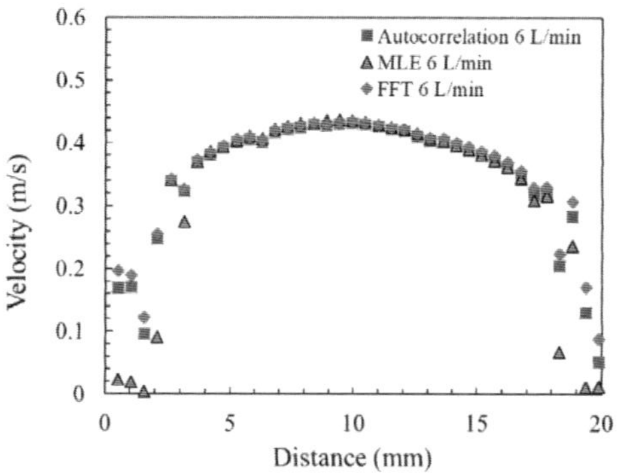

(b)

Figure 5. Measurement result of average single-phase liquid velocity profile (1,000 instantaneous profiles) at (a) 3 L/min and (b) 6 L/min in each Doppler frequency estimation techniques.

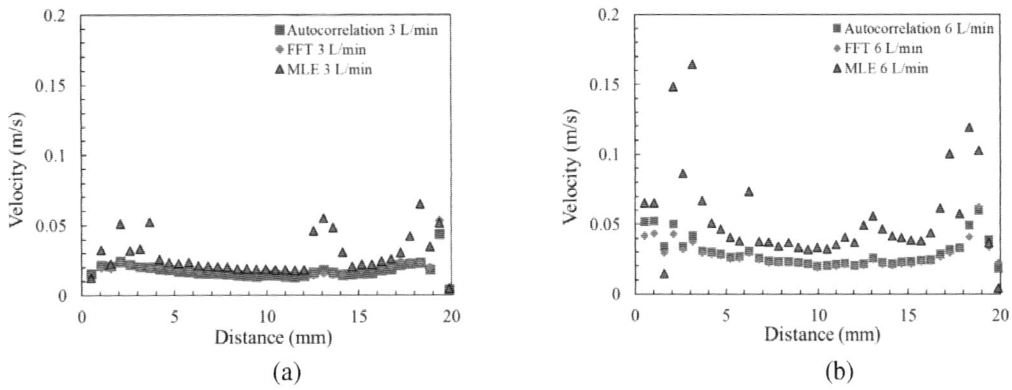

Figure 6. Standard deviation of average single-phase liquid velocity profile (1000 instantaneous profiles) at (a) 3 L/min and (b) 6 L/min in each Doppler frequency estimation techniques.

4.2. Separation of two-phase bubbly flow velocity

The second experiment was carried out in the two-phase bubbly flow condition. The bubbles dispersed into the liquid water through bubble injector as shown in figure 4. Figure 7 shows the result of the instantaneous velocity profile which is measured using FFT-based frequency estimator (without separation), henceforth, it is called as "Original FFT" and our proposed technique, it is called as "Proposed FFT." It can be seen that the instantaneous velocity profile of liquid and bubble is obviously separated when Proposed FFT technique is applied. Instantaneous bubble velocity measurement of the Proposed FFT technique is evaluated by separation velocity profile of bubbly flow. In figure 7(a), at 3 L/min, there are two moving bubbles appear at measurement distance 3~4 mm and at 13~14 mm and their diameter is approximately 2 mm. Their velocity is about 0.228 m/s and 0.292 m/s respectively. In figure 7(b), at 6 L/min, two moving bubbles obviously appear at measurement distance 5~7 mm and at 15~17 mm and their diameter is about 3 mm. Their velocities are about 0.396 m/s and 0.350 m/s respectively. The bubble rise velocity measurement of UVP-Proposed FFT was experimentally compared with the measurement result of Image processing method by the average of 100 bubble velocity. The comparison was specified at bubble diameter about 0.5~1 mm and 2~3 mm and was only demonstrated in far wall region (at position 15~20 mm of measurement distance inside the pipe).

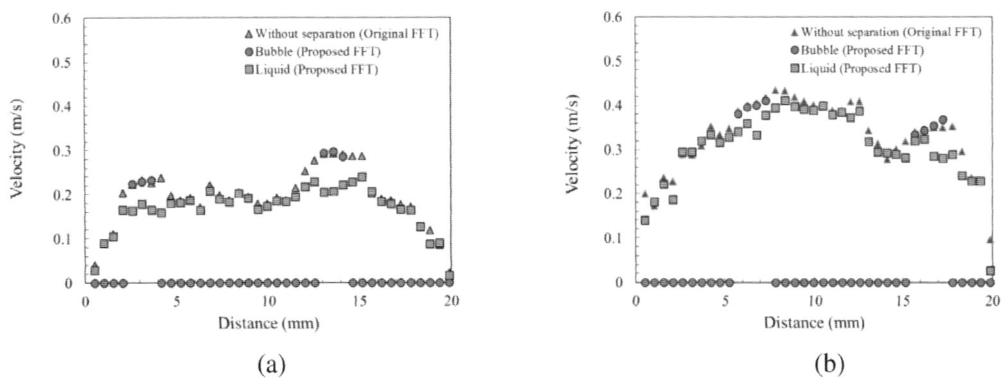

Figure 7. Instantaneous velocity profile of two-phase bubbly flow with separation and without separation; (a) at 3 L/min, (b) at 6 L/min.

Table 5 shows that measurement result of bubble velocity for both systems. The discrepancy is 5.55% at flow rate 3 L/min and 10.95% at 6 L/min when the bubble diameter is about 0.5~1 mm. On the other hand, the discrepancy increases to 34.50% at flow rate 3 L/min and 19.89% at 6 L/min when bubble diameter is about 2~3 mm. Furthermore, it is confirmed that the bubble velocity changes relative to the bubble diameter; bubble velocity increases as increasing the bubble diameter.

Table 5. Comparison of average bubble velocity measurement (100 value) at far wall zone (position 15~20 mm of measurement distance inside the pipe) between the UVP using the Proposed FFT and the image processing method.

Bubble diameter	0.5~1 mm			2~3 mm		
Method	UVP	Image processing	Discrepancy	UVP	Image Processing	Discrepancy
Flow rate 3 L/min	0.210 m/s	0.222 m/s	5.55 %	0.247 m/s	0.350 m/s	34.50 %
Flow rate 6 L/min	0.308 m/s	0.276 m/s	10.95 %	0.344 m/s	0.420 m/s	19.89 %

In the original FFT, liquid and bubble distribution in the bubbly flow cannot be separated instantaneously. It means that the separation technique is needed. Therefore, without separation technique, the behavior of bubble distribution is not obtained by original FFT. Moreover, because bubble motion cannot be obtained instantaneously, a measurement result in liquid-phase is calculated inaccurately due to bubble data is remained in the raw velocity profile data. Nonetheless, the Proposed FFT measure liquid and bubble velocity instantaneously, and it can distinguish two-phase bubbly velocity profile. Therefore, the velocity profile of liquid-phase is calculated accurately due to the bubble data is removed in the raw velocity profile data.

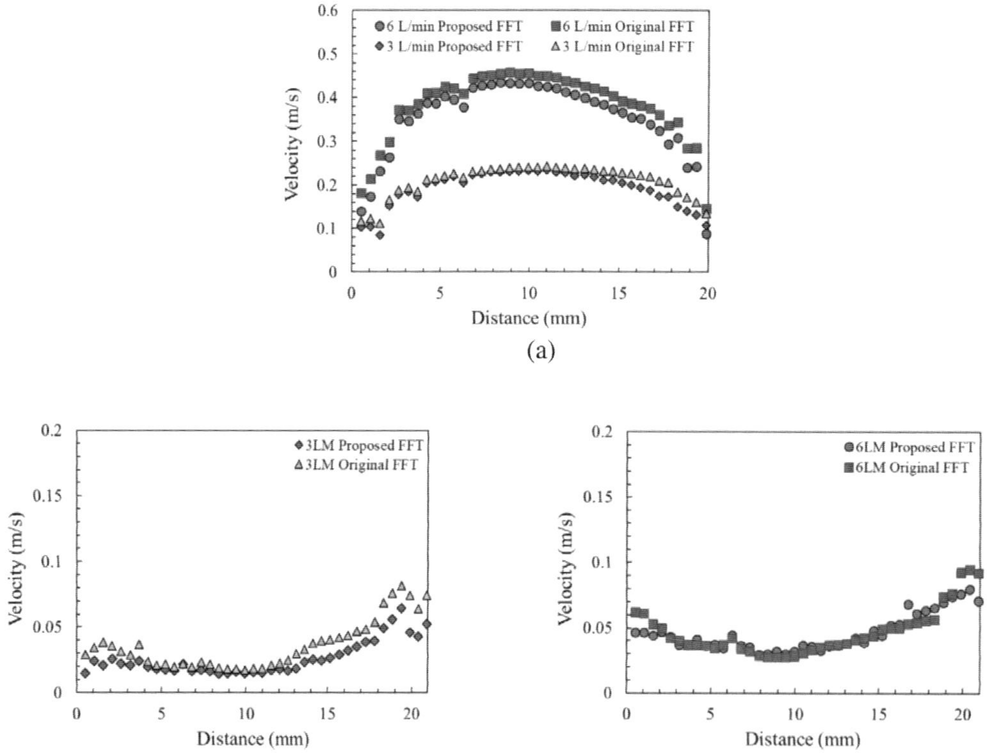

Figure 8. Comparison of 1000 average liquid velocity profile between the Proposed FFT (separated profile) and the Original FFT (without separation).

Table 6. Accuracy evaluation of liquid velocity profile measurement in two-phase bubbly flow at 3 L/min and 6 L/min.

EMF	UVP-Original FFT		UVP-Proposed FFT	
Flow rate (L/min)	Flow rate (L/min)	Accuracy (% of reading)	Flow rate (L/min)	Accuracy (% of reading)
3 L/min	3.547	18.23	3.176	5.86
6 L/min	6.372	6.20	5.753	4.11

Figure 8(a) and table 6 shows a comparison of liquid-phase velocity profile measurement between the Original FFT and the Proposed FFT. It can be summarized that the accuracy of the result that obtained from Proposed FFT after separation is better than the Original FFT both at the flow rate 3 L/min and 6 L/min. Moreover, the standard deviation of the Original FFT is higher than the value of the Proposed FFT at near wall and far wall zone especially as shown in figure 8(b) and 8(c). It can illustrate that the measurement uncertainty by using the Original FFT is higher than the Proposed FFT.

4.3. Number of bubble data

When bubble velocity, bubble diameter, and their position are estimated, it is possible to obtain a number of bubble data along the measurement line. Figure 9 illustrates the number of bubble data, which is obtained from 1,000 bubble velocity profiles after separation. At the flow rate of 3 L/min, the number of bubble data at the far wall region is higher than other zones. It can be interpreted that most of the bubble passes through the far wall zone. At flow rate 6 L/min, the number of bubble data appears on both near wall and far wall zone significantly. It can be concluded that most of the bubble passes through both near wall and far wall regions. In this experiment, the range of bubble diameter was 0.5~4 mm, which is categorized to be a small bubble, it can be summarized that most of the small bubbles existed in the region that is near to a pipe wall. This result also has similarity with the previous study [11].

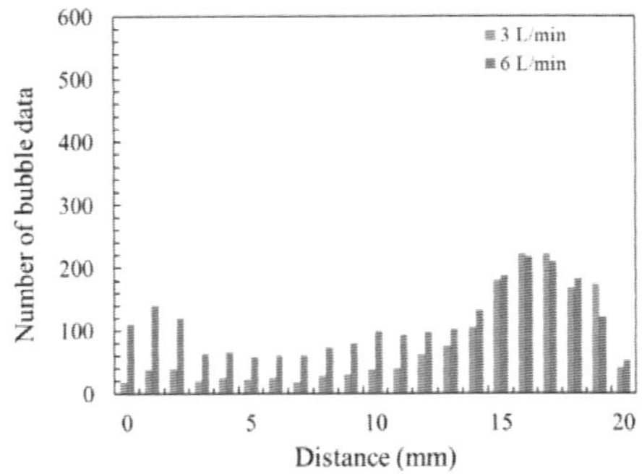

Figure 9. A number of bubble data at flow rate 3 L/min and 6 L/min.

5. Conclusion

FFT-based technique collaborated with Windowing and Signal Enveloping for measuring two-phase bubbly flows was proposed. The measurement system employed 4 MHz transducer and Doppler pulse repetition method that was modified by proposed technique for measuring liquid and bubble velocity distributions. This technique analyzes the behavior of Doppler signal (frequency and amplitude), which are reflected by small particle and bubble. The main advantage of this method was not only liquid, and bubble velocity distributions could be obtained separately, but also this method was capable of acquiring the simultaneous of two-phase velocity at same measurement channel. Furthermore, only one resonant frequency was required. Therefore extra equipment was not a necessity. The performance of FFT-based frequency estimator was experimentally compared with other signal processing techniques for the measurement of the velocity profile in the single-phase (liquid) flow. From the result, it was confirmed that FFT showed a good agreement with other techniques. In bubbly flow, the velocity profile of liquid and bubble was obtained separately. The evaluation was done in the both liquid and bubble phases. Firstly, liquid velocity profile was verified with an electromagnetic flow meter. Secondly, bubble distribution was compared with the image processing method. It was confirmed that the proposed technique could separate both liquid and bubble distributions and the velocity data had a good agreement with another instrument. Moreover, number of bubble data were estimated experimentally. Practically, the diameter of the measured bubble must be smaller than the active diameter of the transducer. The active diameter of 4 MHz ultrasonic transducer is 5 mm, therefore, the maximum size of the bubble that system could detect is limited. The bubble should be smaller than 5 mm. As this limitation, future work should focus on applying lower basic frequency with bigger active diameter transducer to extend maximum limit of bubble diameter.

References

[1] Wang T, Wang J, Ren F, and Jin Y 2008 Application of Doppler ultrasound velocimetry in multiphase flow *Chemical Engineering Journal* vol. 19 pp 205-213
[2] Albrecht H.-E, Borys M, Damaschke N and Tropea C 2003 Laser Doppler and Phase Doppler Measurements Techniques *Springer.*
[3] Chen R C, and Fan L S 1992 Particle image velocimetry for characterizing the flow structure in three- dimensional gas-liquid-solid fluidized beds *Chem. Eng. Sci.* vol. 47 pp 3615-3622
[4] Takeda Y 1986 Velocity profile measurement by ultrasonic Doppler shift method *Int. J. Heat Fluid Flow* vol 7 pp 313-318
[5] Aritomi M, Zhou S, Nakajima M, Takeda Y, Mori M and Yoshioka Y 1996 Measurement system of bubbly flow using ultrasonic velocity profile monitor and video data processing unit *Journal of Nuclear Science and Technology* 33:12 pp 915-923
[6] Suzuki Y, Nakagawa M, Aritomi M, Murakawa H, Kikura H, Mori M 2002 Microstructure of the flow field around a bubble in counter-current bubbly flow *Exp Therm Fluid Sci* vol 26 pp 221–227
[7] Yamanaka G, Kikura H and Aritomi M 2002 Study on the development of novel velocity profiles measurement method using ultrasonic time-domain cross-correlation *In: Proc. 3rd symposium on ultrasonic doppler methods for fluid mechanics and fluid engineering* pp 109–114
[8] Murakawa H, Kikura H and Aritomi M 2008 Application of ultrasonic multi-wave method for two-phase bubbly and slug flow *Flow Measurement and Instrument Journal* vol 19 pp 205-213
[9] Murakawa H, Sugimoto K, Takenaka N 2014 Effects of the number of pulse repetitions and noise on the velocity data from the ultrasonic pulsed Doppler method with different algorithms *Flow Measurement and Instrumentation Journal* vol 40 pp 9-18
[10] Thong-un N, Wongsaroj W, Treenuson W, Chanwutitum J and Kikura H 2017 Doppler Frequency Estimation using Maximum Likelihood Function for Low Ultrasonic Velocity Profile *Acoust. Sci. & Tech. Journal* vol 38 pp 1-4
[11] Wada S, Kikura H, Aritomi M 2006 Pattern recognition and signal processing of ultrasonic echo signal on two-phase flow *Flow Measurement and Instrumentation journal* vol 17 pp 207–224

IOP Conf. Series: Materials Science and Engineering **249** (2017) 012012 doi:10.1088/1757-899X/249/1/012012

Aerodynamic effects by cooling flows within engine room of a car model

T Sawaguchi [1]and Y Takakura[1]

[1] Course of Mechanical Engineering, Graduate school of Tokai University, Japan
E-mail: takakura@tokai-u.jp

Abstract. The purpose of this research is to clarify the change of characteristics of aerodynamic drag and lift of a car by the engine loading system (engine arrangement) and the air inlet system (opening area and position) with and without a radiator in wind-tunnel experiments. A simplified car model with 1/5 scale is generated with reproduction of the engine room covered with the transparent acryl externals for visualization. In the wind-tunnel experiments, the moving-belt ground board is adopted to include ground effects with force measurements by use of load cells. The flows are visualized by the smoke method. As results, with enlargement of the opening area, the drag increased overall although depending largely on the engine loading system and the inlet opening position, the front lift increased and the rear left decreased; the effect of the radiator was to relieve the change of the drag and lift.

1. Introduction

Recently improvements of car fuel efficiency becomes an important topic due to aggravation of the global warming and the energy problem, and therefore reduction of the aerodynamic drag that greatly influences fuel efficiency is proceeded[1,2]. Basic classification of the aerodynamic drag is pressure and friction drags, and as for cars, more concrete classification is made as a shape drag (mainly pressure drag), an inner-flow drag, an induced drag, an interferential drag and a friction drag. Since the shape drag occupies 75% of aerodynamic drag of whole car, it is the optimization for car configuration that has been regarded as most important in aerodynamic developments of cars[2,3,4]. History of improvements of the car shape dates back to the 1920s'. The then average for coefficient of drag (C_D) is about 0.8 but nowadays passenger cars with C_D less than 0.3 are produced, which shows the evolution of car shapes during about 100 years. To realize further reduction of aerodynamic drag, not only the shape drag but also the inner-flow drag should be considered. The inner-flow drag is the drag that is generated by taking the traveling wind from the front inlet into the engine room for cooling of a radiator and an condenser for air conditioning. As another one, there is a drag due to traveling wind passing through the cabin for using air conditioning facilities, with very small drag values. Therefore, as countermeasures for the inner-flow drag, those in the engine room region are mainly conducted. The inner-flow drag within an engine room occupies about 10% of the whole aerodynamic drag of car[2], which is not small as for today circumstances of car aerodynamics, and the reduction of the inner-flow drag in the engine room becomes important[2,5,6,7]. Furthermore, with regard to the engine-room aerodynamics, there exist some reports about the drag and the cooling efficiency, but there are very few documentations about the stability performance of cars such as reduction of lift and side-wind stability, and therefore further research is considered to be necessary.

The final goal of this research is to reduce of undesirable influences in aerodynamics due to airflows within the engine room. A simplified car model with 1/5 scale was generated with reproduction of the engine room. By this car model it is possible to change the placement of the engine loading system and the opening area and position of the air inlet system and to attach and remove a radiator. In the wind tunnel with a moving-belt ground board, the drag and lift forces are measured with flow visualization, and aerodynamic effects by the engine loading system and the air inlet system with and without a radiator are investigated.

2. Experiments

2.1. Test model

The 1/5 scale car model (Fig.1, Table.1) was generated by main frames with surface treatment, pasted exteriors with surface treatment except for front parts, and acrylic plates as a bonnet and a fender for visualization. The air inlet can be chosen from 6 types of opening area including "without opening" (Fig.2) and each dimension of the opening is described in Table 2. As the air inlet system, totally 12 patterns were reproduced with being larger from the lower (Fig.2) and from the upper by reversing these air inlets (Fig.3). As the engine loading system, 2 types of arrangements (Fig.4), length placement and width placement, were reproduced. Further, a radiator model (Fig.5) was generated, so that the loss of pressure coefficient (C_p) be equivalent to that of an actual car radiator. By combination of the air inlet system, the engine loading system, and existence/nonexistence of a radiator, totally 48 patterns were experimentally investigated.

Figure 1. 1/5 scale car model.

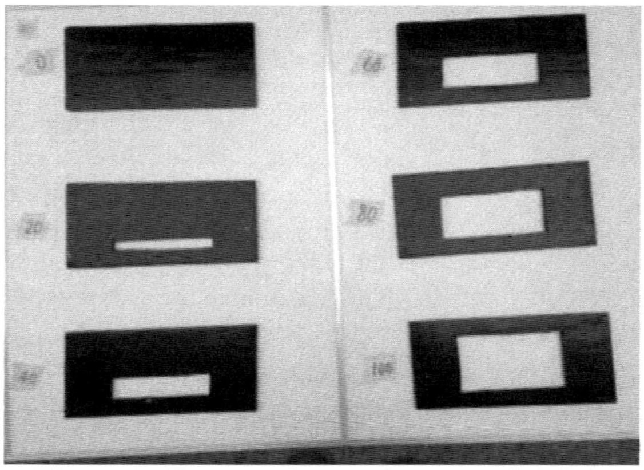

Figure 2. Six types for opening area of air inlet.

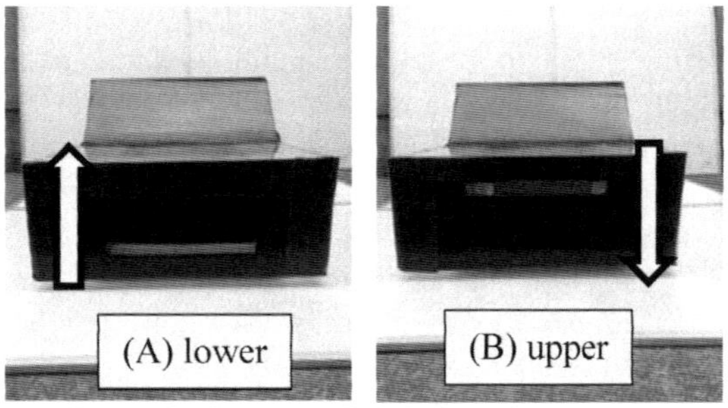

Figure 3. System of air inlet enlargement.

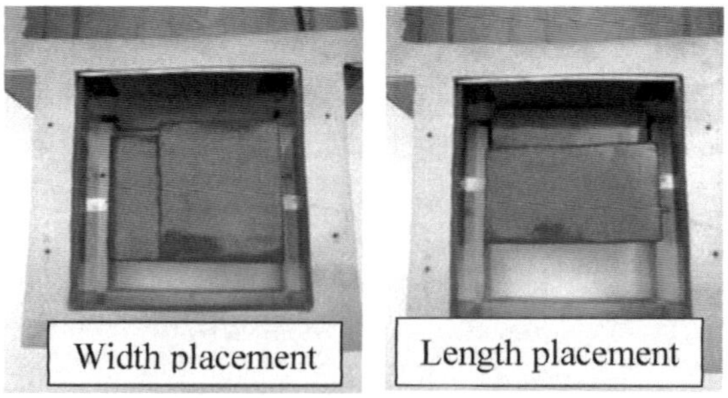

Figure 4. Engine loading system.

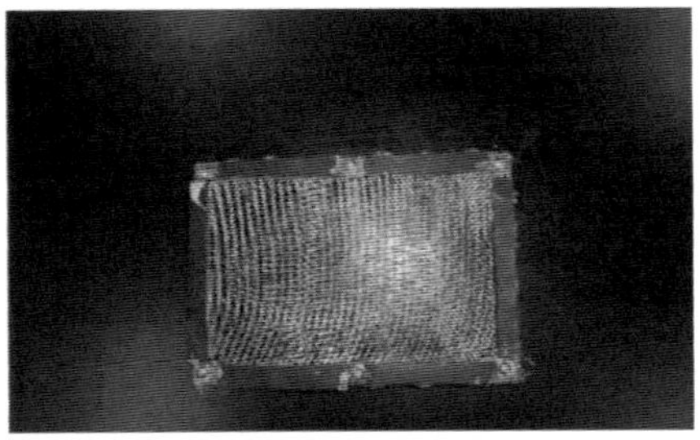

Figure 5. Radiator for test model.

Table 1. Dimensions of test model.

reduced scale	1/5
full length	980 mm
full width	340 mm
full height	315 mm
wheel base	670 mm

Table 2. Dimension of air inlet

width of air inlet	height of air inlet
180mm	0mm
180mm	20mm
180mm	40mm
180mm	60mm
180mm	80mm
180mm	100mm

2.2. Experimental setup

The drag and lift forces were measured by using the large-scale low-speed wind tunnel of Tokai University (nozzle exit: 1.0m [height] × 1.5m [width])[8] and the moving belt ground board, where the test model was supported by rods and wires connected to load cells. From the load cells the electric signals were sent through the strain amplifier and A/D converter to a computer, with sampling period of 10ms and sampling number 1024. Experiments have been carried out with speed of 20 m/s for both the wind and the moving-belt, and road clearance of 30 mm.

3. Results and discussion

Transforming the voltage values obtained in experiments to the coefficients of drag and lift, C_D and C_L, how C_D and C_L are changed for each experimental pattern is evaluated. Figures 6, 7 and 8 show changes of C_D, front C_L and rear C_L in a vertical axis, respectively, for opening heights of the air inlet in a transverse axis, where (A) and (B) indicate the cases without radiator and with radiator, respectively. In the graph legends, "width placement" and "length placement" indicate the engine arrangement shown in Fig. 4, and "lower" and "upper" in ( ) indicate the lower and upper opening position shown in Fig. 3. In Fig. 6, with enlargement of the opening height, the drag increases overall, regardless of existence of a radiator, and even for same height of the air inlet, difference is brought to the drag by differences of opening position and engine loading system. For example, at 20 mm height, the case with the width placement engine and the upper opening position records the lowest drag and the case with the length placement engine and the lower opening position does the highest drag. Figure 9 shows smoke visualization for the former case, where the smoke smoothly flows above the transmission device, while figure 10 shows the latter case at the same opening height but opposite position, where the smoke impinges against the engine and forms eddies in the engine room. From these it is considered that the drag becomes lower as the inner flow within the engine room moves more smoothly without impingement. And Figure 6 also indicates that with the radiator equipped, increase of the drag for enlargement of the opening height and also for each legend pattern is made small. In Fig. 7 (A) the front lift increases with enlargement of opening height. The reason is considered that increase of the inflow volume rate through the opening area by running wind would cause increase of downward momentum of air exhausted from the engine room to the outside. Further from Fig. 7 (B) with the radiator, compared with cases without the radiator, the lift increases for the

IOP Conf. Series: Materials Science and Engineering **249** (2017) 012012 doi:10.1088/1757-899X/249/1/012012

smaller opening height, the range of the lift for each legend pattern becomes narrow, and the lift decreases at 80mm and 100mm of the opening height in the length placement. The reason for the lift increase in the cases with the smaller opening area is considered that the airflow moving from the upstream side of the radiator to the underbody would increase since the radiator interferes with the flow going straight. And the reason for the lift decrease at 80mm and 100mm height in the length placement is considered that the momentum of airflow moving from the downstream side of the radiator to the underbody would decrease by equipping the radiator. In Fig. 8 (A) the rear lift decreases with enlargement of the opening height, which is considered that due to enlargement of the opening height, the flow rate and velocity of the underbody would increase. Further from Fig. 8 (B) with the radiator, the tendency is same as in (A) but the lift value becomes minute, which is considered that due to the equipment of the radiator, the flow rate and velocity of the underbody would decrease, and therefore the effects to the rear lift is limited.

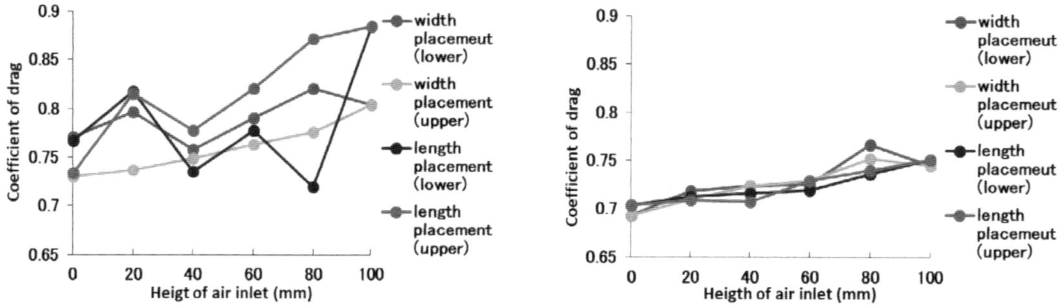

(A)Without radiator. (B)With radiator.

Figure 6. Relation of Drag and height of air inlet.

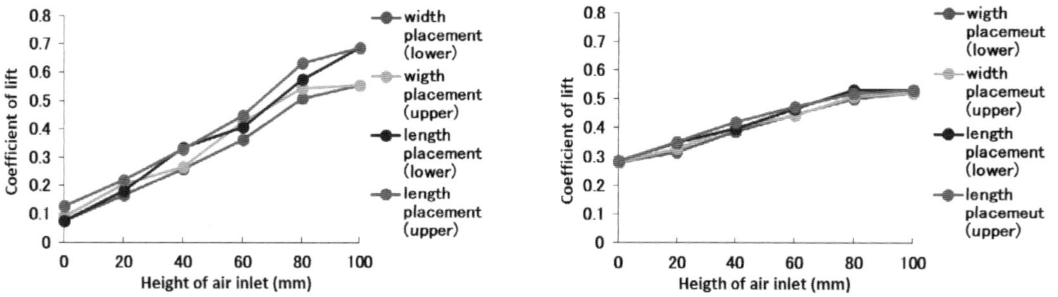

(A)without radiator. (B)with radiator.

Figure 7. Relation of front Lift and height of air inlet.

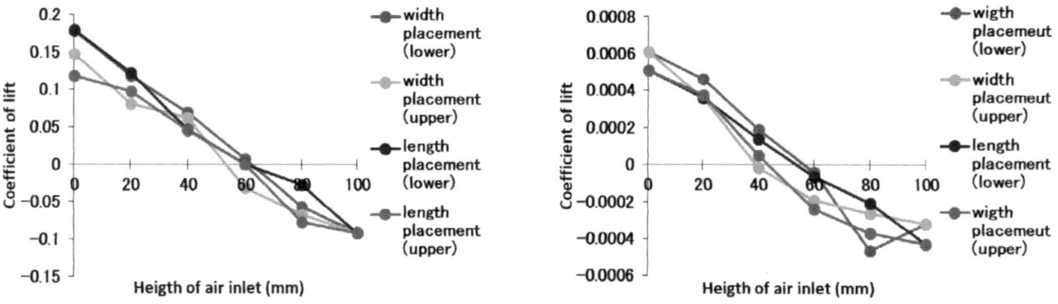

(A)without radiator. (B)with radiator.

Figure 8. Relation of rear Lift and height of air inlet.

Figure 9. Smoke visualization in engine room with width placement and upper position of air-inlet height 20mm.

Figure 10. Smoke visualization in engine room with length placement and lower position of air-inlet height 20mm.

The above findings can be summarized as follows:

1) With enlargement of the air-inlet height, the drag increases overall and slightly.
2) Even for same height of the air inlet, difference is brought to the drag by differences of opening position and engine loading system, since the drag is influenced by states of engine-room inner flows based on the air inlet system and the engine loading system.
3) The front lift increases with enlargement of the air inlet height.
4) The rear lift decreases with enlargement of the air inlet height and enters into a downforce region with border height of 40~60mm.
5) Equipment of a radiator relieves the change of drag and lift:
   1. the increase of drag due to enlargement of the air-inlet height is relieved, and difference of drags due to position of the air inlet and the engine arrangement is made small.
   2. the front lift increases for the smaller air-inlet height, and decreases at 80mm and 100mm height in the length placement.
   3. the rear lift value becomes minute.

4. Conclusion

With enlargement of the air-inlet height, the drag tends to increase slightly but depends remarkably on the engine arrangement and the air-inlet position, since the drag is affected by the inner flows within the engine room. On the other hand, the lift depends on the height of the air inlet. These changes of drag and lift are relieved by equipment of a radiator.

As future study, verification by suitable visualizations is necessary for inner flows within the engine room and outflows to the underbody. On the basis of these results, methods to reduce undesirable aerodynamic influences will be made clear.

References

[1] Kobayashi T and Nouzawa T (Society of Automotive Engineers of Japan, Inc.), eds., 1999 Car design and aerodynamic technology (Asakura Publishing Co., Ltd.)

[2] Wolf-Heinrich Hucho, ed., 1998 *Aerodynamics of Road Vehicles* (Society of Automotive Engineers, Inc.)

[3] Sumitani K, Maeda K and Ichinose K 2004 Development of vehicle fluid dynamics: Flow around the vehicle and aerodynamic characteristics, *Journal of Japan Society of Fluid Mechanics*, Vol. 23, pp.445-454

[4] Aizawa S, Matsumoto R, and Takakura Y 2013 Aerodynamic Characteristics and Wake

Structures on Car Models Changed by Configuration Parameters, *Proc. FLUCOME 2013 (The 12th International Symposium on Fluid Control, Measurement and Visualization)*, Paper No. 0237

[5] Tanaka A, et al. 2015 Aerodynamic Development for New Demio and CX-3, *Matsuda Technical Report*, No. 32, pp.85-92

[6] Ohshima T, Hamatani K, Ninoyu M and Nakagawa K 1998 Influence of the Cooling Air Flow Outlet on Aerodynamic Characteristics, *Transactions of Society of Automotive Engineers of Japan*, Vol.29, No.3, pp.81-86

[7] Nouzawa T, Hiasa K, and Yoshimoto M 1989 Influence of Engine Cooling Air Flow on Aerodynamic Drag, *Transactions of Society of Automotive Engineers of Japan*, No.40, pp.76-84

[8] Kameoka K and Takagi T 1994 *Journal of the Visualization Society of Japan*, Vol. 14, suppl. No. 3, pp.102-103

IOP Conf. Series: Materials Science and Engineering **249** (2017) 012013 doi:10.1088/1757-899X/249/1/012013

Measurement of corner separation zone response on a compression ramp to plasma actuation within the hypersonic boundary layer

Brock E Hedlund, Alec W Houpt, Stanislav V Gordeyev and Sergey B Leonov

Institute for Flow Physics and Control, University of Notre Dame, Notre Dame, IN, 46556, USA

bhedlund@nd.edu

Abstract. This study was performed to characterize the dominant frequencies present in the boundary layer uptsream of and in the corner separation zone of a compression surface in Mach 4.5 flow and to determine a control effect of transient plasma actuation on the boundary layer. Schlieren imaging was used to distinguish the corner separation zone for 20°, 25°, and 30° compression ramps mounted on flat plates. Spectra of the natural disturbances present in the boundary layer and separation zone were gathered using a high-speed Shack-Hartmann wavefront sensor and surface mounted PCB™ pressure sensors while varying flow parameters by adjusting total pressure, temperature, and ramp angle. Shallow cavity discharge plasma actuators were used as a high-frequency localized thermal forcing mechanism of the boundary layer. The plasma effect was negligible for forcing frequencies (50 kHz) below the natural dominant frequency (~55-80 kHz). High frequency perturbations that can promote the transition to turbulence were amplified when the plasma forcing frequency (100 kHz) was higher than the natural dominant frequency (~55-80 kHz). This technique can potentially be used for active control of hypersonic boundary layer transition and the supersonic flow structure on the compression surface.

1. Introduction

Scramjet engines provide a reusable and more efficient form of supersonic propulsion for atmospheric flight as compared to rocket engines. An air-breathing supersonic engine (scramjet) is expected to be used to power high-supersonic and hypersonic aircraft aimed at transport missions in the prospective applications. These vehicles are expected to operate at high altitudes, where the air density is low enough to achieve these speeds. This is an advantage for reducing the drag on the vehicle but can cause issues with scramjet operation, one of which is the maintaining of operational stability of the air inlet. Turbulent flow is desired at the inlet of a scramjet so as to reduce flow separation and total pressure loss as the flow interacts with the compression surface and helps prevent inlet unstart [1]. The low air density associated with high altitudes leads to a relatively low Reynolds number that means the transition from a laminar to a turbulent boundary layer does not always occur naturally on the length scales encountered. Since the incoming flow generally has a low magnitude of freestream disturbances, which could promote turbulence, the flow needs to be preconditioned by an aerodynamic surface before being ingested. It has been suggested that the most effective mechanism of tripping the flow is the formation of streamwise vorticity within the boundary layer [2].

There are multiple ways that the flow can be tripped to promote the transition to a turbulent boundary layer. The most employed methods include local barriers, discrete surface roughness, or cross-flow jets.

These mechanical methods can augment the instabilities but have slow response times (orders of magnitude larger than the gasdynamic time scale) and, in general, are only capable of generating a stationary pattern of flow forcing. It can be assumed that unsteady forcing of the boundary layer at the appropriate time scales would be more effective than the steady forcing methods at generating and growing unstable disturbances.It has been shown that thermal methods, such as plasma, can be used to promote the laminar-to-turbulent transition [3]. Plasma based flow control uses electron-molecular interactions that transfer energy from an applied electric field to gas acceleration, heat release and pressure elevation via collisional momentum transfer and plasma-chemical reactions. Plasma actuators provide many advantages to the aforementioned methods; faster response time, no moving parts, a robust installation flush with the surface, adjustable frequency to match flow characteristics and to generate flow disturbances. Being able to adjust the frequency and power deposition of the plasma actuators is a key advantage over the mechanical methods which provide a steady forcing and are optimized for only one flow condition. Plasma flow control has been studied extensively focusing on AC or pulse driven Surface Dielectric Barrier Discharge (SDBD) plasmas [4]. The power deposition and power density are typically too low for SDBD to have much of an effect on high-speed flows. There have also been studies done with plasma flow control in high-speed applications [5].

2. Test Approach

2.1 Experimental Facility Description

Testing took place at the University of Notre Dame in the high enthalpy arc tunnel, ACT-1, in Mach 4.5 flow. Figs. 1 and 2 below illustrate the test section arrangement and pictures of the model within the test facility respectively. The test conditions consisted of: a range of the unit Reynolds number $Re_L \approx 3-14\cdot10^6\ m^{-1}$ and $T_0 = 300\ K$ (cold flow) stagnation temperature; unit Reynolds number $Re_L \approx 4\cdot10^5\ m^{-1}$and $800\ K - 1250\ K$ (hot flow) stagnation temperature; stagnation pressure $P_0 \approx 0.8 - 4\ bar$; flow enthalpy $h\cdot\dot{m} \approx 10\ kW$ (cold flow) and $50\ kW$ (hot flow). "Cold flow" and "hot flow" will be used to distinguish between low-enthalpy and high-enthalpy tests respectively. The experimental model consisted of a flat plate with a sharp leading edge at an angle $\alpha = 15^{\circ}$. One of three plates were mounted on top of the flat plate to form the compression ramp with a variable angle β. In total, the model measured: length $L = 229\ mm$, width $W = 102\ mm$, and height $H = 19\ mm$. $62\ mm$ upstream of the ramp, a row of electrodes was mounted spanwise through the model. Insulated by a ceramic insert, the electrodes were used to generate a plasma discharge on the top surface of the model. This placement is pointed out in Figs. 1 and 2.

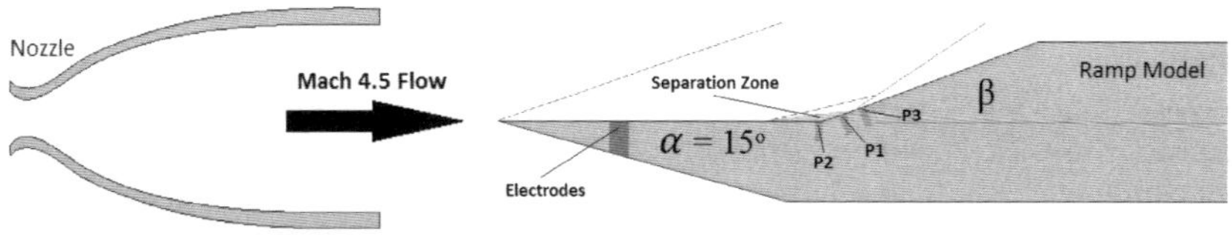

Figure 1. Side view of experimental arrangement.

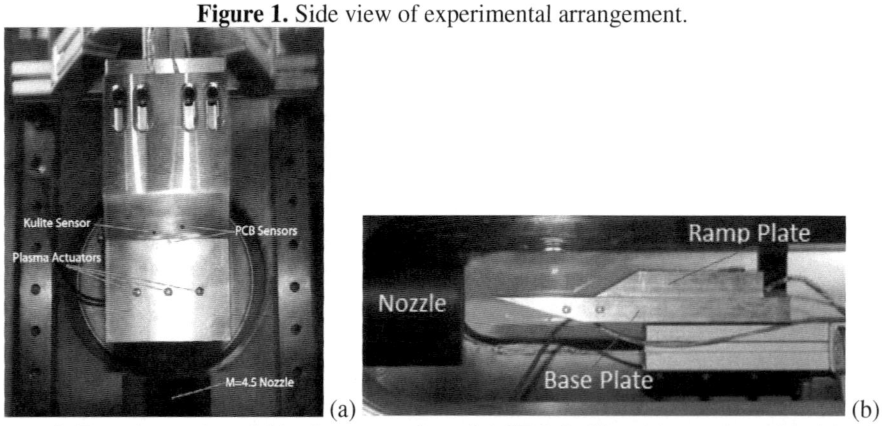

Figure 2. Experimental model in the test section of ACT-1 facility. (a) top view; (b) side view.

Tests lasted up to 1 second, during which the arc heater operated for 0.3 s (for hot flow operation) and plasma generation lasted 0.1 s. The model allows for variations in ramp angles $\beta = 20° - 30°$ in order to vary the corner separation conditions. To establish a baseline and consistency in flow characteristics from one test to the next, data was collected before, during, and after plasma actuation. Instrumentation in the sequence of data collection included a schlieren imaging system, high-speed aero-optical wave front sensor, high speed camera, high sensitivity pressure transducers, and probes to measure current and voltage of the electrical discharge.

Both Kulite™ and PCB™ pressure sensors were selected to provide measurements of the flow pressure and perturbation frequencies within the corner separation zone. A Kulite XCQ-080 miniature pressure transducer provided an absolute pressure measurement within the separation zone at location P1 in Fig. 1. This sensor however, has a limited range of frequency capabilities and measurements are limited to 15 kHz and below, allowing the measuring of the average surface pressure dynamics during the test. A sample of the Kulite sensor measurements is shown in Fig. 3. Pressure amplitude, measured on the ramp at P1 by a Kulite sensor provides pressure profiles for both cold and hot flow tests below.

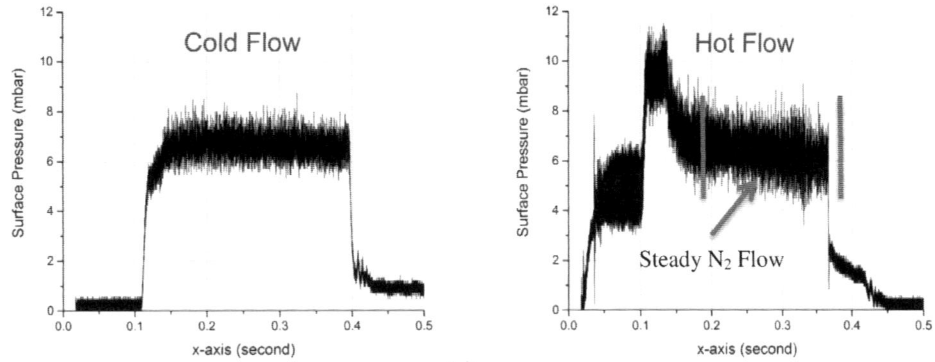

(a) (b)

Figure 3. Pressure data on ramp surface for (a) cold and (b) hot flow tests.

In Fig. 3a, the cold flow profile reflects the steady flow of nitrogen gas for 0.3 seconds. Fig. 3b shows the time sequence for hot flow, beginning with an initial flow of argon gas for approximately 80 ms in order to initiate the arc breakdown. Nitrogen is then introduced prior to the flow of argon cutting out at 0.125 s transitioning the arc to operating in nitrogen only. This data shows the extreme case for the highest pressure to expect in the separation bubble and acts as a guide for the other test conditions.

Two PCB™ 132 series microsensors located at P2 and P3 in Figs. 1 and 2 (only for the $\beta = 30°$ case) provided high frequency pressure measurements in the corner separation zone. The locations of these sensors (on the flat plate and ramp respectively) were chosen to read the traces of perturbation on both surfaces.

Prior to performing the tests with the ramp model, a test was performed to directly measure the freestream speed from the nozzle for both the cold and the hot flow cases using Shack-Hartmann wavefront sensing (principles and setup described later in Section II D). The laser beam was sent normal to the flow direction immediately downstream of the nozzle assembly near the centerline. Deflection angles were gathered with a sampling frequency of 530 kHz on the rectangular grid with 10 points in the streamwise direction and 5 points in the cross direction. A multi-point cross-correlation technique [6] was used to calculate the deflection angle power spectra, phase difference between adjacent streamwise points and the corresponding convective speeds at several points away from the nozzle. It should be emphasized that these are *direct* measurements of the convective speeds of the optically-active small-scale turbulence structures naturally occurring in the freestream.

The extracted deflection angle power spectra at different streamwise points for the cold flow case are presented in Fig. 4. The spectra decay in the streamwise direction, consistent with the isotropic turbulence decaying process. Between frequencies 10 kHz and 100 kHz all spectra exhibit the power slope of "-4/3", indicated as a dashed line in Fig. 4. In Ref [7] it was shown that this slope corresponds to the Kolmogorov-type spectra of turbulence in the self-similar energy transfer region.

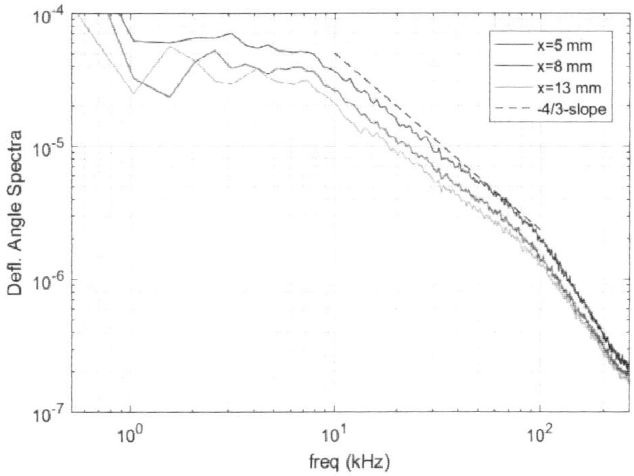

Figure 4. Characterization of the flow disturbances near the nozzle: power spectra

The freestream convective velocities, gathered with the aero-optical technique for the cold flow case at different streamwise points are presented in Fig. 5. The speed is found to be approximately 700 *m/s* immediately after the nozzle. An isentropic calculation of the freestream speed for M = 4.5 and the total temperature of 300 *K*, gives the consistent value of 700 *m/s*. As the flow expands into a vacuum chamber, it forms an under-expanded round jet with a slightly increasing speed further downstream of the nozzle; this effect is also present in Fig. 5. For the hot flow case, the density drops by approximately a factor of 6, compared to the cold flow case, so the optical distortions are reduced by the same factor. Nevertheless, an analysis of the experimental data, performed for the hot flow case, had given the value of the freestream speed between 1,100 and 1,500 *m/s*, with the estimate of the flow stagnation temperature between 800 and 1,250 *K*.

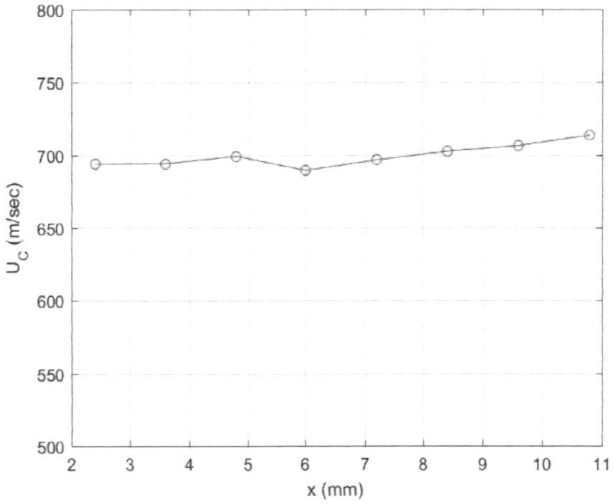

Figure 5. Freestream velocity at streamwise locations for cold flow.

2.2 Estimation of Natural Instability Frequency

The most-amplified naturally occurring disturbances in the test article boundary layer were estimated from Linear Stability Theory results for a flat plate as follows. Fig. 3h in Ref. [8] presents the non-dimensionalized frequency *F* as a function of $(Re_{x,e})^{1/2}$ for an adiabatic flat plate at Mach 4.5. The non-dimensional frequency *F* is defined as:

$$F \equiv \frac{2\pi f}{U_e Re_e} \tag{1}$$

where *f* is the disturbance frequency, U_e is the boundary-layer edge velocity, and Re_e is the unit Reynolds number at the boundary-layer edge.

For a flat plate at zero angle of attack, the unit Reynolds number at the boundary-layer edge equals the freestream unit Re; for these test conditions ($P_0 = 0.8\ bar, T_0 = 300\ K$), $Re_e = 3.0 \cdot 10^6\ /m$. The Reynolds number based on the distance from the leading edge to the ramp ($x = 0.10\ m$) and boundary-layer edge conditions, $Re_{x,e}$, for this test condition is thus $3.0 \cdot 10^5$. According to Fig. 3h in Ref. [8], for $(Re_{x,e})^{1/2} = 539$, $F \approx 7.8 \cdot 10^{-5}$ then the dominant frequency to be expected is $f \approx 25\ kHz$. The low-enthalpy case has an approximately adiabatic wall. In the high-enthalpy case, the wind tunnel run time was too short to reach an adiabatic wall, so the relatively cold wall would result in a thinner boundary layer and somewhat higher disturbance frequency [9].

It should be noted that this approximation is extrapolated from Fig. 3h in Ref. [8], so it is not 100% reliable. It also must be noted that the Reynolds number used in the experiments are significantly lower than the studies done, so the disturbances may not yet have enough time to develop and grow to become well-observable.

When the total pressure is increased to $2.2\ Bar$, the unit Reynolds number increases to a more reasonable value of $Re_e = 8.0 \cdot 10^6\ /m$. This increases the predicted most amplified frequency of perturbations to be $f \approx 55\ kHz$. Neutral stability curves for this case are presented in Fig. 10.5 of Ref. [9]. The point corresponding to $(Re_{x,e})^{1/2} = 894$ and $F \approx 6.3 \cdot 10^{-5}$ is well within the unstable region for first-mode (Tollmien-Schlichting) waves. The maximum stagnation pressure tested was $4\ Bar$ which corresponds to a unit Reynolds number of $Re_e = 14 \cdot 10^6/m$, $(Re_{x,e})^{1/2} = 1200$, and $F \approx 4.7 \cdot 10^{-5}$. This is again well within the first-mode unstable region. The most amplified frequency for this max case is $f \approx 78\ kHz$.

Due to the extremely low Reynolds number while testing with stagnation pressures under $1\ Bar$ and the linear stability prediction, the perturbations seen in these cases are not believed to be first mode (Tollmien-Schlichting) waves. Alternatively, it follows that these could be second mode acoustic vibrations in nature. Assume the measured separation zone thickness of $\delta = 1.8\ mm$ is one-quarter of the wavelength of some acoustic wave ($\lambda = 7.2\ mm$). Assuming the temperature of the boundary layer is approximately equal to the total temperature, $T_0 = 300$, the local sonic velocity is $a = 353\ m/s$. Using the equation $f = a/\lambda$ the estimated frequency of this acoustic wave is $49\ kHz$. Using the same method for hot flow, the estimated frequency is $127\ kHz$.

2.3 Approach to the Plasma Actuator Design

The fast development of a gas dynamic instability can be promoted by the introduction to the system of an oscillation of pressure or gas density, with a proper frequency, to the boundary layer. The main method for doing so is a modulation of the electrical power deposited to the plasma. This method might not be fast enough because the response time is limited by the gas dynamic time scale, related to a characteristic length of a few centimeters. At the same time, a modulation of electrical parameters of the plasma might initiate much faster kinetic processes in the gas flow.

Over the last few years, kinetic mechanisms of "rapid heating" resulting from energy transfer from internal molecular energy to translational/rotational temperature, on a sub-microsecond time scale, have been analyzed in great detail [10]. This suggests that rapid heating, caused primarily by the quenching of N_2 excited electronic states by oxygen, $N_2^* + O_2 \rightarrow N_2 + O_2$, would result in significant pressure overshoot in the plasma zone [11]. The process of gas dissociation may deliver even faster pressure elevation, although with a lower overall efficiency. In a multicomponent gas (air), the gas pressure can be represented by the sum of partial pressures of each component. The equation takes into account the ionization with degree α and the dissociation of a molecular gas with degrees δ_k for the k-component (N2, O2, at least). For the dissociation of multiatomic molecular gases the dissociation degree is defined as $\delta_k = \frac{\sum_i N_{ki} - N_k}{N_k}$, where i-species are originated due to k-species dissociation. Thus the gas pressure is

$$P = k_B \cdot \sum_{k=1}^{m} N_k \cdot T_k \approx\approx k_B \cdot (N_{N2} + N_{O2}) \cdot \left((1 + (1-\chi) \cdot \delta_{N2} + \chi \cdot \delta_{O2}) \cdot T_g + \alpha \cdot T_e\right) \quad (2)$$

Here k_B is the Boltzmann constant and χ the molar fraction of molecular oxygen in the gas. Nitrogen dissociation is negligibly small at these conditions. If only fast processes are considered (frequency $f \geq 10^5\ Hz$), for the rough estimation the gas density can be posed as a lesser variable than the gas temperature and pressure. After simplification the pressure dynamics can be expressed by the following formula:

$$\frac{dP}{dt} \approx k_B \cdot (N_{N2} + N_{O2}) \cdot \left((1 + \chi \cdot \delta_{O2}) \cdot \frac{\partial T_g}{\partial t} + \chi \cdot \frac{\partial \delta_{O2}}{\partial t} \cdot T_g\right) \quad (3)$$

Note, that the derivative of the dissociation degree can be calculated through a dissociation reaction rate coefficient k^d_{ki} (2nd order). :

$$\frac{\partial \delta_k}{\partial t} = \frac{1}{N_k} \cdot \sum_i \frac{\partial N_{ki}}{\partial t} = \sum_i (k^d_{ki} \cdot N_k - k^r_{ij} \cdot \frac{N_{ki} \cdot N_{kj}}{N_k}) \tag{4}$$

where the recombination processes with rate coefficients k^r_{ij} could be neglected with fast temperature elevation.

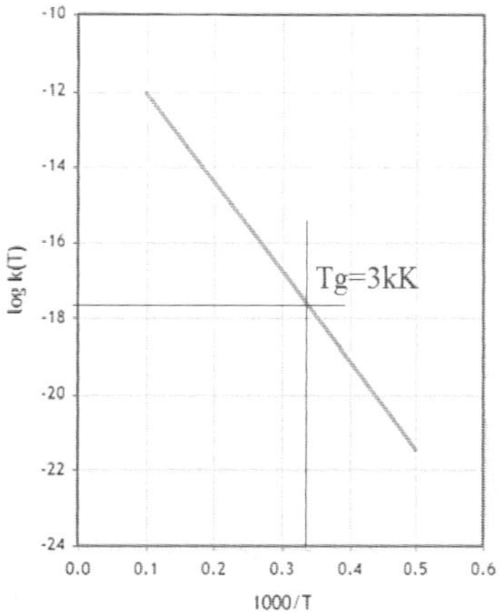

(a)

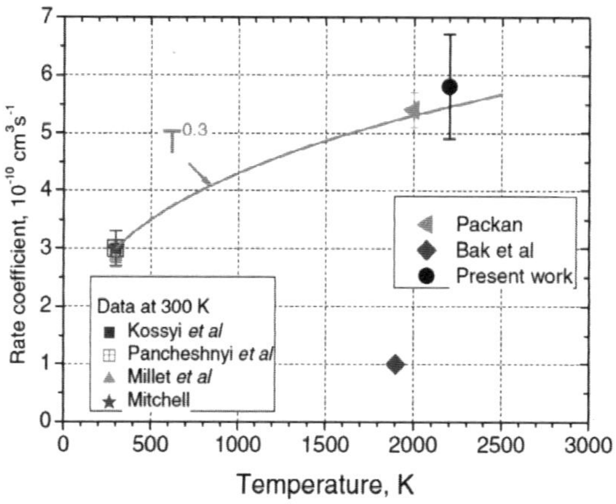

(b)

Figure 6. Thermal dissociation rate constant. (a) reaction O2 (thermal) = O + O [k]=m³/s [12]; (b) reaction N2*+O2=N2+O+O - quenching rates of N2(C) by O2 [13].

The measurements performed in a nanosecond duration discharge [13] confirm that the ultrafast dissociation of molecular oxygen is mainly due to the quenching reaction of N2(B) and of N2(C) by oxygen. The fraction of dissociated oxygen reaches about 50% after 20 ns. Both the gas temperature evolution and oxygen density are in good agreement with the ultrafast mechanism simulations of Popov [10, 14]. The fraction of energy spent on dissociation of oxygen is about 35%, and the fraction of energy spent on gas heating is about 21%. Figure 6 presents data on reaction rate constants for thermal dissociation of oxygen and oxygen dissociation in low-temperature pulsed plasma. It is well seen that a thermal dissociation of oxygen does not play a significant role in pressure elevation even at gas temperature $T_g \approx 3kK$, compared to the non-equilibrium dissociation in plasma. The effect of the plasma-based dissociation might be $>10^3$ times higher than one due to the thermal mechanism. To realize such a mechanism, the plasma should be organized in a proper way. It is also important that in the case of a plasma kinetic mechanism the upper boundary of the frequency of pressure oscillations is not limited by a slow, purely thermal, process.

Existing data [15] allows suggestion that the most effective frequency of artificial disturbances for tripping the boundary layer for excitation of the first mode of BL disturbances at $M = 5$ is $f = 100 - 200\ kHz$. For a proper optimization of the system, the frequency of plasma operation has to be adjustable over a wide range. Previously, data was collected with isolated roughness elements installed in one line in the spanwise direction with 5 to 25 mm spacing [2]. Based on suggestion to produce the local instant gas temperature increase $\Delta T \approx 10^2 - 10^3\ K$ the average electrical power of each actuator should be in the range of $P = 100 - 500\ W$. The duration of each plasma pulse should be of microsecond or submicrosecond scale to produce pointwise disturbances. A low level of gas density in a hypersonic boundary layer flow may cause an issue because in a low gas density environment the discharge strives to transit to a glow type discharge where the power density becomes too low for any observable effect. Preliminary tests [16] indicate that it is possible to generate filamentary plasma at the molecular concentration of $n > 2\times10^{17}\ cm^3$ using a simple two-electrode scheme.

The Shallow Cavity Discharge actuator [16] has been used in these tests. It is flush mounted, with a size of a few millimeters, and can be installed on the metallic surface of the model. It has the proper discharge geometry, reasonably low applied voltage, and a sufficient level of the disturbances excited. A time resolved image of the SCD discharge in $M = 4.5$ flow taken by the Andor iStar ICCD camera is shown in Fig. 7. At $f = 50 - 100\ kHz$, the discharge works as a push-pull plasma mini-jet. At lower frequencies of repetition a second operation mode was observed, currently considered as a cathode sheath pattern, where a rather thin layer of plasma covers most of the model surface. The cathode sheath is where a major portion of the electric power is released [17] that is potentially beneficial for BL control.

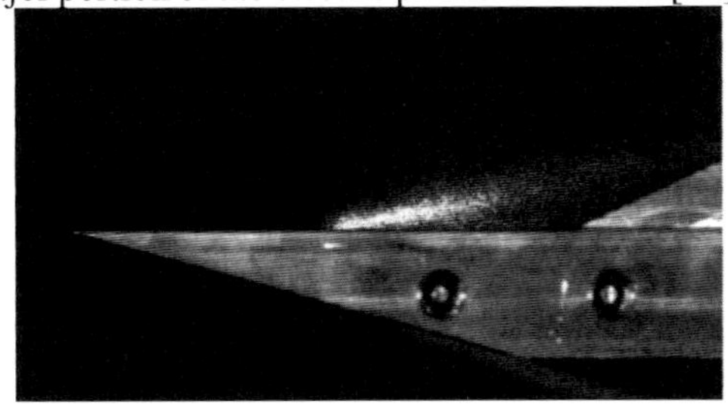

Figure 7. SCD in $M = 4.5$, $P = 4\,mbar$ flow in plasma mini-jet operation mode. Exposure 1μs, delay time 3 μs (within the electric pulse).

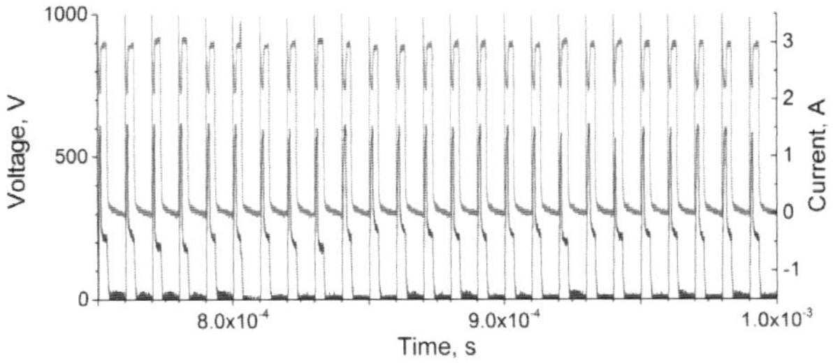

(a)

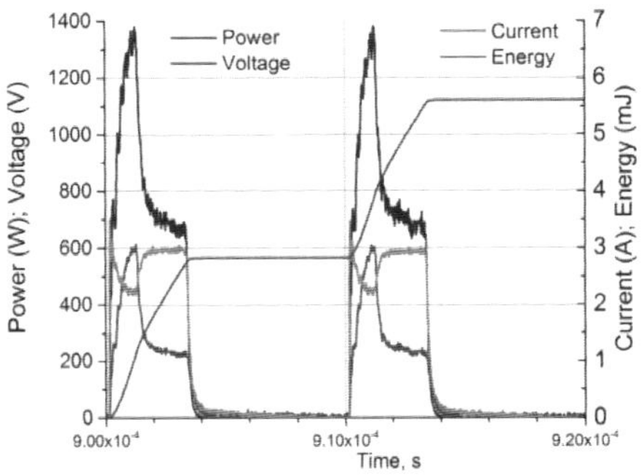

(b)

Figure 8. Voltage –current time series of the SCD operation at $f = 100\,kHz$ (a); calculation of the discharge pulse power and energy for 3 SCDs (b).

Typical records of the electric parameters in M=4.5 flow are shown in Fig. 8a for a frequency of repetition $f = 100\,kHz$. The pulse energy and average power were calculated based on this data, as it is shown in Fig. 8b. The discharge parameters were as follows: frequency of repetition $f = 10 - 100\,kHz$; pulse duration $t < 4\,\mu s$; voltage $U < 1\,kV$; pulse energy $E = 0.8 - 1.2\,mJ\,/unit$; average power $W_{av} < 400\,W$. Three units of SCD have been arranged in the model.

2.4 Aero-Optical Measurements

Aero-optical diagnostic measurements were performed using a high-speed Shack-Hartmann wavefront sensor [18]; a schematic of the experimental setup is shown in Fig. 9. The laser beam was expanded to a 50-mm-diameter collimated beam and passed along the spanwise direction over the corner region of the model mounted in the test section. The spanwise beam propagation was chosen for two reasons. First, the flow is expected to be primarily spanwise-uniform. Second, as the beam traverses the 4-inch-long region of the flow, aero-optical distortions become stronger thus improving the signal-to-noise ratio [19]. After exiting the test section, the beam is reflected off the return mirror, which sends the beam back along the same path. This so-called double-path setup further amplifies the aero-optical signal by a factor of two, as the beam traverses through the flow of interest twice, and also simplifies the optical setup. The returning beam is split off using a cube beam splitter, sent though a contracting telescope, which reduces the beam size to 12.5 mm in diameter, and recorded by a Phantom v1611 high-speed digital camera. The camera had a 38 mm focal length, 70×60 lenslet array with 0.3-mm-pitch with 100% fill ratio, attached to it. After passing through the lenslet array, the beam was split into subaperture beams and focused on the camera sensor, creating a series of dots. To achieve the high, 531 kHz, sampling rate, only a small portion of the image (128×64-pixel) was acquired for the full duration of the tunnel run. In-house software was used to extract the temporal motion of the dots and to convert this into a time series of the deflection angles.

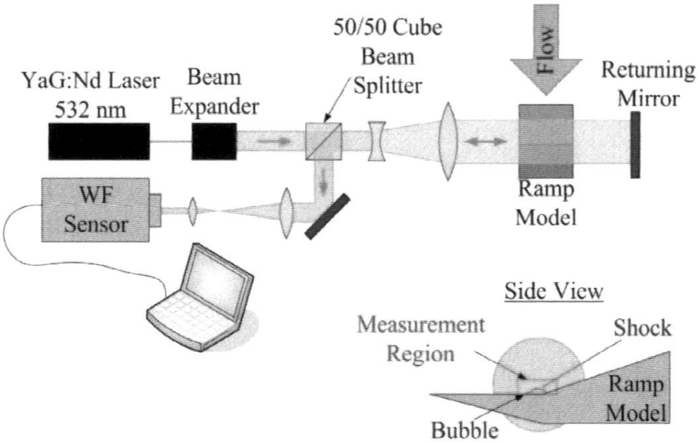

Figure 9. Schematic of set-up for aero-optical measurements.

It is important to note that all aero-optical measurements are non-intrusive by nature, sensitive to the density field only, provide good spatial resolution and limited in temporal resolution only by the digital camera technology. Comparing to other optical diagnostics, like schlieren or shadowgraph techniques, wavefront measurements provide additional advantages. The light from each lenslet is focused on a digital sensor, increasing the dot intensity, so wavefront measurements can be done with less-powerful lasers. A location of the dot is typically extracted with better than 0.1 pixel resolution using centroiding algorithms [20], providing accurate measurements of the deflection angle over the subaperture, since the displacement of each dot is proportional to the subaperture-averaged deflection angle. The time series of the deflection angles carry information about the temporal characteristics of the density field over different spatial locations across the laser beam. The wavefront sensor simultaneously measures both streamwise (X-) and wall-normal (Y-) components of the deflection angle; this is in contrast to Schlieren, where wavefront changes are measured along only one fixed direction. The 2-component deflection angle is a local gradient of the wavefront, so knowing both components of the deflection angle the wavefront (or equivalently, Optical Path Difference or OPD) can be computed. OPD is proportional to the spanwise-integrated density field, $OPD(x,y,t) = K_{GD} \int \rho'(x,y,z,t)dz$ [21], where K_{GD} is the Gladstone-Dale constant, and knowing OPD, the amplitude of the density fluctuation over a subaperture can be estimated [19]. In contrast, time-resolved Schlieren images provide good spatial and temporal information, but density information is not easily extracted from them. Finally, the wavefront measurements using the Shack-Hartmann sensor are fairly insensitive to mechanical vibrations, imposed on the test section, the model and different components of the set-up [21].

3. Test Results

For all tests it was important to distinguish the formation, stability, and size of the corner separation zone. Schlieren images were made using a pulsed NIR laser diode with a pulse duration of $0.15\ \mu s$, essentially showing the flow for one moment in time to examine these separation zone characteristics in $M = 4.5$ flow. It was observed that the corner separation zone forms and stabilizes within $15\ ms$ of the flow starting. Fig. 10a notes the important features of the flow field on a schlieren image. The low Reynolds number of the flow in these experiments causes the laminar boundary layer, resulting in the flow separation for all three ramp angle cases, as it is seen in the 25° and 30° in Figs. 10c and 10d. There is a shallow separation zone, much longer than it is deep, for the 20° case in Fig. 10b. This shallow separation zone with laminar boundary layers was also found numerically in Ref. [22].

IOP Conf. Series: Materials Science and Engineering **249** (2017) 012013 doi:10.1088/1757-899X/249/1/012013

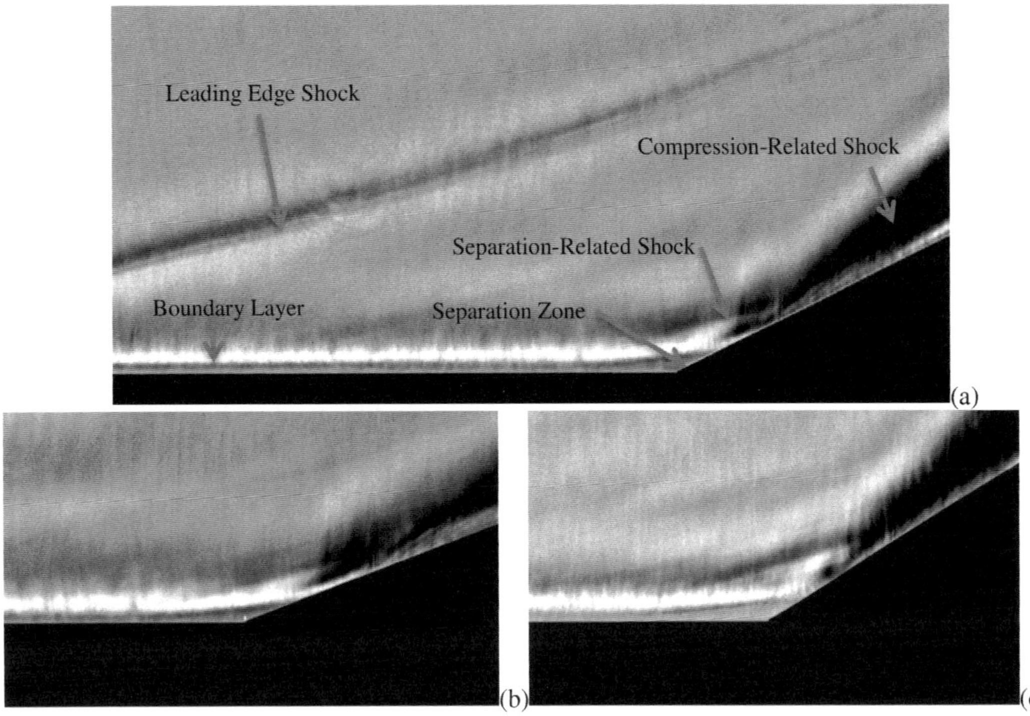

Figure 10. Schlieren images of the ramp corner and the full flow field with ramp angle: (a) 25° and enlarged separation zones for: (b) 20° and (c) 30°.

To gather basic information about the flow field, both a Shack-Hartmann wave front sensor and surface mounted PCB™ sensors were used to collect data on the perturbations in the flow. The aero-optical method measures the gas density gradients at multiple points in the flow field while gathering data at the surface with pressure sensors can provide more information about the boundary layer conditions. The ramp angle and total pressure, P_0, were varied to observe the effects of each on the amplitudes of flow perturbations present. The two PCB™ sensors, mounted at P2 and P3 (see Fig. 1) detected flow perturbations on the flat plate and ramp portions of the model surface. The data collected on the flat plate with varying total pressure and ramp angle is shown in Fig. 11 and the ramp data collected for the 30° case is shown in Fig. 12.

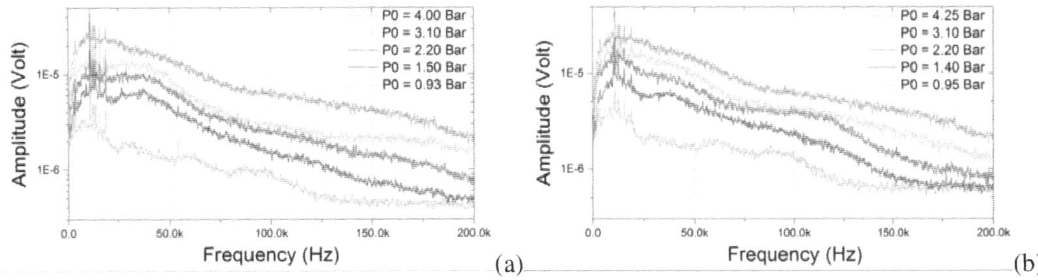

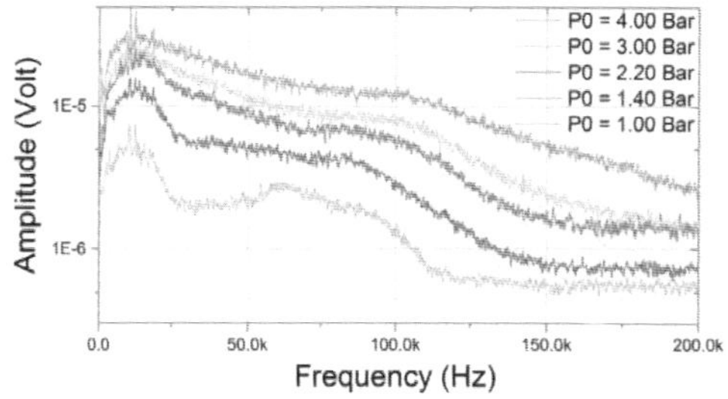

(c)

Figure 11. Spectra of flow disturbances measured on the flat plate by a PCB™ pressure sensor with ramp angle: (a) 20°, (b) 25°, and (c) 30°.

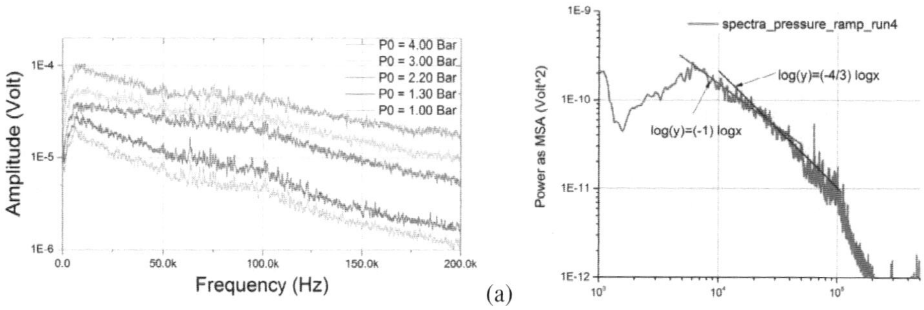

(a) (b)

Figure 12. (a) Spectra of flow disturbances measured on the ramp surface near the separation zone on a 30° ramp and (b) log-log plot of pressure spectra

It can be seen that the total pressure has a significant effect on the spectra of the pressure perturbations. The increase in total pressure results in a higher unit Reynolds number flow with a significantly higher magnitude of initial perturbations in the flow; this is seen as a higher magnitude of oscillations across the frequency spectrum. A higher total pressure also shifts the peak of the dominant frequencies higher. In Fig. 11c the dominant peak for $P_0 = 1\ bar$ is approximately $F = 60\ kHz$. The peak broadens and shifts to the right such that for $P_0 = 2.2\ bar$ the peak is $F > 75\ kHz$. This shifting effect can also be attributed to a change in the unit Reynolds number, which has a significant impact on the development of the boundary layer and the perturbations present. The dominant peak structure tends to become less dominant and all frequencies experience an increase in amplitude as the stagnation pressure increases to $P_0 = 4\ bar$. The same effect of increasing magnitude is observed in the data collected on the ramp surface in Fig. 12a, where the spectra are nearly an order of magnitude higher. Fig. 12b is a sample of ramp data plotted on a log-log scale. Here, the "-1" power slope in the mid-range frequencies is associated with eddy motion in the logarithmic layer of the boundary layer [23]. The "-4/3" power slope of the higher frequencies suggests that the flow is turbulent. Flow tends to transition to turbulence quickly after the reattachment point with corner separation zones.

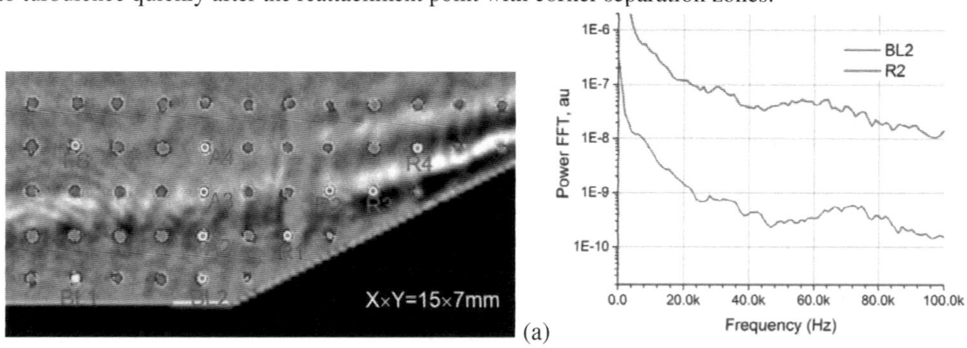

(a) (b)

Figure 13. (a) shawdowgraph and measurement locations; and (b) FFT spectra of optical density perturbations.

To better visualize the relation between the spatial distribution of the deflection angles, measured by the wavefront sensor, and the flow structure in the vicinity of the flow separation zone, the same optical setup was used to obtain high-speed shadowgraph videos. The lenslet array was removed and distorted intensity patterns were recorded at the same sampling speed of 531 kHz over the same 64×128 area. This image size corresponds to a 15×7 mm measurement region over the model. The representative shadowgraph image with an overlaid dot pattern (with 1.2 mm spacing between the dots) from time-averaged Shack-Hartmann wavefront sensor images is shown in Fig. 13a. The flow regions, indicated by yellow circles in Fig. 13a, were selected to investigate the temporal dynamics and the flow structure sensitivity to the plasma generation, specifically: point BL2 is inside the separation zone, A2 and A3 correspond to the unsteady shock over the separation zone, and A4 is in the outside portion of the flow downstream of the plasma generators.

Figure 13b demonstrates the optical density fluctuation spectra taken by the aero-optical technique to be compared to the pressure sensor dataset. Two lines of observation have been chosen to be close to the locations of the pressure sensors. The major frequency of disturbances, measured by the two techniques is similar, thus initially verifying both techniques and proving that the perturbations are indeed flow related and not due to some sensor characteristic. The aero-optical technique was used to determine the perturbation frequencies present in the boundary layer upstream of the corner separation zone. Fig. 14 below shows this data for cold and hot flows on a 30° ramp and Fig. 15 shows the corresponding measurement locations in this test series. Figure 14a indicates that as the boundary layer approaches and enters the corner separation zone that the peak frequency shifts lower. This effect is less apparent in Fig. 14b. It is difficult to optically determine the oscillations of the boundary layer prior to the separation zone due to the extremely low density of the gas. Here the appearance of the local maximum at $F\sim 110\ kHz$ at BL8 and BL9 correspond to the measurement points being in the corner separation zone.

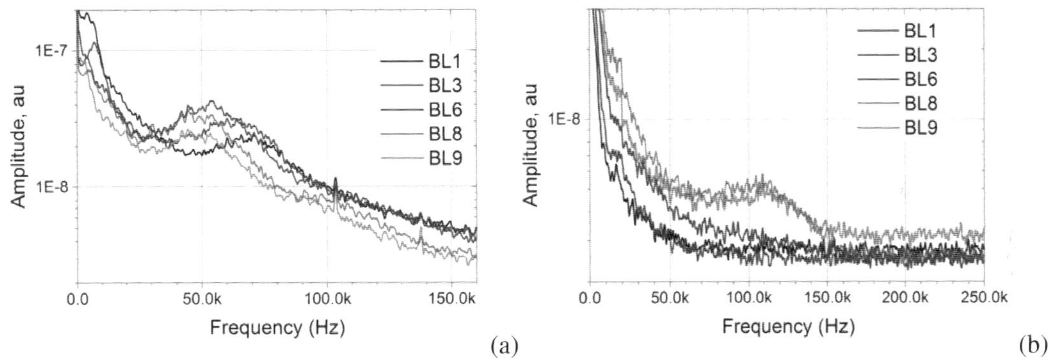

Figure 14. Wavefront oscillation spectra for different positions along the plate: (a) cold flow; (b) arc-heated tests.

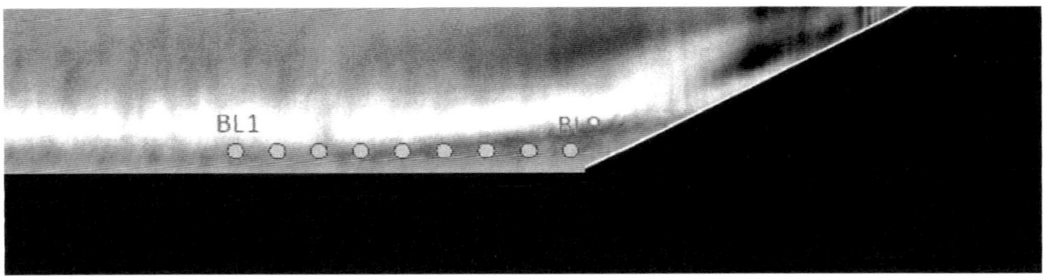

Figure 15. Shack-Hartmann data points overlaid on a schlieren image used for measurements of the BL condition.

Figure 16 compares the data collected with the Shack-Hartmann to pressure sensor data for a 30° ramp. The pressure sensor data was normalized to equalize the magnitude of the signal in the two methods. The datasets appear to agree quite well. Some shift in the peak frequency for Figs. 16a and 16b as well as an increase in amplitude for Fig. 16b is observable which can be attributed to a few factors. The pressure sensor is measuring pressure oscillations whereas the Shack-Hartmann measures density oscillations. These two properties are indeed related but not necessarily the same. Also, there is a difference in sensitivity to what is being measured. The pressure sensors are more sensitive than the optical measurements, hence why the data had to be normalized after transforming to the frequency domain. This factor can account for the larger amplitude of the peak frequency in Fig. 16b.

None of the results follow the linear stability theory predictions. Recall that the peak frequencies seen in Fig. 16 are not believed to be T-S waves, but instead are trapped acoustic disturbances. The estimated $49\ kHz$ and $127\ kHz$ match well with the data in Fig. 16 thus improving the validity of the origin of the perturbations. The peak dominant frequency differs between the cold and hot flow cases. It is expected for the dominant frequency to be higher in the hot flow case since the sonic velocity and flow speed are higher, affecting the acoustic resonance frequency. In addition, the run time is so short that the model walls do not have time to reach adiabatic conditions resulting in a cooling effect on the boundary layer. This can also cause the boundary layer to be thinner, resulting in an extra shift to a higher dominant frequency.

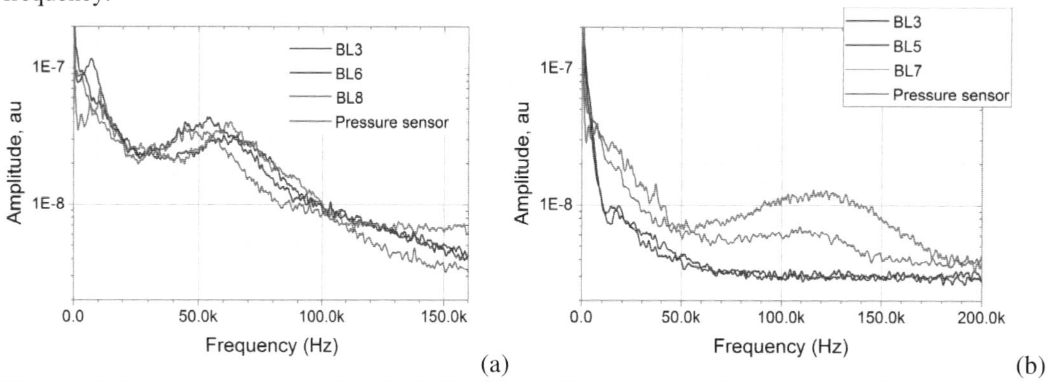

Figure 16. Spectral data comparison including Shack-Hartmann wavefront sensor and pressure sensor: (a) low enthalpy test; and (b) high enthalpy test.

With the dominant natural frequencies present in the boundary layer determined under these test conditions it was then possible to study the effect of pulse-repetitive SCD operation on the spectra of flow perturbations in and near the corner separation zone. Deflection-angle spectra obtained using the high-speed Shack-Hartmann wavefront sensor were analyzed at the locations indicated in Fig. 13a. Note, the use of the pressure sensors is challenging under these conditions due to a high level of the electromagnetic noise associated with the plasma operation. Early results showed [24] that SCD operation affects the spectra only slightly if the repetition frequency is less than the dominant frequency of the first-mode instability, $F_1 = 60 - 80\ kHz$. Operating at a frequency greater than F_1 produces significant changes in the spectra of disturbances.

As mentioned above, the wavefront sensor measures horizontal (X) and vertical (Y) components of the wavefront gradient. Isotropic flows generally display close to equal magnitudes of the wavefront gradient in the horizontal and vertical directions. The near-wall flow considered in this study is expected to have larger magnitudes of deflection angles in the wall-normal (Y) direction compared to the streamwise (X) direction as a result of the larger density gradients in the wall-normal direction. This in turn amplifies the effect that the plasma has on exciting the higher frequency disturbances and results in larger magnitudes of deflection angles.

Figures 17a-d show data for plasma actuation at a frequency of $100\ kHz$ for two points in the incoming boundary layer and two points near the compression ramp (refer to Fig. 13a for the corresponding measurement points). Each test included measurements of $0.1\ s$ prior to plasma actuation, and $0.1\ s$ during plasma operation. There appears to be a small effect due to the plasma in the boundary layer at point BL1 (see Fig. 13a). At the point BL2, Fig. 17b, the amplitude of perturbations is higher and the amplitude of high-frequency disturbances larger than $F = 80\ kHz$ increases. However, looking at Figs. 17c and 17d gathered for the ramp locations, it can be seen that these high-frequency disturbances were excited at the cost of a significant reduction of the peak frequency oscillations.

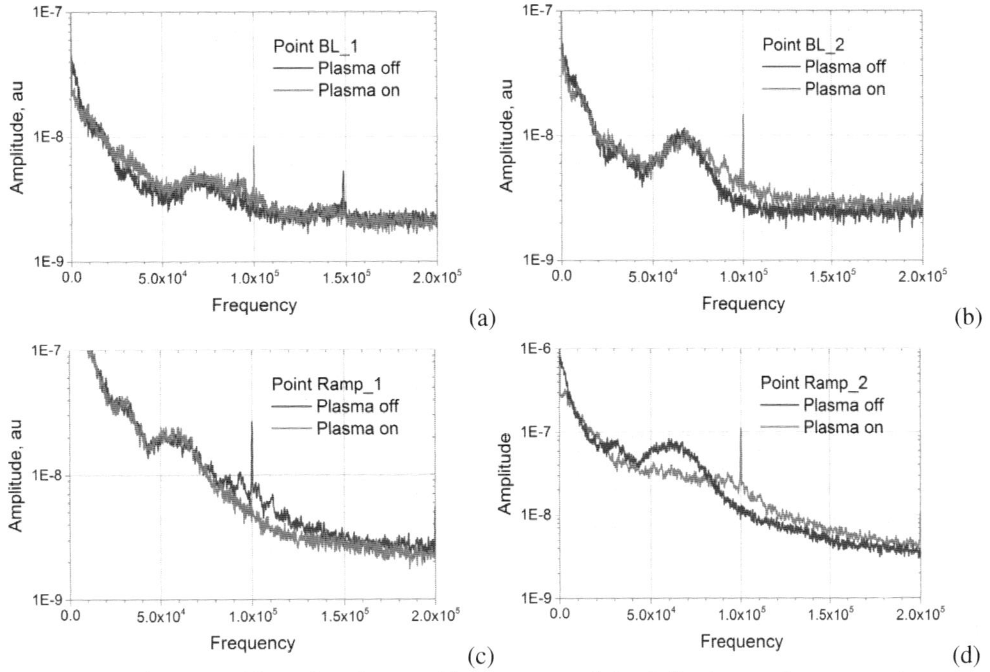

Figure 17. Spectra of flow disturbances. Y-component of the deflection-angle spectra, plasma excitation $f = 100\ kHz$. (Fig. 13a for reference)

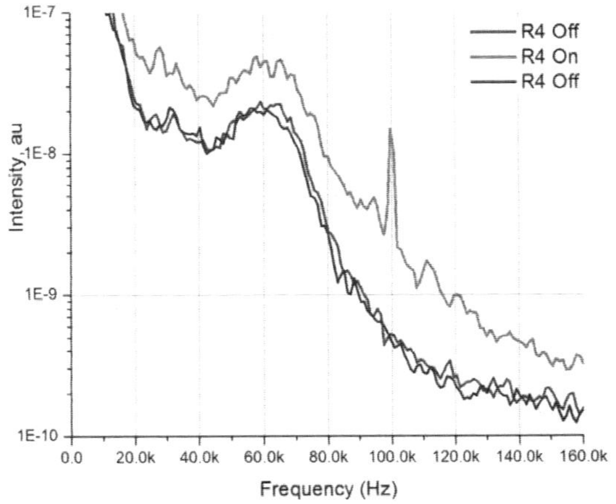

Figure 18. Spectra of flow disturbances on the ramp, point R4 (Fig. 13a for reference).

Figure 18 displays the spectra of flow perturbations at the point Ramp_4 with the largest observed plasma effect. Here the data is presented for three time periods: before the SCD plasma actuation, during SCD actuation, and after the SCDs are turned off. This was done to ensure that flow parameters did not change during operation. Point R4 lies on the compression ramp and is very near the reattachment point of the boundary layer to the ramp. The magnitude of the flow disturbances was increased for all observed frequencies and even extends beyond $100\ kHz$. The resulting amplification of flow perturbations at R4 due to the operation of the SCD actuators is $A/A_0 = 2 - 8$. It may reflect the earlier transition of the flow to a turbulent state on the compression ramp after the separation reattachment.

4. Conclusion

A canonic geometrical configuration has been tested in $M = 4.5$ high-enthalpy flow consisting of a flat surface at zero angle of attack and a compression ramp with inclinations of 20, 25 and 30 degrees in order to vary the size of the corner separation zone from a negligibly small one to a well-detected area characterized by the appearance of an oblique shock wave originating from a wedge of the separation zone. The spectra of gas pressure perturbations were gathered from surface mounted pressure sensors and compared to the gas density perturbations measured non-intrusively by a high-frequency Shack-Hartmann sensor (aero-optical technique). The dominant frequency in the boundary layer was detected in the range of $F = 55 - 70\ kHz$ in cold flow ($T_0 = 300K$) and $F = 90 - 120kHz$ in hot flow ($T_0 = 800 - 1250K$). There is a shift to lower frequencies as the boundary layer approaches the corner separation zone. It is believed that this is due to the increasing thickness of the boundary layer, especially once it reaches the corner separation zone. This increase in boundary layer thickness increases the wavelength of the trapped acoustic waves, thus decreasing the frequency. Varying the ramp angle had little effect on the dominant frequencies present.

Employing pressure sensors with high-frequency capabilities allowed for the verification of the results obtained from the aero-optical techniques and vice-versa. The Shack-Hartmann sensor collected data about the flow field density perturbations, while the pressure sensors collected pressure oscillation data at the surface. The aero-optical techniques employed have high potential for use in high-speed flows due to the ability to gather data in a non-intrusive way at high-frequencies (as high as $1\ MHz$), with a high spatial resolution, in the range of $1 - 2\ mm$. Utilizing optical methods and surface mounted pressure sensors it is possible to determine the flow perturbations in the flow field, on the surface and their cross-correlation.

An attempt to control the flow field pattern in the compression configuration of the flow has been performed with application of a highly-transient plasma generator arranged midway between the test model leading edge and the compression ramp. The generation of a constricted plasma in a low-density gas, characteristic of high-speed boundary layers, is of utmost importance in actually being able to affect the flow structures on the appropriate time scales. The shallow cavity discharge has been shown to fit this need and is still being studied to define the full limits of its operation. The hope is to be able to push its operating frequency to $f > 200\ kHz$ and beyond while maintaining its power density such that it can be utilized in a wide range of high-speed flows into the hypersonic range. The specific geometry of the SCDs used, a one-dimensional array of plasma elements in the spanwise direction with individual control of each element, has demonstrated its effectiveness in exciting high-frequency disturbances within the boundary layer. These tests have also demonstrated the feasibility of high repetition rate plasma operation in low-density, high-speed flow. Looking at all of the data collected, it is reasonable to conclude that hypersonic boundary layers are sensitive to highly transient plasmas. Active tripping of the boundary layer by electrical discharges can be done for a wide range of flow conditions so long as the forcing frequency is higher than the dominant frequency of perturbations present in the boundary layer; $f > F$.

Going forward, it is proposed to gather more data for varying flow conditions, most importantly an increased Reynolds number. Linear stability theory predicted for a flow case with $P_0 = 2.2\ bar$, the dominant first mode frequency would be $55\ kHz$ and is well within the unstable region for first mode waves as should be when $P_0 = 4\ bar$. It is also considered to study the geometry of the model and how the location of the plasma actuators in relation to the leading edge can affect the receptivity to perturbations that will grow nonlinearly. If the ramp location is moved further away from the leading edge, it is possible to have second mode waves of acoustic nature play a role in transitioning to turbulence which could also be utilized.

5. References

[1] Valdivia A, Yuceil K B, Wagner J L, Clemens N T and Dolling D S 2014 Control of Supersonic Inlet-Isolator Unstart Using Active and Passive Vortex Generators *AIAA Journal* **52** 6 pp 1207-1218

Wu Y, Yi S, He L, Chen Z and Zhu Y 2015 Flow Visualization of Mach 3 Compression Ramp with Different Upstream Boundary Layers *Journal of Visualization* **18** pp 631-644

[2] Berry S A, Nowak R J and Horvath, T J 2004 Boundary Layer Control for Hypersonic Airbreathing Vehicles *34th AIAA Fluid Dynamics Conference* AIAA-2004-2246 Portland Oregon

Schneider S, "Effects of Roughness on Hypersonic Boundary-Layer Transition," *Journal of Spacecraft and Rockets,* **45**, no. 2, pp. 193-209, 2008.

Reshotko E and Tumin A, "Role of Transient Growth in Roughness-Induced Transition," *AIAA Journal,* **42**, no. 4, pp. 766-770, April 2004.

Choudhari M, Li F and Edwards J, "Stability Analysis of Roughness Array Wake in a High-Speed Boundary Layer," in *47th AIAA Aerospace Sciences Meeting, AIAA-2009-170,* Orlando, Florida, 5-8 January, 2009.

[3] Yan H and Gaitonde D, "Effect of Thermally Induced Perturbation in Supersonic Boundary Layers," *Physics of Fluids,* **22**, no. 064101, pp. 1-17, 2010.

[4] Adamovich I V, Little J, Nishihara M, Takashima K and Samimy M, "Nanosecond Pulse Surface Discharges for High-Speed Flow Control," in *6th AIAA Flow Control Conference, AIAA 2012-3137,* New Orleans, Louisiana, 25-28 June, 2012.

Corke T, Enloe C and Wilkinson S, "Dielectric Barrier Discharge Plasma Actuators for Flow Control,," *Annual Review of Fluid Mechanics,* **42**, pp. 505-529, 2010.

Kotsonis M, Correale G, Michelis T, Ragni D and Scarano F, "Nanosecond-Pulsed Plasma Actuation in Quiescent Air and Laminar Boundary Layer," *Journal of Physics D: Applied Physics,* **47**, no. 10, p. 105201, 2014.

Kriegseis J, Duchmann A, Tropea C and Grundmann S, "On the Classification of Dielectric Barrier Discharge Plasma Actuators: A Comprehensive Performance Evaluation Study," *Journal of Applied Physics,,* **114**, no. 053301, 2013.

Leonov S, Opaits D, Miles R and Soloviev V, "Time-Resolved Measurements of Plasma-Induced Momentum in Air and Nitrogen Under Dielectric Barrier Discharge Actuation," *Physics of Plasmas,* **17**, no. 113505, 2010.

Moreau E, "Airflow Control by Non-thermal Plasma Actuators," *Journal of Physics D: Applied Physics,* vol. 40, pp. 605-636, 2007.

[5] Macheret S, Shneider M and Miles R, "Magnetohydrodynamic and Electrohydrodynamic Control of Hypersonic Flows of Weakly Ionized Plasmas," *AIAA Journal,* **42**, no. 11, pp. 1378-1387, 2004.

Meyer R, Palm P, Ploenjes E, Rich J W and Adamovich I V, "The Effect of a Nonequilibrium RF Discharge Plasma on a Conical Shock Wave in a M=2.5 Flow," *AIAA Journal,* **41**, no. 5, pp. 465-469, 2003.

Minucci M, Toro P, Oliveira A, Ramos A, Chanes J, Pereira A, Nagamatsu H and Myrabo L N, "Laser-Supported Directed Energy "Air Spike" in Hypersonic Flow," *Journal of Spacecraft and Rockets,* **42**, no. 1, pp. 51-57, 2005.

Leonov S, Yarantsev D, Kuryachii A and Yuriev A, "Study of Friction and Separation Control by Surface Plasma," in *42nd AIAA Aerospace Sciences Meeting and Exhibit, AIAA-512,* Reno, Nevada, 2004.

Adelgren R, Elliot G, Crawford J, Carter C, Donbar J and Grosjean D, "Axisymmetric Jet Shear-Layer Excitation Induced by Laser Energy and Electric Arc Discharges," *AIAA Journal,* **43**, no. 4, pp. 776-791, 2005.

Samimy M, Adamovich I, Webb B, Kastner J, Hileman J, Keshav S and Palm P., "Development and Characterization of Plasma Actuators for High Speed Jet Control," *Experiments in Fluids,* **37**, no. 4, pp. 577-588, 2004.

Falempin F, Firsov A, Yarantsev D, Goldfeld M, Timofeev K and Leonov S, "Plasma control of shock wave configuration in off-design mode of M = 2 inlet," *Experiments in Fluids,* **56**, no. 54, 2015.

[6] Smith A, Gordeyev S, Ahmed H, Wittich III D and Paul M, "Shack-Hartmann Wavefront Measurements of Supersonic Turbulent Boundary Layers in the TGF," AIAA Paper 2014-2493.

[7] Gordeyev S, Smith A, Cress J and Jumper E 2014 Experimental Studies of Aero-Optical Properties of Subsonic Turbulent Boundary Layers *Journal of Fluid Mechanics* **740** pp 214-253

[8] Mack L M 1975 Linear Stability Theory and the Problem of Supersonic Boundary- Layer Transition *AIAA Journal* **13** 3 pp 278-289

[9] Anderson J D 2000 Hypersonic and High Temperature Gas Dynamics *Viscous Flow: Basic Aspects, Boundary Layer Results, and Aerodynamic Heating* Ch 6 Reston VA

[10] Aleksandrov N L, Kidysheva S V, Nudnova M M and Starikovskii A Y 2010 Mechanism of Ultra-Fast Heating in a Non-Equilibrium Weakly Ionized Air Discharge Plasma in High Electric Fields *Journal of Physics D: Applied Physics* **43** 255201

Popov N A 2011 Fast Gas Heating in a Nitrogen-Oxygen Discharge Plasma: I. Kinetic mechanism *Journal of Physics D: Applied Physics* **44** 285201

Shkurenkov I and Adamovich I 2016 Energy Balance in Nanosecond Pulse Discharges in Nitrogen and Air *Plasma Sources Science and Technology* **25** 015021

[11] Nishihara M, Takashima K, Rich J and Adamovich I 2011 Mach 5 Bow Shock Control by a Nanosecond Pulse Surface Dielectric Barrier Discharge *Physics of Fluids* **23** 066101

[12] NIST Chemical Kinetics Database National Institute of Standards and Technology [Online] Available: http://kinetics.nist.gov/kinetics/

[13] Rusterholtz D L, Lacoste D A, Stancu G D, Pai D Z and Laux C O 2013 Ultrafast Heating and Oxygen Dissociation in Atmospheric Pressure Air by Nanosecond Repetitively Pulsed Discharges *Journal of Physics D: Applied Physics* **46** 464010

[14] Popov N A 2011 Kinetic Processes Initiated by a Nanosecond High-Current Discharge in Hot Air *Plasma Physics Reports* **37** pp 805-817

Popov N A 2001 Investigation of the Mechanism for Rapid Heating of Nitrogen and Air in Gas Discharges *Plasma Physics Reports* **27** 10 pp 886-896

[15] Mack L M 1984 Boundary Layer Linear Stability Theory *Special Course on Stability and Transition of Laminar Flow* AGARD Report 709 pp 1-81

Fedorov A, Ryzhov A and Soudakov V 2013 Effect of Local Volume Energy Supply on High-Speed Boundary Layer Stability *43rd AIAA Fluid Dynamics Conference* San Diego California

[16] Leonov S, Houpt A and Falempin F 2015 Control of Hypersonic BL Transition by Electrical Discharge (feasibility study) *20th AIAA International Space Planes and Hypersonic Systems and Technologies Conference* Glasgow Scotland

[17] Raizer Y P 1991 Gas Discharge Physics, Berlin: Springer

[18] Tyson R K 1997 Principles of Adaptive Optics, New York: Academic

[19] Sontag J and Gordeyev S 2015 Non-intrusive Velocity and Density Measurements in Subsonic Turbulent Boundary Layer *46th AIAA Plasmadynamics and Lasers Conference* Dallas Texas

[20] Nightingale A and Gordeyev S 2013 Shack-Hartmann Wavefront Sensor Image Analysis: A Comparison of Centroiding Methods and Image Processing Techniques *Journal of Optical Engineering* **52** 071413

[21] Wang M, Mani A and Gordeyev S 2012 Physics and Computation of Aero-Optics *Annual Review of Fluid Mechanics* **44** pp 299-321

[22] Novikov A, Egorov I and Fedorov A 2016 Direct Numerical Simulation of Wave Packets in Hypersonic Compression-Corner Flow *AIAA Journal* **54** 7 pp 2034-50

[23] Beresh S, Henfling J, Spillers R and Pruett B 2010 Measurement of Fluctuating Wall Pressures Beneath a Supersonic Turbulent Boundary Layer AIAA-2010-305 Orlando Florida

[24] Houpt A W, Gordeyev S, Juliano T J and Leonov S B 2016 Optical Measurement of Transient Plasma Impact on Corner Separation in M=4.5 Airflow AIAA-2016-2160

IOP Conf. Series: Materials Science and Engineering **249** (2017) 012013 doi:10.1088/1757-899X/249/1/012013

Acknowledgments
The current work is supported by the FlowPAC Institute at the University of Notre Dame. A portion of this work (design of the SCD plasma actuator [16]) was previously supported by MBDA-France (Mr. Francois Falempin supervision). The authors would like to thank Dr. Alexander Fedorov for useful discussion pertaining to the nature of the disturbances present in our test cases.

IOP Conf. Series: Materials Science and Engineering **249** (2017) 012014 doi:10.1088/1757-899X/249/1/012014

Effect of Opposing Multiphase Jet on Hypersonic Flow Around Blunt Body

Zhong Ce[1], Rei Tomita[1], Kojiro Suzuki[2] and Yasumasa Watanabe[3]

[1]Graduate Student, Department of Advanced Energy, Graduate School of Frontier Sciences, The University of Tokyo, 5-1-5 Kashiwanoha, Kashiwa, Chiba 277-8561, Japan

[2]Professor, Department of Advanced Energy, Graduate School of Frontier Sciences,, The University of Tokyo, 5-1-5 Kashiwanoha, Kashiwa, Chiba 277-8561, Japan

[3]Assistant Professor, Department of Aeronautics and Astronautics, Graduate School of Engineering, The University of Tokyo, 7-3-1 Hongo, Bunkyo, Tokyo 113-8656, Japan

E-mail: kjsuzuki@k.u-tokyo.ac.jp

Abstract. An experimental research has been conducted to investigate the flow control method using opposing jet with air and water as a fluid spike on a bunt body in hypersonic freestream at Mach number 7, using the Kashiwa hypersonic wind tunnel. The dynamic characteristics and instability of opposing jet were clarified. It can be seen from the comparison between the air opposing jet and the water opposing jet that a liquid jet seems more effective in generating fluid spike for reduction in the drag force and the heating rate on the body. Details of the flow field and the jet oscillation have been captured by Schlieren system with high-speed cameras. The experimental data reveal the interaction of the jet oscillation with the shock wave fluctuation. Such interaction has strong impact on the whole flow field structure and should be taken into consideration in the design of the active flow control system using the opposing jet method.

1. Background

The opposing jet has been attracting our research interest since it has been regarded as an effective flow control method for cooling, drag reduction, flow field modification and so on. Lots of the experimental and numerical studies have been conducted so far since the early 1950s.

Opposing jet, which is a kind of active flow control method, has advantages for reusability in comparison with a "physical" solid spike on a blunt body, which suffers from severe aerodynamic heating at its tip. The opposing jet works as a "virtual" spike, which extends forward into the incoming stream and forms the oblique shock wave at the tip of the spike. The pressure and temperature behind that oblique shock wave are much lower than those behind the detached bow shock wave on the blunt body without jet flow. As a result, significant reduction in the drag force and the thermal load are expected for a vehicle flying at hypersonic speeds. In addition, by switching off the opposing jet, the "virtual" spike instantaneously disappears. Such controllability cannot be expected for the solid spike.

Meyer and Nelson conducted the research on the active flow control by the opposing jet and the results showed that large reductions in heat transfer can be obtained by upstream injection[1].

Experimental study of Warren investigated the effectiveness of opposing gas jet from a spherical cone model in the Mach 5.8 wind tunnel. They found that the thermal load is effectively reduced by the gas injection[2].

The efficiency and performance of the flow control using opposing jet are explained in terms of the change in the flow field characteristics. The penetration length of the jet is one of the characteristic features of the jet flow. Daso, Preitchett and Wang conducted numerical and experimental studies on the opposing jets in the freestreams at Mach 3.48 and 4.0[3]. The results showed that the penetration length changes with the flow rate of the opposing jet. In some cases, however, the flow becomes unsteady and long penetration mode (LPM) and short penetration mode (SPM) were observed from the experimental data.

In most of the researches, gas is mainly considered for the injected fluid . When the length of the spike is not sufficient, the oblique shock wave formed from the tip of the spike hits the body surface, the severe peak pressure and peak heating are produced. To avoid such situation, the penetration length of opposing jet must be long enough. To realize a long virtual spike, the jet reservoir pressure must be high enough. Liquid or particle-laden gas seems promising for generating a long opposing jet, because its density is much larger than the gas and the strong inertia elongates the jet even in the presence of the free stream against it. Also, water jet can achieve a considerable cooling effect and protect the models from thermal damages, as shown in the Rashis' experiment[4].

Though the flow control by the opposing jet seems promising, Daso and Woods pointed out in their experimental study that the complexity and instability of the jet flow can have significant effect on the flow field structure, cooling effectiveness and the aerodynamic drag[5]. Experimental investigation on the cross-flow injection of liquid in Mach 6 flow by Beloki Perurena showed a coupling between the jet breakup and shock fluctuation. For the opposing jet, more complicated phenomena including jet oscillation, shock wave reflection and oblique shock wave/bow shock wave occurred in hypersonic flow[6].

The above review shows the researches on the use of liquid jet injection for flow field modification. However, few studies reported the details of the opposing jet flow dynamics and instability and our understanding on those is not satisfactory especially for the opposing jet of liquid in hypersonic flows. Thus, there remains a paucity of work on characteristics of opposing jet

In this paper, experimental studies were conducted to obtain the effect of opposing jet on hypersonic flow and details of the flow dynamic characteristics of opposing jets in order to gain a better understanding of its flow mechanism. This information should be helpful to advance the knowledge and application of opposing jets as a feasible method of flow control. The major objectives in the present study are 1) to visualize the opposing jet around the blunt body in hypersonic flow and capture the details of the flow field structure, 2) to analyze the flow dynamics, instability of opposing jet and its effect on hypersonic flow, 3) to compare the opposing air jet and opposing water jet .

2. Models and experiment apparatus

The test model is a blunt body imitating the Apollo capsule, as shown in Fig. (1a). The diameter of the model is 30 mm. The jet hole of 1mm diameter was drilled at the stagnation point. A tube was connected to a cavity inside the body from the back of the model. A flexible PVC tube was used to provide air or water from the reservoir outside of the test section at the atmospheric pressure. The water tank was set at the same height as the jet exit to avoid further acceleration of the water jet injection due to the gravitational force. The 3-D schematic of the model is shown in Fig. (1b). The photograph of the model installed in the test section is shown in Fig. 2. A spherical model was used to compare the flow structure of opposing jet around models with different shapes. The diameter of the spherical model is 30mm.

A Schlieren flow visualization system equipped with normal (30 frames per second) and high-speed cameras (60000 frames per second) was used in this experiment. For the cross-section imaging of the flow and the PSP (pressure sensitive paint) measurement, a laser light source (450nm, power Max. 2W) was used . The schematic of the experimental setup is shown in Fig. 3. Figure (3b) shows the front view of the setup of the laser cross-section imaging, the laser light source was placed on the upside of the test section.

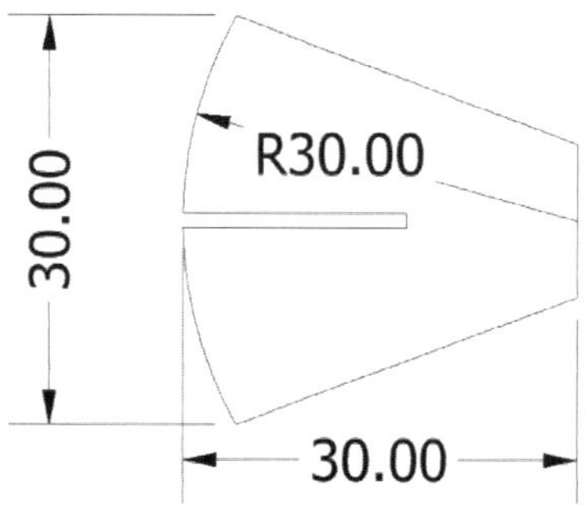

a) Sketch of the model design

b) 3-D model

Fig. 1. Schematic of the Apollo capsule model

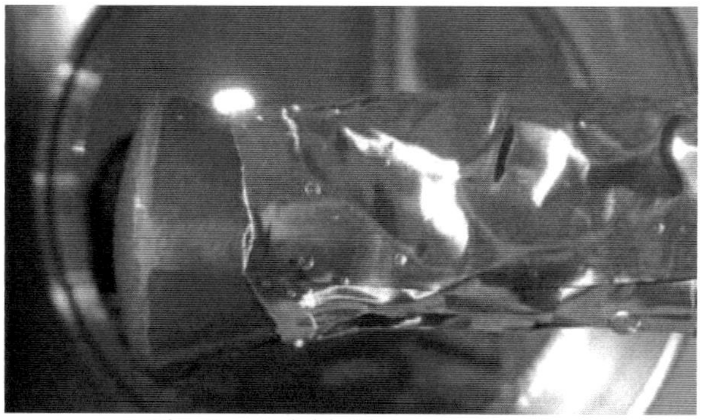

Fig. 2. Model in the test section

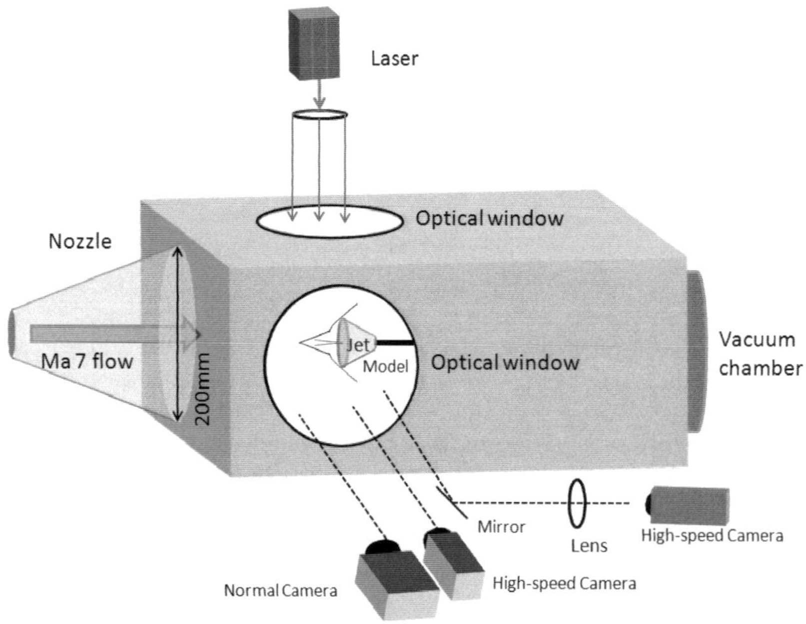

Fig. 3. Schematic of the experimental setup

3. Facility

The test facility used for the experiments is the Kashiwa Hypersonic and High-Temperature Wind Tunnel in the University of Tokyo[7]. It is a blow-down tunnel that operates at Mach number of 7 and 8. In the present study, the Mach 7 nozzle was used. Figure 4 shows the schematic description of the wind tunnel. The test model is injected after the steady hypersonic flow has been obtained. The tunnel conditions for this experiment are given in Table 1.

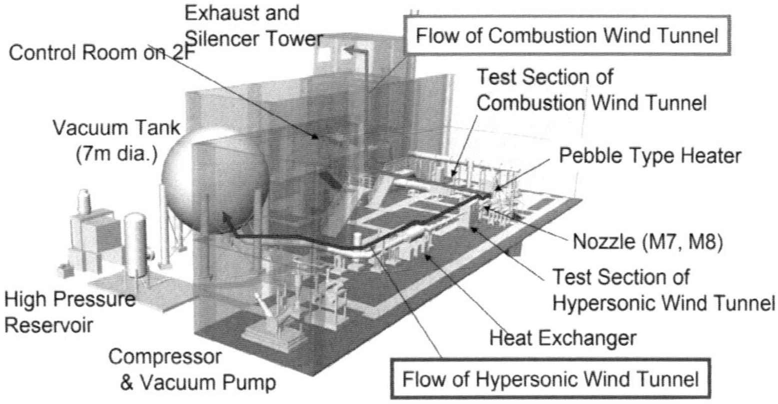

Fig. 4 Hypersonic and High Enthalpy Wind Tunnel at UT Kashiwa Campus (http://daedalus.k.u-tokyo.ac.jp/wt/info/UTHYP.pdf)

Table 1. Tunnel operation conditions in present experiment

Mach Number	7.0
Unit Reynolds No. (1/m)	1×10^5
Stagnation Temperature	540-670 K
Stagnation Pressure	950 kPa
Test Duration	Max. 60s
Nozzle Exit	200mm dia.
Uniform Flow Core Size	120mm dia.
T_0 of Jet Injection	299.15K
P_0 of Jet Injection	101 kPa

4. Result and Discussion

The discussion is divided into two parts. The first part is the test results of opposing air jets, the second part is results of opposing water jets..

4.1. Air opposing jet

Figure 5 is the Schlieren images of the flow field when the model was set in the Mach 7 freestream without the jet injection. The bow shock wave is clearly seen. This picture was taken by a normal video camera and the knife-edge is placed vertically. The separated flow was seen behind the edge of the shoulder of the model.

IOP Conf. Series: Materials Science and Engineering **249** (2017) 012014 doi:10.1088/1757-899X/249/1/012014

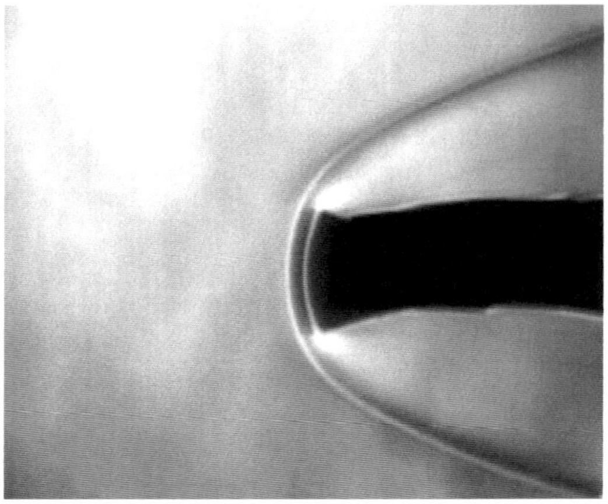

Fig. 5. Schlieren image of the flow field with no jet injection

Figure (6a) and (6b) shows snapshots of the Schlieren pictures in the case of the longest virtual spike obtained by the air jet injection. Here the penetration length is defined as the maximum distance between the pointed end of the jet and the surface of the model. The schematic of opposing jet flow filed is shown in Fig. (6c). The interface of the detached shock wave and bow shock can be seen at the position where the bow shock and the jet induced shock interact. The jet stream impinges the shear layer and the shock wave reflection can be seen in (6b), the high-speed Schlieren image Figure.

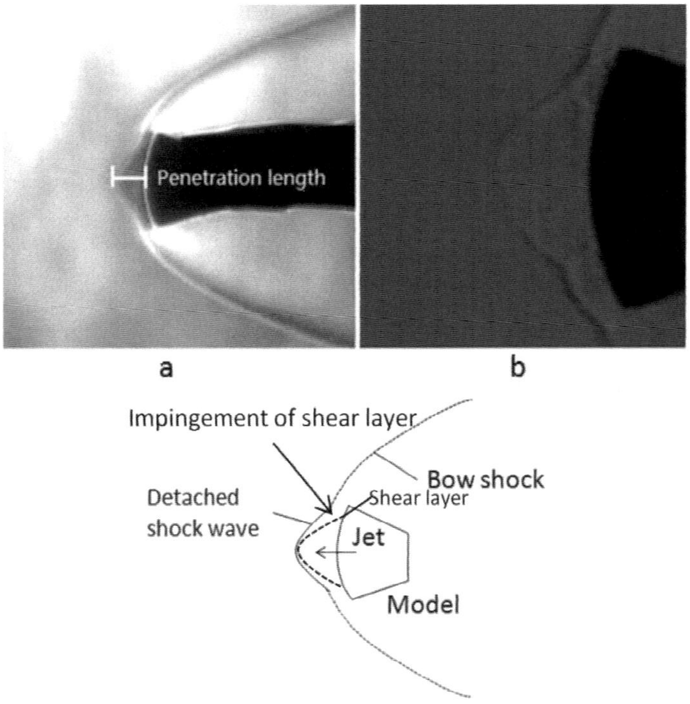

Fig. 6. Penetration length of opposing jet
from normal-speed Schlieren system(left) and high-speed Schlieren system(right)

Figure 7 is a comparison of the Schlieren images from normal-speed camera(30 frames/sec) and high-speed camera(60,000 frames/sec, 256×128 pixels). The exposure time of high-speed camera was

1μs. These images were taken in the same blow and show the time history of the flow structure for the flow field with air jet injections at the attack angle α=0°. From high-speed Schlieren images, it is noticeable that there exist significant jet oscillation and unsteadiness at the front of the nose of the model. The phenomenon is one of the most important dynamic characteristics to show the instability of the opposing jet in hypersonic freestream. The mechanism or explanation of oscillation and instability remains unclear. One possible mechanism is the transition of the boundary layer flow on the nozzle wall from laminar to turbulent [8].

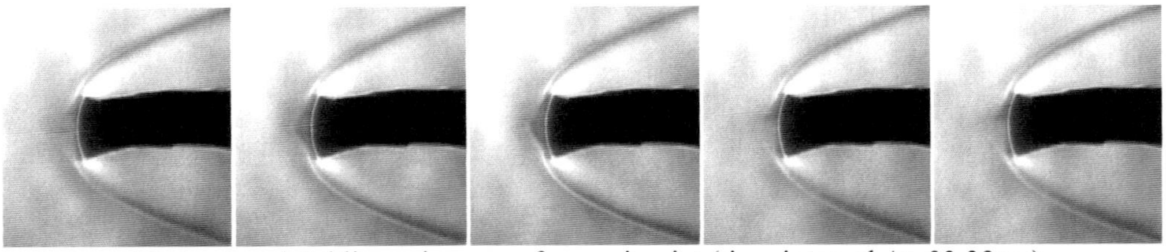

a) Low-speed Schlieren images of opposing jet (time interval Δt=33.33ms)

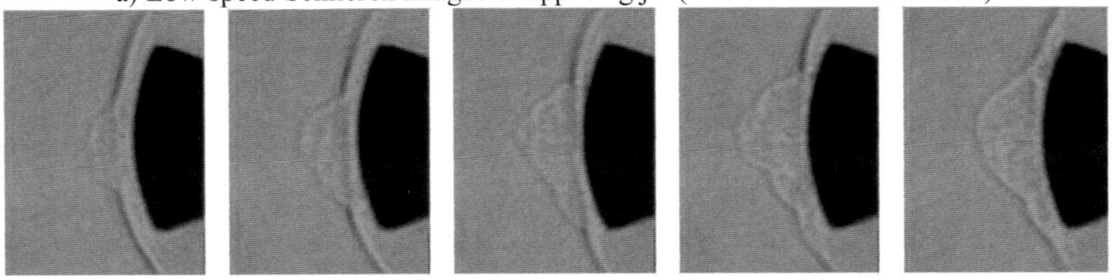

b) High-speed Schlieren images of opposing jet(Δt=16.67μs)

Fig. 7. Comparison of Low-speed and High-speed Schlieren images

However, such oscillations cannot be captured by the normal-speed Schlieren video. As can be seen in high-speed Schlieren images, the irregular compression wave generated by jet-free stream interaction appears in different positions during its oscillation. The high-speed Schlieren images at 60,000 frames/s successfully captured the oscillatory motion with the period of 250μs.

Figure 8 shows the penetration length fluctuate with time. Though the jet penetration is not necessarily symmetric with respect to the centerline as shown in Fig. 7, we define the most upstream point as the pointed end of the jet. From Fig. 8, which shows the part of the results used for measuring the distance between the pointed end of the jet and the surface of the model for 1000 frames of the Schlieren video, it can be seen that the amplitude of the fluctuation is about 10mm and the frequency is estimated at about 3963 Hz though the fluctuation is not perfectly periodic.

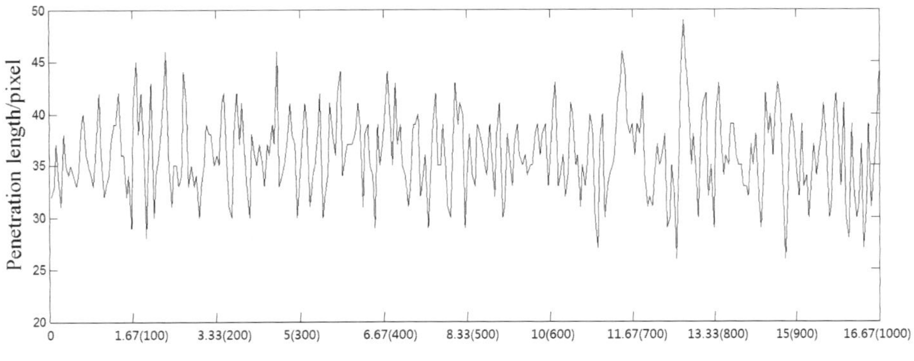

Fig. 8. Penetration length of air opposing jet fluctuation

Figure 9 is a series of the images of the jet/shock interaction, it can be seen from the figure that the jet oscillation interacts with the bow shock and the intersection of compression wave and bow shock locates on the model surface. The penetration length changes, and the feature of the jet expansion changes from one to another. The intersection of both waves kept moving reciprocally on the surface and the contour of bow shock fluctuated in a synchronizing manner. This indicates the presence of a correlation between the jet oscillation and the shock fluctuation. Such fluctuation may change the pressure distribution on the body and may cause unfavorable forces and/or moments on the body. For the design of the flow control system by the opposing jet, such unsteady behavior of the jet must be taken into account to avoid the unexpected effects on the body.

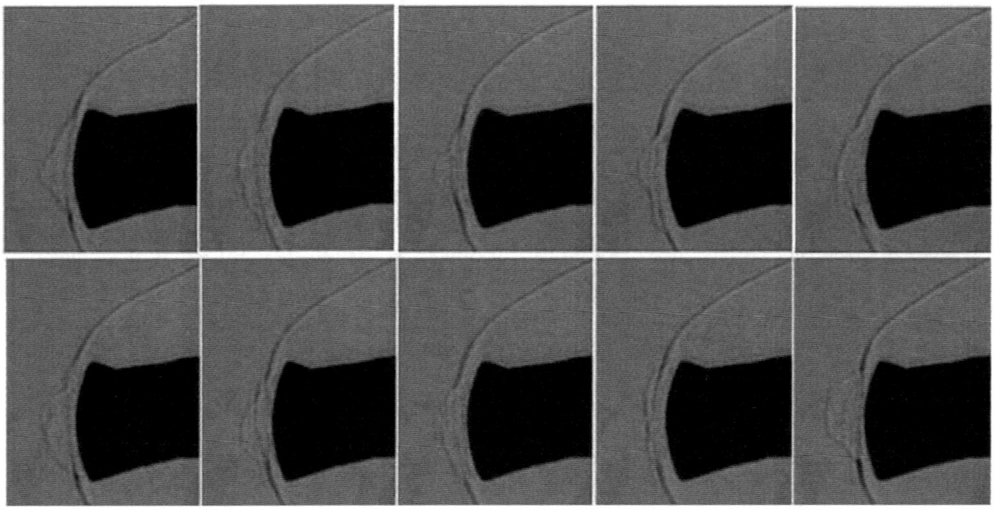

Fig. 9. Schlieren images of jet/shock interaction

The experiment for a spherical model was also conducted for comparison. Figure 10 is a time series of Schlieren images of the flow field. The penetration length for the spherical model is longer than that of capsule model. These results suggest that shape of model, in other words, the nose radius has obvious impact on the jet dynamic characteristics and oscillation.

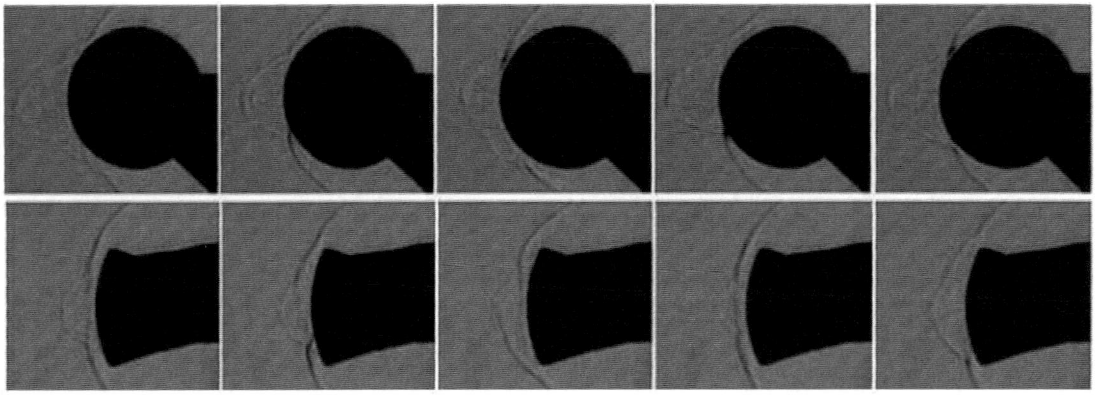

Fig. 10. Flow field of opposing jet on spherical model and capsule model

4.2. Water opposing jet

In this case, the opposing jet is injected into the stagnation pressure behind the normal shock wave, which is 14.58kPa not 950 kPa. The jet stagnation pressure of 101 kPa is sufficiently high to generate the jet. However, once the tip of the opposing jet goes out of the shock layer over the body, it is exposed to the freestream stagnation pressure. Consequently the air opposing jet can change the shape of the detached shock wave but cannot make a long penetration. As mentioned above, it needs high jet reservoir pressure for air opposing jet to realize a long penetration length to avoid impingement of the shear layer (see Fig. 6c), which can be danger for the surface of hypersonic vehicles. It seems reasonable to use liquid such as water for generating a long spike of the opposing jet. Since it is easier to obtain higher jet momentum with its density much larger than the gas, the penetration length of liquid jet is expected to be longer than that of gas jet. The dynamic characteristics and instability of liquid jet should be investigated in the same way as in the case of the air jet.

The flow field of water injection into Mach 7 flow is shown in Fig. 11. The attack angle for each run was 0°. As expected, longer virtual spike was formed for the water opposing jet and sharper conical shock wave was formed at the tip of the water jet. That oblique shock wave did not hit the surface of the blunt body when the penetration length reaches its maximum. The dark spots and regions in the Schlieren images indicate the presence of the liquid water. The water jet is well confined by the conical shock wave without spreading. The cone angle and the contour of conical shock wave in the Schlieren images seem relatively stable. Interaction between conical shock wave and bow shock can be seen near the rims of model. However, it can be seen clearly that the water jet bent downward due to both the gravity force and the aerodynamic force. Once the water spike inclines downward by the gravitational force, the asymmetric aerodynamic force bends it further.

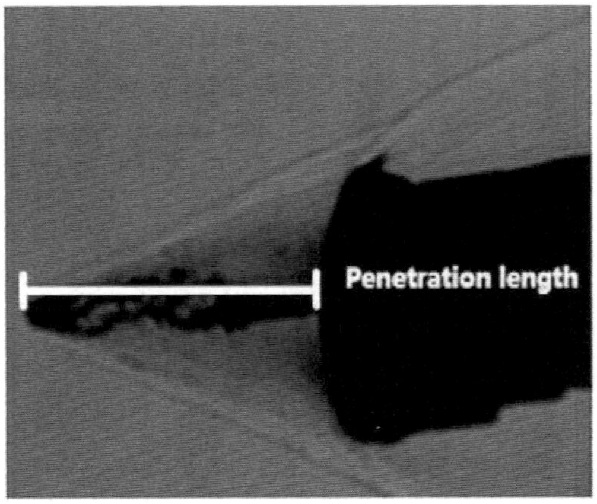

Fig. 11. Schlieren image of opposing jet and its penetration length

From the Schlieren images, the penetration length of water opposing jet is measured to be about 1.05 D, where D is the diameter of model. Comparison is made between water jet and air jet in Fig. 12. We can see the penetration length of water jet is much longer than that of air jet (about 0.3D). For the water tank was set at the same height as the jet exit, and the reservoir pressure of water jet was the same as that of the air jet. These results suggest that a liquid jet may be more effective than the gas jet from a viewpoint of the virtual spike effect. In the present experimental setup, the sonic jet was only tested as the opposing air jet. The effect of the jet exit Mach number should be investigated as well as the air jet stagnation pressure in the future.

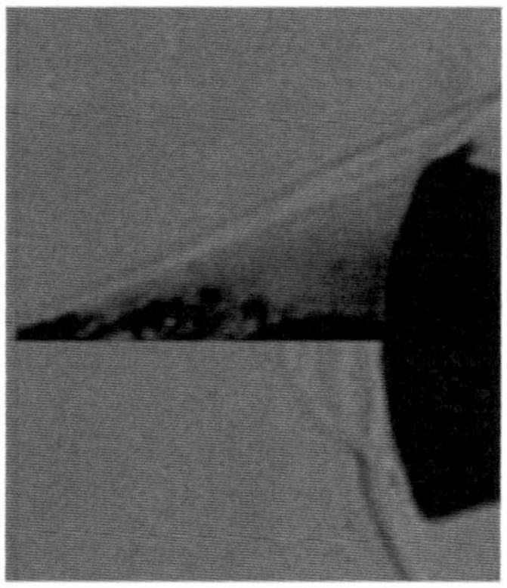

Fig. 12. Comparison of penetration length of water(upper) and air opposing jet

Figure 13 is the time history of the flow field of water jet captured by high speed camera, the snapshot was taken each 40 frames(Δt=666.67μs). It can be seen from the images that the water jet is unstable and oscillates at a relatively low frequency. The length and width of jet change along with time. Fig. 13(e) shows that when the penetration length was shorter than a certain value, the jet impinged on the model surface and the oblique shock wave hit the model, inducing the shock reflection. Complex interactions exist among the oblique shock wave, reflection shock and bow shock. Strong impact of the jet oscillation on the shock separation area can be seen from the enlarged view in Fig. 14. In addition, the frequency of the water jet oscillation is estimated at 66.65Hz, which can be calculated from the Fig. 15. Break up and bending of the water jet caused the fluctuation indicated as the second peak in Fig. 15.

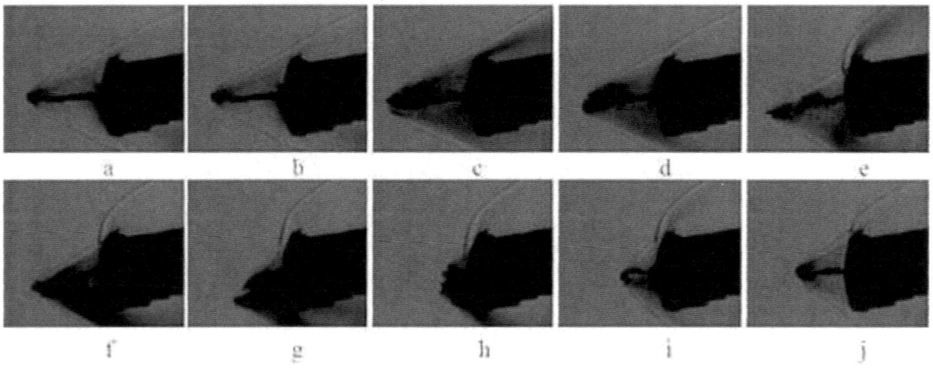

Fig. 13. Schlieren images of water opposing jet(interval of 40 frames, Δt=666.67μs)

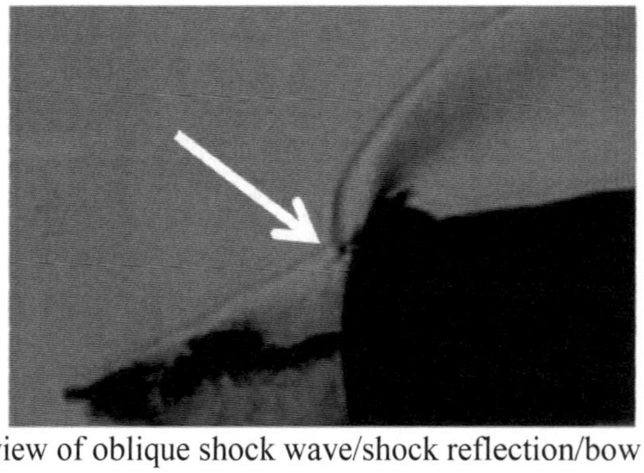

Fig. 14. Zoom in view of oblique shock wave/shock reflection/bow shock interactions

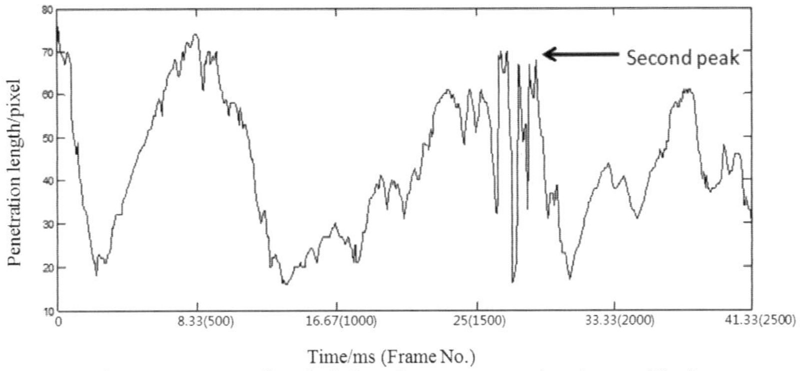

Fig. 15. Penetration height of water opposing jet oscillation

The enlarged view also shows that water flew over the rim of the capsule model and entered the wake flow area. The behavior of the injected water itself seems to be complicated and no explanation has been given up to this present time. In addition, it is reported by Rashis[4] that water jet is promising for cooling the re-entry vehicles. However, such uneven cooling can cause extra problem in the heating environment over the vehicle. Detailed investigation needs to be done on the water jet individually. The boundary of water jet is indiscernible in the Schlieren images due to the interaction with shock wave. On the other hand, the information of the cross-sectional image of the flow with jet is expected to be helpful for understanding the water jet behavior.

Flow visualization method using the laser sheet to obtain cross-section distribution, called as the laser cross-section imaging hereinafter, was performed. The schematic of laser cross-section imaging is shown in Fig. 16. It can be seen that the water spike forms an oblique shock wave at the tip and there is a recirculation region behind the oblique shock wave. Laser beam passing through a lens to fan out into a line, creating a cross-section through the flow. The water scatters the laser light towards the camera. Figure 17 is a time series of the jet cross-section taken by the high speed camera(5000 frames/sec, exposure time=198μs). The contour of water opposing jet can be seen from Fig. 17 clearly, the conical water jet well covered the model surface. The scattering of laser lights indicates the zone, where the liquid water or ice pieces exist. Most of the water stayed in liquid and solid state without turning into vapor, even the pressure in the test section is extremely low.

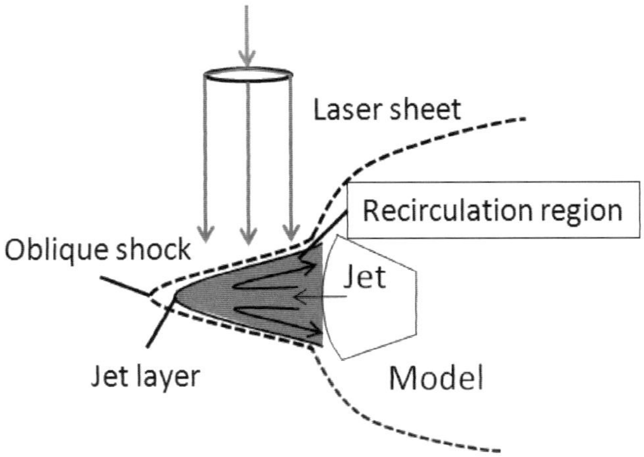

Fig. 16. Schematic of laser cross-section imaging

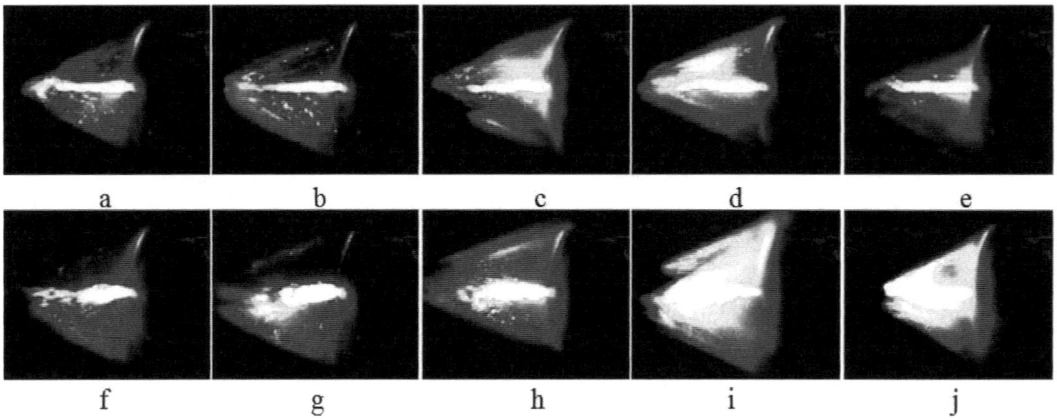

Fig. 17. Laser cross-section images of water jet(Δt=200μs)

This observation suggests that phase change cooling can be also available for thermal protection. In the Fig. 17(c), at the moment after the penetration height reached its maximum and began to back to the jet outlet, part of the water seems to be 'peeled' off the conical surface and no longer confined by the water. The contour of oblique shock was unstable and began dispersing. Such breakup process of liquid jet was found in hypersonic cross flow by Beloki Perurena[6]. The results of Beloki's research indicates the correlation between jet breakup and shock fluctuation.

4.3. Visualization of air spike using Pressure-sensitive Paint technology

The above work shows that different kinds of visualization methods have been used in this research including Schlieren photography and laser cross-section imaging. In order to obtain detailed and quasi-quantitative/quantitative results of opposing jet, a trial experiment of using binder-less Pressure sensitive paints technology[9] was conducted to realize the measurement of the pressure distribution of opposing jet. The dye probes was absorbed by some small inorganic spherical porous particle (Godd ball, B-25C, Suzuki Yushi Kougyo, Japan). Such particles were seeded into the jet. It can be seen from the Fig. 18 that the signal of the luminescence of dye probe molecules was captured near the tip of the air 'virtual' spike. Figure (18b) shows the enlarged view of PSP signal.

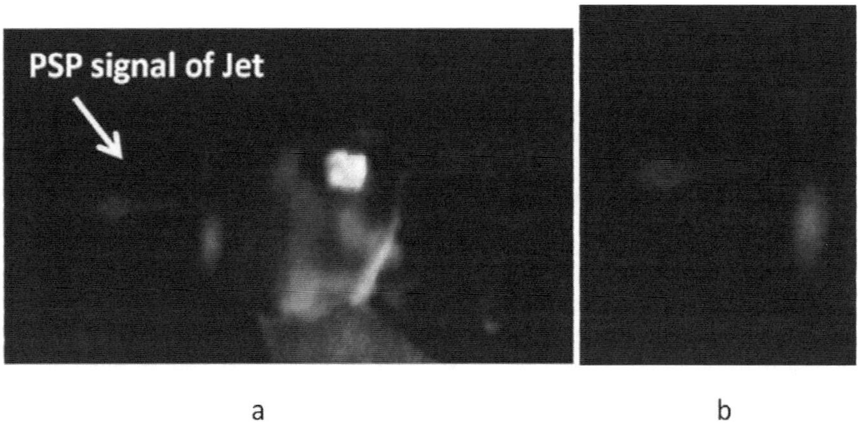

Fig. 18. PSP signal of PSP particles in the air jet

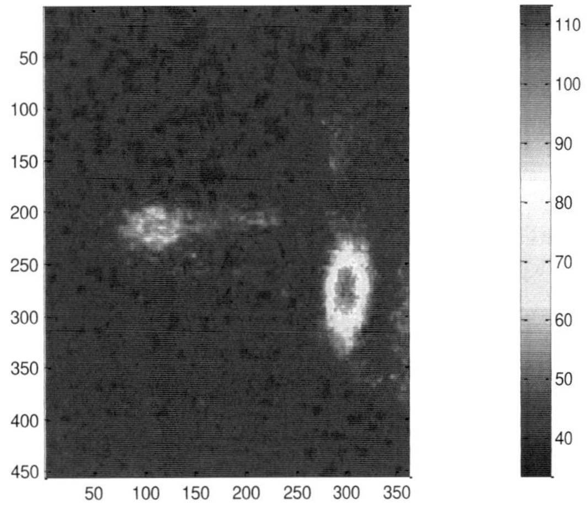

Fig. 19. Post-processed image of PSP signal

Figure 19 is the post-processed image of PSP signal. The light blue area is the PSP signal. A piece of tape coated with PSP powder was used as the object of reference for camera focusing. However, it can be seen from Fig. 19 that the intensity of PSP signal is low and the signal/noise ratio is not sufficient for flow visualization of opposing jet and calculation of pressure distribution. The PSP system will be improved in the near future for further research.

5. Conclusion

Experimental research was conducted to investigate the effect of opposing jet on hypersonic flow around the blunt body. The major conclusions are as follows:

1) The flow structure of opposing jet around the blunt body in hypersonic flow is visualized in this research. With the aid of high speed camera, the details of opposing jet-free stream interaction and jet instability have been captured. As a result of analyzing the Schlieren data, it was revealed that the jet oscillation and instability has a strong impact on the shock separation area.

2) Comparison of air opposing jet and water opposing jet was done. The penetration length of air jet is shorter than that of water jet. Frequency of air jet oscillation is larger than that of water jet oscillation. These results suggest that a liquid jet may be more effective than the gas jet from a viewpoint of the virtual spike effect.

3) Flow visualization methods including laser cross-section imaging and Pressure-sensitive paint technology were explored and new information about opposing jet was obtained in this research. Results of the laser cross-section imaging reveals the complicated water jet behavior. Trial experiment for binder-less PSP measurement of opposing jet was performed. The image of PSP signal indicates that PSP technology has potential to be used for quantitative measurement of opposing jet in hypersonic flow.

References

[1] Meyer, Benjamin, Harlan F. Nelson, and David W. Riggins. "Hypersonic drag and heat-transfer reduction using a forward-facing jet." Journal of Aircraft 38.4 (2001): 680-686.

[2] Warren, C. Hugh E. "An experimental investigation of the effect of ejecting a coolant gas at the nose of a bluff body." Journal of Fluid Mechanics 8.3 (1960): 400-417.

[3] Daso, Endwell O., et al. "Dynamics of shock dispersion and interactions in supersonic freestreams with counterflowing jets." AIAA journal 47.6 (2009): 1313-1326.

[4] Rashis, Bernard. "Preliminary Indications of the Cooling Achieved by Injecting Water

Upstream from the Stagnation Point of Hemispherical, 80 Conical, and Flat-Faced Nose Shapes at a Stagnation Temperature of 4000 F." NACA RM L 57103 (1957): 10.

[5] Woods, William C., Kenneth M. Jones, and Noah N. Genzel. "Preliminary Investigation of Stagnation Point Liquid Injection Influence on Blunt Body Aerodynamics." (2002).

[6] Perurena, J. Beloki, et al. "Experimental investigation of liquid jet injection into Mach 6 hypersonic crossflow." Experiments in fluids 46.3 (2009): 403-417.

[7] Kojiro Suzuki, Kashiwa Wind Tunnel Working Group. " Introduction of Hypersonic Wind Tunnel at UT Kashiwa Campus ." http://daedalus.k.u-tokyo.ac.jp/wt/info/UTHYP.pdf

[8] Romeo, David J., and James R. Sterrett. Exploratory investigation of the effect of a forward-facing jet on the bow shock of a blunt body in a Mach number 6 free stream. National Aeronautics and Space Administration, 1963.

[9] Abe, Shinkichi, Koji Okamoto, and Haruki Madarame. "B234 Development of PIV-PSP hybrid system using Pressure Sensitive Particle." Journal of the Visualization Society of Japan 23.Supplement2 (2003): 179-182.

IOP Conf. Series: Materials Science and Engineering **249** (2017) 012015 doi:10.1088/1757-899X/249/1/012015

Visualization of Projectile Flying at High Speed in Dusty Atmosphere

Chihiro MASAKI[1*], Yasumasa WATANABE[2], and Kojiro SUZUKI[1]

[1] Department of Advanced Energy, Graduate School of Frontier Sciences, The University of Tokyo, 5-1-5 Kashiwanoha, Kashiwa, Chiba 277-8561, Japan.
[2] Department of Aeronautics and Astronautics, Graduate School of Engineering, The University of Tokyo, 7-3-1 Hongo, Bunkyo, Tokyo 113-8656, Japan.

E-mail: masaki@daedalus.k.u-tokyo.ac.jp

Abstract. Considering a spacecraft that encounters particle-laden environment, such as dust particles flying up over the regolith by the jet of the landing thruster, high-speed flight of a projectile in such environment was experimentally simulated by using the ballistic range. At high-speed collision of particles on the projectile surface, they may be reflected with cracking into smaller pieces. On the other hand, the projectile surface will be damaged by the collision. To obtain the fundamental characteristics of such complicated phenomena, a projectile was launched at the velocity up to 400 m/s and the collective behaviour of particles around projectile was observed by the high-speed camera. To eliminate the effect of the gas-particle interaction and to focus on only the effect of the interaction between the particles and the projectile's surface, the test chamber pressure was evacuated down to 30 Pa. The particles about 400μm diameter were scattered and formed a sheet of particles in the test chamber by using two-dimensional funnel with a narrow slit. The projectile was launched into the particle sheet in the tangential direction, and the high-speed camera captured both projectile and particle motions. From the movie, the interaction between the projectile and particle sheet was clarified.

1. Introduction

Particle-laden environment is not unusual in space. In deep space missions, a spacecraft may encounter a particle-laden environment during an interplanetary flight, atmospheric flight or travel on the ground. Such encounter may cause serious damage on the spacecraft surface. Generally speaking, the particle-laden environment can be categorized into three situations: the dusty atmosphere, the regolith as packed particles, and the cloud of sparsely scattered particles in the vacuum.

A typical example of the dusty atmosphere will be found on Mars. When an atmospheric probe enters the Martian dusty atmosphere, the flow around the body is significantly influenced by the gas-particle two-phase flow effect [1]. In such case, the aerodynamic heating may be augmented and the damage on the surface may become more severe due to the augmented aerodynamic heating as well as the collision of the dust particles. On those problems, the experimental studies and the numerical studies have been conducted [1]-[3]. The regolith on planets, satellites and asteroids is typical example of the packed particles. For investigation on the interior structure under the ground, impact probe missions have been considered in the past [4]. The behavior of the regolith at the impact will be well described by the soil dynamics and some numerical analysis methods have been developed [5].

The most well-known example of a cloud of sparsely scattered natural particles will be the Saturn's ring. However, such environment appears by an artificial effect. When a spacecraft is approaching to the ground of an asteroid, lots of dust particles must fly upward at high speed, receiving the jet of the landing thruster of the spacecraft [6],[7]. If the same or another spacecraft is flying in the dust cloud, it may encounter the particle cloud at high relative speed. The behavior of the particle cloud is expected to be described by the dynamics of the granular flow. The granular flow has been extensively studied so far [8]. However, most of them consider relatively low speed regimes. The phenomena in high-speed regimes, which may be encountered by a spacecraft, have not been clearly understood, because the granular flow dynamics in high-speed regimes must become much complicated. It includes the crash and abrasion of particles due to high-speed impact among the particles and/or between particles and the spacecraft surface.

Therefore, it is necessary to experimentally clarify the fundamental characteristics of the high-speed granular flow. Specifically, the case that a projectile flies at high-speed in the particle-laden environment must be studied for consideration on possible hazardous situation of a spacecraft.

The experimental method of high-speed inrush into the particle-laden environment was established and changes in the flow field due to differences in the material of the projectile and particles have been observed [9]. In the present study, the shape of the projectile is changed to sphere and observed flow field. In addition, the image processing method for a gray image obtained by experiment is established. The objectives of the present study are as follows:

1) To establish the experimental technique to simulate the high-speed flight of sphere into particle-laden environment by using the ballistic range,

2) To establish the image processing method, to reveal the influence zone of the destruction of particles due to collisions around the projectile, from snapshots of the high-speed camera,

3) To compare the results of image processing with different projectile shapes.

2. Method of experiment

2.1. Ballistic range and optical system

The schematic view of ballistic range is shown in Fig. 1, and the concept of the present experimental setup is shown in Fig. 2. As shown in Fig. 2(a), a sheet of particles is formed in the test chamber, to capture the high-speed flight of a projectile in particle-laden environment from the glass window (Fig.2 (b)). In the case of thick cloud, a recorded image will become integrated one in the viewing direction and the analysis of the image will become difficult to obtain the cross section of the phenomena. A projectile was launched in the tangential direction to the sheet as shown in Fig. 2 (a). The behavior of particles around a projectile was captured by using a high-speed camera, whose viewing direction was normal to the sheet. Such thin planar particle sheet is convenient for observation of the phenomena.

To launch a projectile at a high speed, the ballistic range facility was used. The facility used in this laboratory can launch a projectile with about 10 g mass up to a speed of about 400 m/s, depending on the charged pressure. This facility was used for observation of the shock wave around a body at a free flight and for high-speed collision experiment of a penetrator on an icy target [10]. As shown in Fig. 1, the facility consists of the high-pressure chamber, the barrel tube, the accelerate tube, the adapter, and the test chamber. A projectile is set from the front of the diaphragm section before the experiment. The injection velocity is controlled by appropriately charging the compressed air in the high-pressure chamber. The maximum available pressure is 1 MPa relative to the atmospheric pressure. To eliminate the aerodynamic effect, the test chamber is evacuated down to about 30 Pa before the experiment by using the mechanical booster pump.

To capture the image in the test chamber through the glass window, a high-speed camera Phantom Miro 310 (Nobby Tech. Ltd.) was used. The specifications and the operation parameters are listed in Table 1. In this research, in order to investigate the high-speed collision phenomena, it is necessary to increase the frame rate and to decrease the exposure time as much as possible. Then the high-intensity metal halide lamp MID-25FC (Lighterrace Inc.) with the maximum power 250 W was used.

As shown in Fig. 2(b), the light enters the test section through the window on the opposite side of the camera. The particles and projectile were illuminated with the backlight configuration. The oil clay in the box of the size 15 × 20 × 20 cm was set ahead of particle sheet. The launched projectile passes through the particle sheet and is finally stopped by the cray.

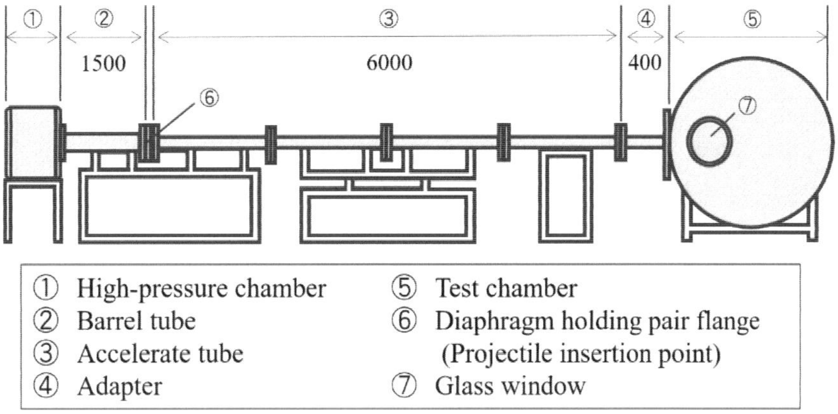

Fig. 1. Photograph and conceptual diagram of ballistic range facility.

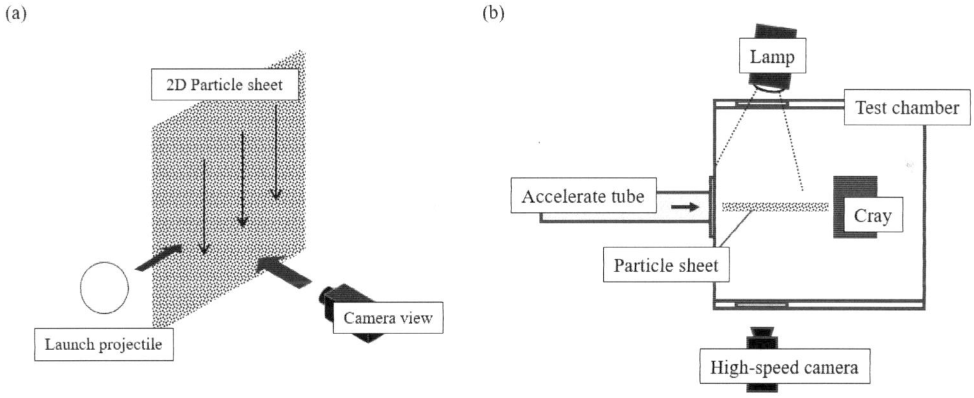

Fig. 2. Experimental concept and arrangement from the top of test chamber.

Table 1. Specifications and filming parameters of high speed camera.

Specifications	
Brightness resolution	Monochrome (12bit)
Sensitivity (ISO / ASA)	16,000
Minimum exposure time	1.0 μs
Filming parameters	
Image acquisition rate	20,000 pps
Exposure time	1.0 μs
Image resolution	512 x 256 pixels

2.2. Projectiles

The requirements of projectile used in the present experiment were the visibility of the projectile shape, the appropriate size for observation, and the strength to withstand the strong acceleration in the tube. In the present study, spherical shape was selected for a projectile. In our previous study [9], a hemisphere-cylinder was used. The hollow opened to the base of the projectile made the observation of the wake difficult. The outer diameter was the same as the inner diameter of the accelerate tube. Such configuration enables us to launch the projectile without the sabot. The launch without the sabot enhances the stability and repeatability of the experiment, because the sabot often causes a significant disturbance in the attitude and motion of the projectile when it is separated from the projectile. Polycarbonate was selected as a material, because it is strong and has good visibility in the present optical system. To reduce the projectile mass and to obtain high launch velocity, the inside was hollowed out so that the thickness of the body was reduced to about 2 mm. The spherical body was composed of two hemispherical shells, which were tightly glued at their spigot joints. Thanks to the thin shell structure, the mass of the spherical projectile was only about 4.3 g, which was sufficiently light to be launched in the present ballistic range. A photograph of projectile is shown in Fig. 3.

Fig. 3. Photograph of projectile, which diameter is 25.75 mm, mass is 4.3 g, respectively.

2.3. Particle sheet

2.3.1. Particles Considering that the objective of this study is to clarify the fundamental properties of the particle behavior around the projectile, the material and the particle size were selected from a viewpoint of the visibility for the present camera system. Glass beads with about 400 μm diameter was used as brittle particles, and is compatible with the optical system of this work. The appearance and 180 times enlarged images is shown in Fig. 4.

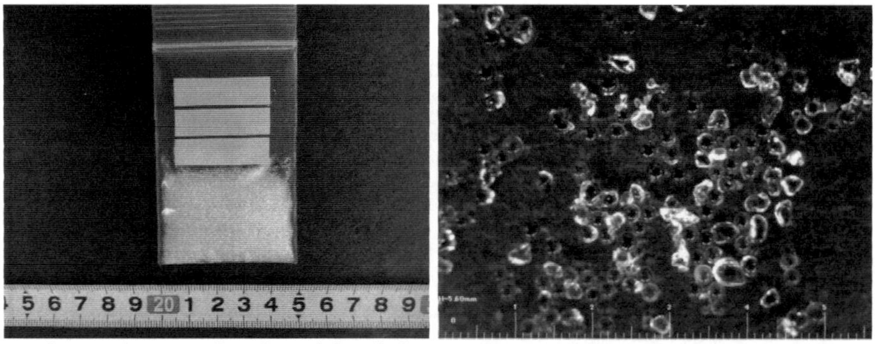

Fig. 4. Photographs of particles (left: appearance, right: 180 times enlarged).

2.3.2. Particle sheet generating device In order to create the particle-laden environment in the test chamber, a device to generate a particle sheet was developed. The uniformity of the particle number density on the sheet is important. Therefore, a mechanism for dropping particle from a slit by the gravitational force was selected. As shown in the Fig. 5 (b), the length of the slit is 100 mm and the gap is 2 mm. Due to the volume of the dust reservoir above the slit, the duration of the particle sheet is about 10 seconds. As shown at the left side of Fig. 5 (a), a shutter at the slit is closed before the experiment, and it is removed by the motor at an arbitrary trigger timing. The terminal speed of the free dropping particles is about 3.0 m/s, which is much smaller than the projectile speed. Consequently, the particles are considered to be floating still in the test chamber in the time scale of the projectile's passage. Using the above device, almost the uniform two-dimensional planar particle sheet was successfully generated. The particle-supply rate and the number density can be calculated. The supply rate is 12.4 g/s. This is calculated from the amount of particles in the funnel and the discharge time. The number density in the particle sheet is estimated about 5×10^8 particles/m^3 from the supply rate and the volume swept by falling particles. These parameters can be changed by particle diameter and material or adjusting slit width.

Photograph of test chamber interior is shown in Fig. 6. Projectile is launched from the right injection port to left. The particle sheet generating system is installed at the top of test chamber. After passing through particle sheet, projectile is stopped by cray.

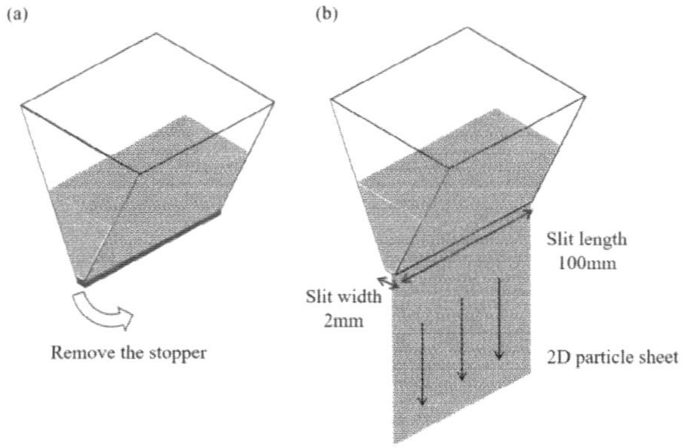

Fig. 5. Illustration of particle sheet generating devise.

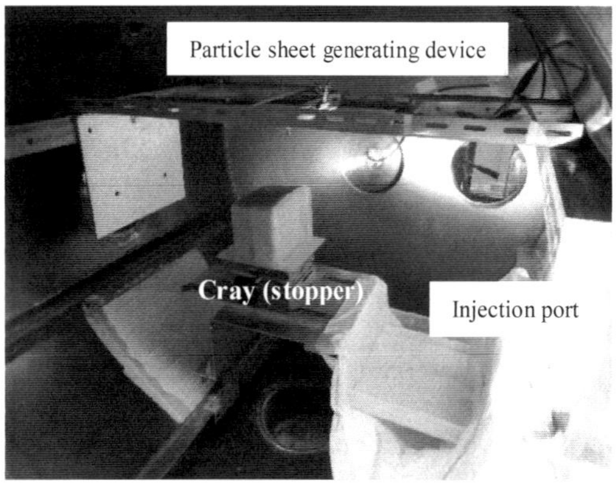

Fig. 6. Photograph of test chamber interior.

2.4. *Operations and experimental conditions*

The experiment was carried out in the following procedure. The test conditions of the present experiments are summarized in Table 2.

a) Set the projectile from diaphragm holding pair flange (shown Fig. 1),

b) Evacuate the test chamber down to 30 Pa,

c) Fill the air into the high-pressure chamber (nominally 0.5 MPa, 1 MPa at the maximum),

d) Remove the stopper of two-dimensional funnel by the remote control switch, and then start particle sheet generating device,

e) Launch the projectile and capture the movie by the high-speed camera.

Table 2. Test conditions.

Parameters	Selected condition
High pressure chamber pressure	0.5Mpa
Test chamber pressure	30Pa
Projectile	Polycarbonate sphere
Particle	Glass beads (diameter = 400 μm)

2.5. Image processing
Due to camera specs, the result obtained is grayscale. For consideration, a method for extracting only the region of interest by colorizing the image will be described below.

At first, the brightness of each pixel with 256 levels from black (= 0) to white (= 255) was extracted. Then, the brightness matrix of 512 × 256 size was obtained from one frame. Next, contour lines with respect to the brightness were drawn. The region with brightness lower than 65 was painted in white to remove the image of the background. Processes and benefits of this image processing are discussed in the next section.

3. Results and discussions
The snapshots of high-speed movie of a projectile flying in a particle sheet are shown in Fig. 7. About 5 frames were extracted from the movie during the period for the projectile to pass through the particle sheet. From these snapshots, the influence of the gas in front of a projectile and that used for acceleration of the projectile in the accelerate tube is thought to be negligible, because the particle sheet does not seem to be disturbed by the air flow. From the angle of the adhesive part of the sphere, the projectile is not rotating. A spherical projectile can ignore posture misalignment more than a hemisphere-cylinder projectile, used in our previous study [9].

In Fig. 7(d)-(f), the particle collision occurs on the projectile surface, and the zone of light emission expands gradually. The emission occurs only in the region where the particles exist and the light is scattered by the presence of particles. If the intensity of the light emission increases with the increase in the local number density of particles, a larger number of particles, some of which may be fine crushed particles, exist in the zone of brighter emission. A black area on the projectile surface is expected to be a shadow of a particle floating between the projectile and the light source.

To estimate the extent of the presence of particles scattered by the projectile, the image processing was made to the original gray scale picture. The process of image processing is shown in Fig. 8. Left picture of Fig. 8 is before subtracting background, right is after subtracted. As seen from these figures, only the particularly bright parts in front of the projectile are extracted.

The snapshots after processing are shown in Fig. 9. From these figures, the influence range is gradually spreading and separated by contour lines. By comparing Fig. 7-(f) and Fig. 9-(f), particles are coming around in the rear part of the projectile. Unlike hemisphere-cylinder projectile, particles are gradually propagated to the back of the projectile. A comparison of images after processing of different projectiles is shown in Fig. 10. In the case of hemisphere-cylinder projectile, the propagation of particles to the wake is stopped at the corner of the cylinder. Also, disorder of posture influences the surrounding propagation. In the case of a sphere, the propagation continues to the rear. As the number of particles going backward increased, the particle density in the front is smaller than that in the case of hemisphere-cylinder projectile.

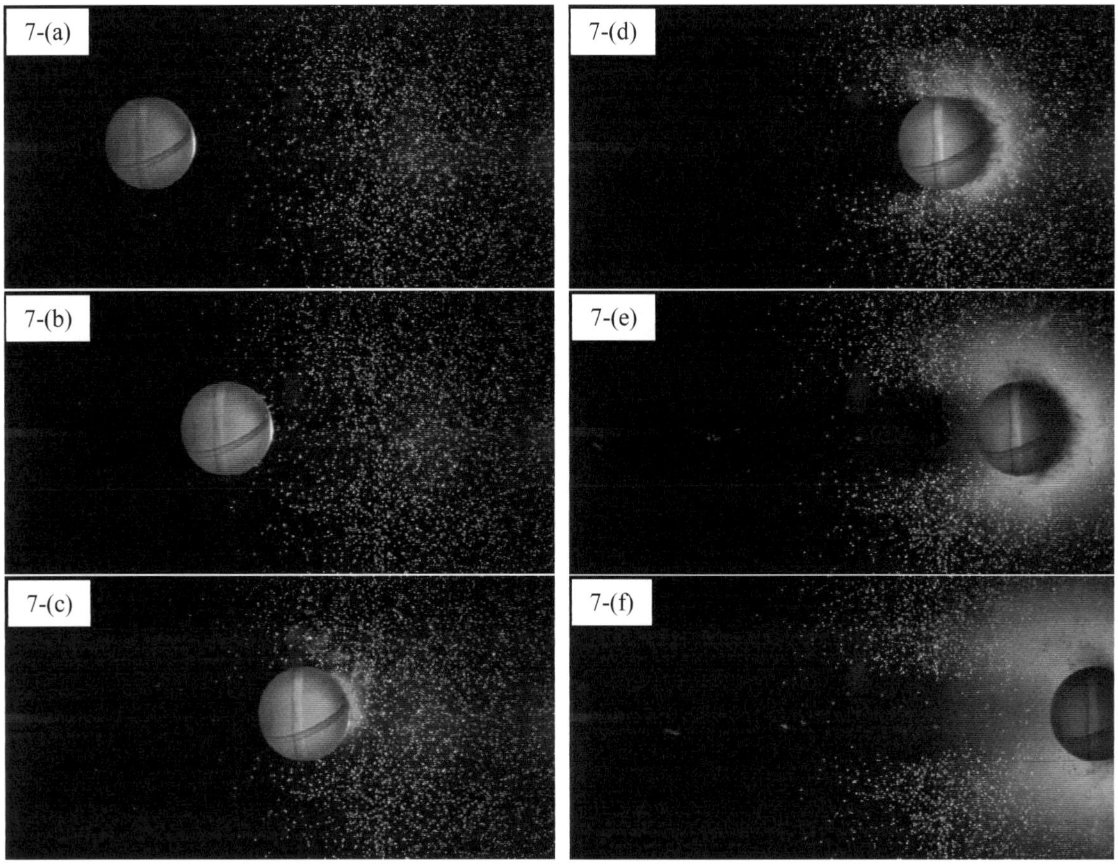

Fig. 7. Snapshots of high-speed movie of a sphere flying in particle sheet.

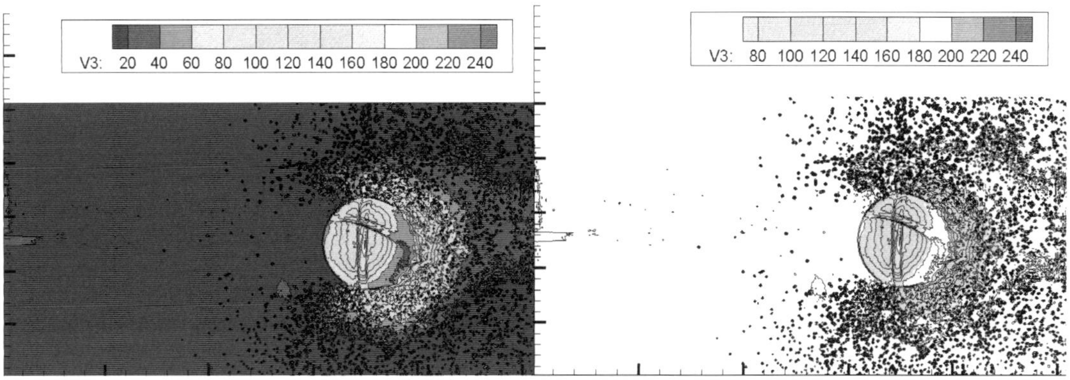

Fig. 8. Process of image processing (left: with background, right: background painted in white).

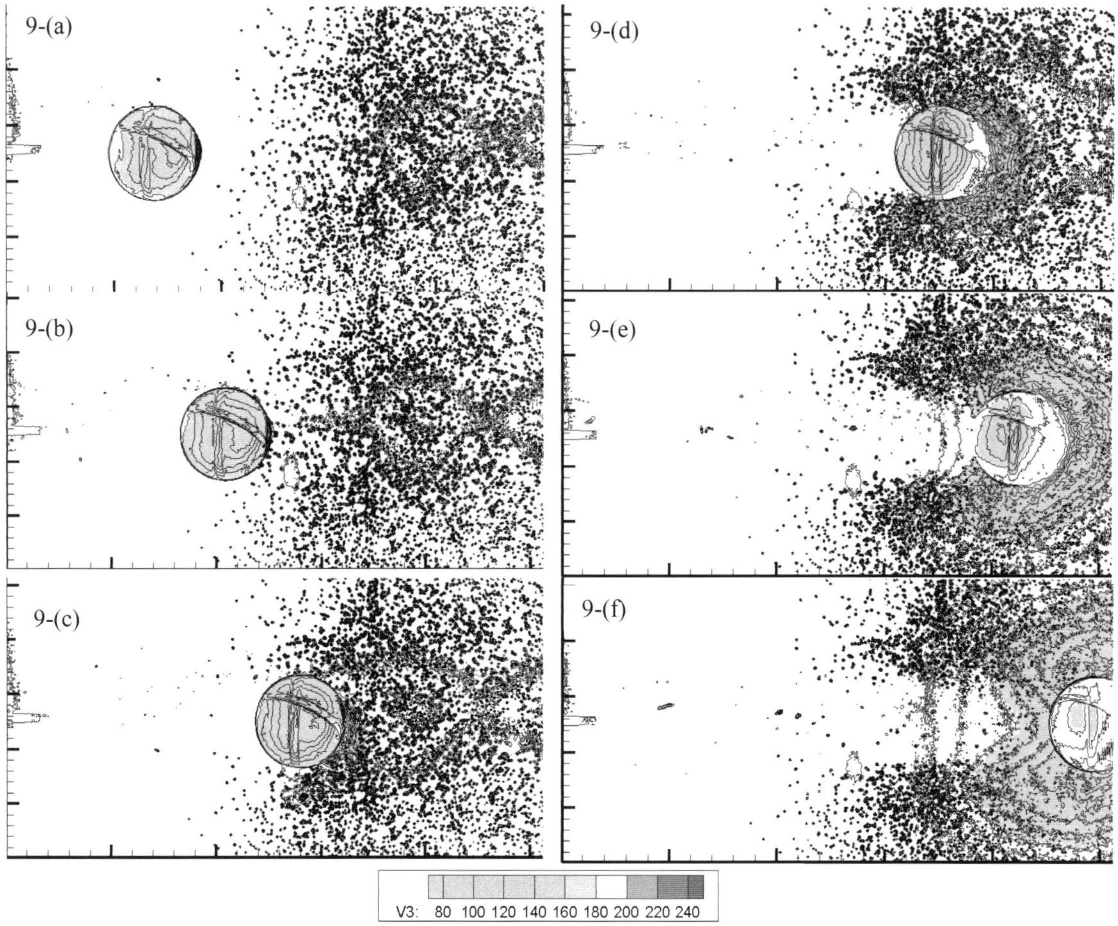

Fig. 9. Snapshots of high-speed movie after processing.

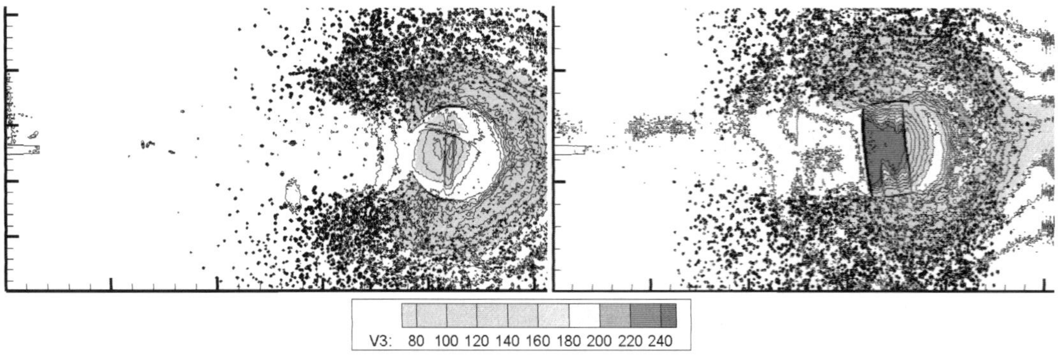

Fig. 10. Comparison of different projectile (left: sphere, right: hemisphere-cylinder[11])

4. Conclusions

Considering the hazardous situation of the spacecraft, high-speed collision experiment with spherical projectile and particles was conducted by using ballistic range. The major conclusions are as follows:

1) The experiment of a high-speed flight of a spherical projectile in particle sheet was successfully performed using a ballistic range facility. In order to enable a spherical and lightweight projectile, the

spherical body was composed of two hemispherical shells, which were tightly glued at their spigot joints.

2) To clarify the influence range of particles, image which captured the pattern of the particle behavior around the projectile was processed. By changing grayscale to color and painting white the background, it was possible to extract only the moving range of particles.

3) The image processing results of different projectile shapes were compared and the state of propagation was discussed. In the case of spherical projectile, as the number of particles going backward increased, the particle density in the front is smaller than that in the case of hemisphere-cylinder projectile.

Acknowledgements
This work is supported by Grant-in-Aid for Scientific Research (B) No. 16H04585 of Japan Society for the Promotion of Science. For image analysis, the authors would like to express sincere thanks to Mr. Zhong Ce, who is a doctor course student in the authors' laboratory, for his technical support on the image processing.

References

[1] Papadopoulos, P., Tauber, M. E. and Chang, I. D.: Heatshield Erosion in a Dusty Martian Atmosphere, J. Spacecraft Rockets, 30(1993), pp. 140-151.
[2] Vasilevskii, E. B., Chirikhin, A. V., and Osiotsov, A. N.: Heat Transfer to a Stagnation Region of a Blunt Body in a Hypersonic Gas Flow with and Admixture of Solid Particles, Aerothermodynamics for space vehicles, ESA SP, 426(1998), pp.301-307.
[3] Migita, S. and Suzuki, K.: Numerical Analysis on Dusty Supersonic Shock Layer Flow over a Blunt Body, Transactions of the Japan Society for Aeronautical and Space Sciences, 49(2001), No.566, pp.63-69 (in Japanese).
[4] Mizutani, H.: Lunar Interior Exploration by Japanese Lunar Penetrator Mission, LUNAR-A, Journal of Physics of the Earth, 43 (1995), No. 5, pp. 657-670.
[5] Suzuki, K., Abe, T., Higuchi, K., Fujimura, A. and Mizutani, H.: Development of Analysis Model of Penetration Dynamics for Lunar-A Mission, 20th International Symposium on Space Technology and Science(1996), Gifu, 96-i-02V.
[6] Hirotani. T, et.al.: Spacecraft Contamination by Propulsion Jet Impinging on Dusty Planetary Surface, 32nd Fluid dynamics conference (2000), pp.363-366 (in Japanese).
[7] Hirotani. T, et.al.: Contamination spacecraft hovering on dusty planet, Symposium on shock wave (2000), pp.589-592 (in Japanese).
[8] Tomas Schwager, Computational Granular Dynamics: Models and Algorithms, Germany, 2014.
[9] Masaki. C, Watanabe. Y, and Kojiro. S: Ballistic Range Experiment of Sphere in Dusty Atmosphere, 31st International Symposium on Space Technology and Science(2017), Ehime, 2017-s-10-e.
[10] Namba. K, and Suzuki. K.: Fundamental Study on Penetration Dynamics and Projectile Shape Effect of the Cryo-Penetrator into Icy Object, master's thesis, Department of Aeronautics and Astronautics, The University of Tokyo, 2014 (in Japanese).
[11] Masaki. C, Watanabe. Y, and Kojiro. S: Experimental Study on Granular Flow around a Body Flying in Dens Dust Cloud at Low Atmospheric Pressure, 49th Fluid dynamics conference (2017), 2D06 (in Japanese).

IOP Conf. Series: Materials Science and Engineering **249** (2017) 012016 doi:10.1088/1757-899X/249/1/012016

Visualization of Supersonic Microjets Using LIF and MTV Techniques

Taro Handa[1], Hideto Kambara[2] and Masaru Harada[2]

[1] Department of Advanced Science and Technology, Toyota Technological Institute, 2-12-1 Hisakata, Tempaku-Ku, Nagoya, 468-8511, Japan
[2] Department of Energy and Environmental Engineering, Kyushu University, 6-1 Kasuga-Koen, Kasuga, 816-8580, Japan

handa@toyota-ti.ac.jp

Abstract. Supersonic microjets issuing from a convergent rectangular nozzle whose height is 500 μm at its exit are visualized using the acetone-based PLIF (planar laser-induced fluorescence) and MTV (molecular tagging velocimetry) techniques. The jet Reynolds numbers are ranged between 200 and 7000 and the fully-expanded jet Mach-number is set to 1.5. The PLIF visualization results reveal that the jets spread abruptly at a certain location. Instantaneous MTV images are captured in the vicinity of the location where a jet spreads abruptly and the process of the jet collapse is successfully visualized. The distance between the nozzle exit and the jet-collapse location is estimated from each PLIF image. The distance is found to be constant for the Reynolds numbers higher than ~700 and to be inversely proportional to the Reynolds number for the numbers lower than ~700. Moreover, the distance changes abruptly at the Reynolds number of ~700.

1. Introduction

Recently, study on supersonic microjets has become important due to increase in engineering applications such as power generation [1], propulsion [2], cooling of MEMS component [3], laser cutting machine [4], micro pump [5], and so on. In order to improve the performance of a supersonic microjet for each engineering application, the detailed characteristics of this flow have to be understood. A relatively large number of studies have investigated supersonic microjets experimentally. However, there are no experimental studies on supersonic "rectangular" microjets beyond the researches of Lempert et al. [6] and Handa et al. [7].

Lempert et al. [6] created microjets with a convergent-divergent rectangular nozzle. They measured the jets' velocity profiles using the MTV (molecular tagging velocimetry) technique for jet Reynolds numbers of ~200. They compared the measured profiles with profiles obtained by a direct simulation Monte Carlo (DSMC) method. Their experimental results agreed well with their simulation results. Using the MTV technique, Handa et al. [7] also measured the velocities in supersonic microjets issuing from a convergent-divergent rectangular nozzle. In their experiments, the Reynolds numbers were ranged between 154 and 5560. Their experimental results showed that the jets abruptly spread at a certain location owing to their collapse.

In spite of the above-mentioned studies on the rectangular supersonic microjets, we have no detailed description of microjet's collapse. In the present study, the supersonic microjets issuing from a convergent rectangular nozzle whose height is 500 μm at its exit are visualized using two techniques of PLIF (planar laser-induced fluorescence) and MTV techniques.

2. Experimental setup

A suction-type wind tunnel was used, and the working gas was nitrogen. Stored in a balloon made of vinyl chloride, the gas was supplied to a micronozzle. The gas was then accelerated by the nozzle and was discharged into the wind-tunnel vacuum chamber, which was evacuated by a vacuum pump. To visualize the microjets using the PLIF and MTV techniques, acetone was evaporated and mixed into nitrogen gas in the acetone seeder. By adding pure nitrogen gas to the nitrogen gas (containing acetone vapor) stored in the balloon, the seeding rate of acetone is set at less than ~3%.

Figure 1 shows the detailed structure of the nozzle, a symmetric convergent rectangular nozzle. The coordinate system used to analyze the results is defined in the left hand side of Fig. 1, where x and y are the streamwise and height coordinates, respectively, and the origin point O is located at the nozzle exit on the nozzle centerline. The nozzle was fabricated using electric discharge machining. The nozzle height at the exit is h_{exit} = 500 μm. The nozzle width is 5000 μm; i.e., the width-to-height ratio (aspect ratio) is 10 at the nozzle exit.

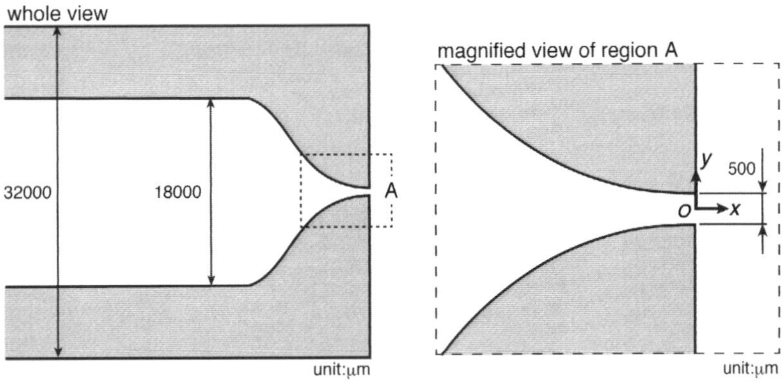

Figure 1. Micronozzle.

The flow in the entrance region with a nozzle height of 18000 μm (Fig. 1) is almost stagnant. The pressures in this region were ranged between p_0 = 3 kPa and 90 kPa. The pressure in the wind-tunnel vacuum chamber was adjusted so that the nozzle pressure ratio (NPR) became 3.67. In general, the jet characteristics are discussed relating to the fully-expanded jet Mach-number M_j [8, 9]. This Mach number is calculated so that the gas is expanded fully from the stagnation pressure p_0 to the back pressure p_b. The ratio p_0/p_b corresponds to the NPR. The value of M_j thus obtained is 1.5. The authors confirmed from the preliminary MTV measurements that all the visualized jets accelerated to the Mach numbers close to 1.5 at x/h_{exit} = 1.0. The nozzle pressure ratio is set within an accuracy of ± 0.5%. The stagnation temperature is maintained at 295 ± 4 K. The Reynolds numbers, which was calculated based on the variables at the nozzle exit, were ranged between Re_j ~200 and ~7000.

In the present study, the flow was visualized by the PLIF technique. The laser was a Nd:YAG laser (Spectra Physics, 170-10) operated with a fourth harmonic mode of 266 nm. Located in the central plane of the nozzle, the laser sheet was irradiated into the flow field parallel to the y-axis shown in the right hand side of Fig. 1. The incident laser energy was ~10 mJ. The resulting LIF signal was collected through a lens and was detected by the CCD camera with an image intensifier (Hamamatsu C7970). The image intensifier's gate was set to be open during 50 ns. In a single experimental run, fluorescence intensity corresponding to 100 laser shots were integrated on the CCD camera (i.e., fluorescence intensity for one image corresponds to 100 laser shots). In a single experimental run, 50 to 400 images were captured and averaged.

The sensitivity of an image intensifier is not uniform over its output phosphor and the laser intensity is not uniform over the laser sheet. To take these into account, the measurement system was calibrated by filling the still nitrogen with a small amount of acetone in the expansion chamber. The captured PLIF image of the still nitrogen was used as a reference image to correct the image data.

Handa et al. [10] investigated the characteristics of acetone fluorescence intensity for pressures and temperatures lower than the atmospheric pressure and temperature, respectively. They determined the empirical constant appearing in the multi-step decay model proposed by Thurber et al. [11] so that it could be used in the low temperature and pressure ranges. According to the multi-step decay model having the empirical constant determined by Handa et al., the fluorescence intensity per molecule is almost constant within ±8%. Therefore, the fluorescence intensity is almost proportional to number density of the jet.

The MTV visualization system was the same as that used by Handa et al. [7]. In this system the acetone molecules seeded in the flow were tagged by a beam-shaped Nd:YAG laser operated with a fourth harmonic mode of 266 nm (Spectra Physics, 170-10) and the resulting fluorescence signal was detected by the CCD camera with an image intensifier (Hamamatsu C7772S). The image intensifier's gate was set to be open during 20 ns. The time delay between tag and interrogation was controlled by the delay pulse generator (Stanford Research System, DG645) and was set to 400 ns.

In general, the spatial resolution of the image intensifier of an intensified CCD camera is lower than that of its CCD array. The spatial resolutions of the image intensifiers used in the present PLIF and MTV visualizations were estimated according to Lempert et al. [12]. The resolutions on the objective planes were 31 μm and 21 μm for PLIF and MTV visualizations, respectively.

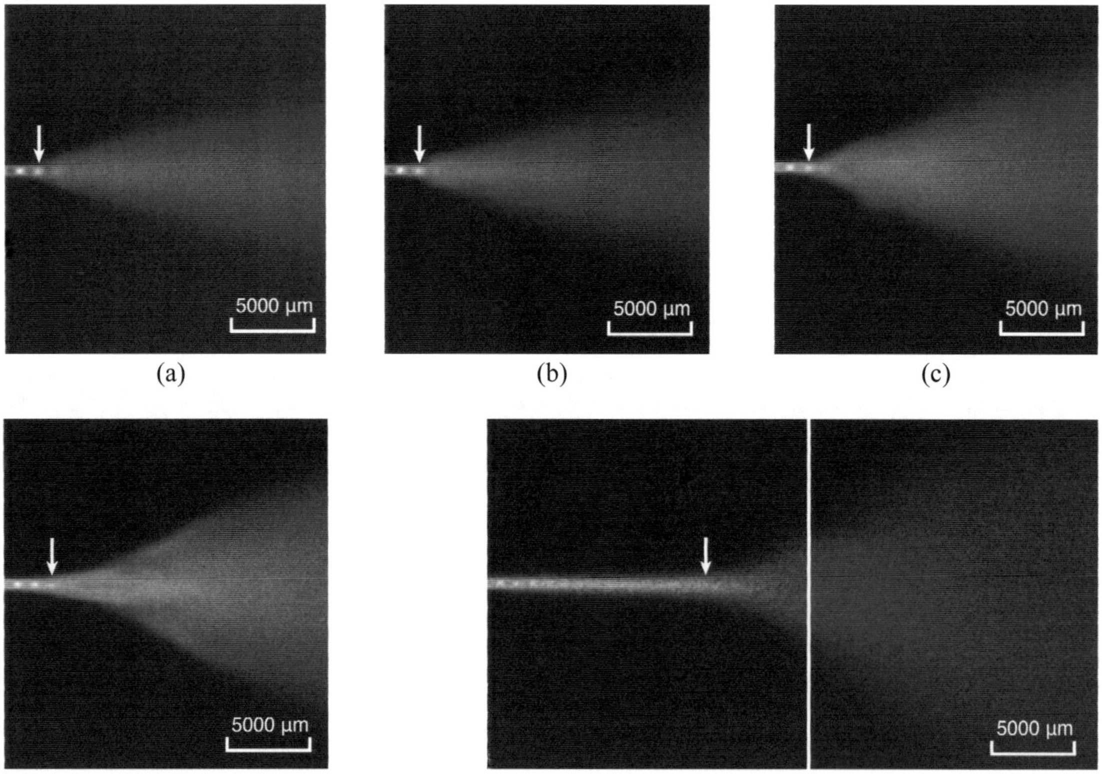

(a) (b) (c)

(d) (e)

Figure 2. PLIF images. (a) $Re_j = 6810$. (b) $Re_j = 3780$. (c) $Re_j = 1510$. (d) $Re_j = 757$. (e) $Re_j = 568$.

IOP Conf. Series: Materials Science and Engineering **249** (2017) 012016 doi:10.1088/1757-899X/249/1/012016

3. Results and discussion

Figure 2 shows the PLIF images of microjets for $M_j = 1.5$. It is found that all the visualized jets in the figure spread abruptly at a certain location. Such abrupt jet spreading is generally seen in incompressible and compressible microjets. Gau et al. [13] visualized incompressible rectangular microjets using the smoke wire technique. They tested three rectangular micronozzles whose heights at the exit were 100 μm, 200 μm, and 500 μm, respectively. They observed that the jet spreads abruptly at a certain location, and this location shifted downstream with decrease in Reynolds number.

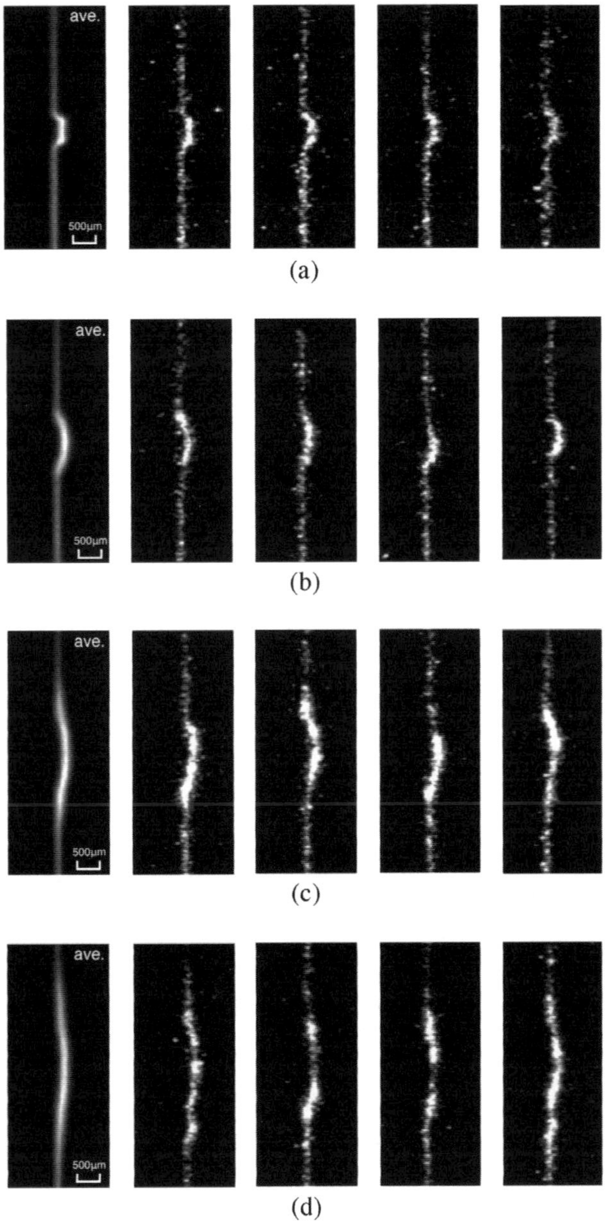

Figure 3. Instantaneous MTV images (Re_j = 6810). (a) x/h_{exit} = 2.0. (b) x/h_{exit} = 4.0. (c) x/h_{exit} = 5.0. (d) x/h_{exit} = 6.0.

By using MTV, Handa et al. [7] obtained the streamwise distributions of jet half-widths for the microjets issuing at M_j = 2.0 from a convergent-divergent rectangular nozzle, whose heights were 500 μm at its exit. In their experiments, the nozzle pressure ratio was set so that the jet was ideally expanded assuming the flow in the nozzle to be inviscid. They also observed that the jets spread abruptly at a certain location and this location also shifted downstream with decrease in Reynolds number. However, in the present visualization results of Fig. 2, the location where the jet spreads abruptly is almost constant for $Re_j \geq 757$ and shifts downstream abruptly between Re_j = 757 and 568.

Corresponding to x/h_{exit} =2.0, 4.0, 5.0 and 6.0, MTV images captured for Re_j = 6810 are shown in Fig. 3(a)–(d). In each figure, the image on the left end is the averaged image. The other four images are representative instantaneous images. Although shot noise degrades instantaneous images, the four interrogated line profiles have almost the same shape for x/h_{exit} =2.0 (Fig. 3(a)). However, the profile fluctuates up and down for x/h_{exit} = 4.0 and 5.0 and the degree of the fluctuation at x/h_{exit} = 5.0 is higher than that at x/h_{exit} = 4.0. These imply that the jet might flap and the degree of flapping motion grows along the flow's path. At x/h_{exit} = 6.0, the profiles of instantaneous interrogated lines are considerably different from the profile of averaged line. This implies that the jet collapses immediately downstream from at x/h_{exit} = 5.0 as a result of the growth in flapping motion. Owing to such jet collapse, large-scale structures might be produced. It is considered that the undulations seen in profiles in Fig. 3(d) might be attributed to such large-scale structures. The process of the jet collapse shown in Fig. 3 is quite similar to that observed by Handa et al. [7] who captured the MTV images of microjets issuing from the convergent-divergent rectangular nozzle although the Reynolds-number dependences of the shift in jet collapse location are different between the presently (convergent nozzle) and previously (convergent-divergent nozzle) visualized microjets.

The jet collapse length, which is defined as the distance between the nozzle exit and the location where the jet abruptly spreads, is estimated from each PLIF image (the location is denoted by an arrow in each image) and plotted against the jet Reynolds number Re_j as shown in Fig. 4. In the figure, the collapse length estimated from the macrojet data measured by Walker and Thomas (M_j = 1.49) [14] is also plotted. The jet tested by them is underexpanded although it issued from the convergent-divergent nozzle. It is found from the graph that the collapse length normalized by the nozzle exit height, i.e., L_c/h_{exit}, is constant for $Re_j > \sim700$ and the length is inversely proportional to Re_j for $Re_j < \sim700$. In addition, the collapse length changes abruptly in the vicinity of $Re_j \sim 700$. The uniformity of the collapse length for $Re_j > \sim700$ intimates that the length depends only on the jet Mach number; i.e., the length seems independent of the nozzle shape as long as the jet is underexpanded.

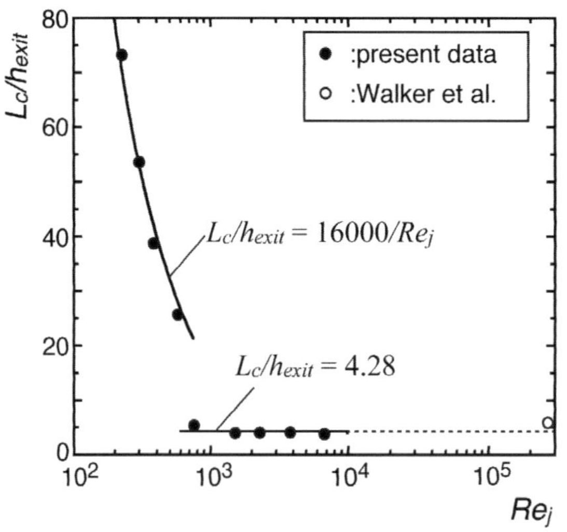

Figure 4. Plots of jet collapse length vs. Reynolds number.

Gau et al. [13] visualized subsonic incompressible microjets by the smoke-wire technique. Their visualization results also showed that the jets spread abruptly at a certain location. They also plotted the jet collapse length (normalized by the nozzle exit height) vs. the jet Reynolds number Re_j. Their plots showed that the jet collapse length was inversely proportional to the Reynolds number over the entire Reynolds number range that they tested; i.e., no abrupt change in the collapse length was seen in their plots. This is why the relationship between the collapse length and Reynolds number shown in Fig. 4 is peculiar to underexpanded microjets.

4. Conclusions

Supersonic microjets issuing from a rectangular convergent nozzle whose height was 500 μm at its exit was investigated using the acetone-based PLIF and MTV techniques. The jet Reynolds numbers was ranged between 200 and 7000 and the jet Mach number was set to 1.5. The PLIF visualization results revealed that the jets spread abruptly at a certain location owing to jet collapse and the jet-collapse process was successfully observed in the instantaneous MTV images. The distance between the nozzle exit and the jet-collapse location was estimated from each PLIF image. This distance was found to be constant for the Reynolds numbers higher than ~700 and to be inversely proportional to the Reynolds number for the Reynolds numbers lower than ~700. Furthermore, the distance changed abruptly at the Reynolds number of ~700.

References

[1] Satoh D, Tanaka S, Yoshida K and Esashi M 2005 *Sens. Actuators A*, **119** 528
[2] Bruccoleri A R, Leiter R, Drela M abd Lozano P 2012 *J. Prop. Power.* **28** 96
[3] Phalnikar K A, Kumar R and Alvi F S 2008 *Exp. Fluids* **44** 819
[4] Mai C-C and Lin J 2002 *Opt. Laser Tech.* **34** 479
[5] Doms M and Muller J 2005 *Sens. Actuators A* **119** 462-467
[6] Lempert W R, Boehem M, Jiang N, Gimelshein S and Levin D 2003 *Exp. Fluids* **34** 403
[7] Handa T, Mii K, Sakurai T, Imamura K, Mizuta S and Ando Y 2014 *Exp. Fluids* **55** 1725
[8] Tam C K W 1995 *Annu. Rev. Fluid Mech.* **27** 17
[9] Raman G 1999 *J. Sound and Vib.* **225** 543
[10] Handa T, Masuda M, Kashitani M and Yamaguchi Y 2011 *Exp Fluids* **50** 1685
[11] Thurber M C, Grisch F, Kirby B J, Votsmeier M and Hanson R K 1998 *Appl. Opt.* **37** 4963
[12] Lempert W R, Jiang N, Sethuram S and Samimy M 2002 *AIAA J.* **40** 1065
[13] Gau C, Shen C H and Wang Z B 2009 *Phy. Fluids* **21** 092001
[14] Walker S H and Thomas F O 1997 *Phys. Fluids* **9** 2562

IOP Conf. Series: Materials Science and Engineering **249** (2017) 012017 doi:10.1088/1757-899X/249/1/012017

Microscopic bubble behaviour in suppression pool during wetwell venting

G Zablackaite, H Nagasaka, H Kikura

Laboratory for Advanced Nuclear Energy, Institute of Innovative Research, Tokyo Institute of Technology, 2-12-1-N1-7, Ookayama, Meguro-ku, Tokyo Japan

zablackaite.a.aa@m.titech.ac.jp

Abstract. During a severe accident PCV failure should be avoided and fission products inside PCV should be confined as much as possible. In order to minimize FPs release, Wetwell venting is conducted by releasing steam-non-condensable gas mixture carrying FPs from the Drywell to Suppression Pool. Steam is condensed by subcooled water in the pool, and most of FPs are retained into water. The removal of FP in the water pool is referred to as "Pool Scrubbing effect". Hydrodynamic parameters of bubbles have impact on pool scrubbing effect. However, there is only few data available to evaluate quantitatively the bubble behaviour under depressurization and/or thermal stratification conditions. Series of experiments were conducted to evaluate the influence of temperature distribution, non-condensable gas content and pressure in the Wetwell on bubble behaviour. Bubbles were visualized using High Speed Camera and adopting shadowgraphy technique. Applying Particle Tracking Velocimetry, bubble velocity and size distribution were obtained from recorded images. Experimental results show that with increasing suppression pool temperature, bubbles reaching the pool surface decreased in size and traveling velocity became slower. In pressurized wetwell, bubble behaviour was similar to that in the heated up suppression pool case, although bubble parameters were similar to the low temperature case. Higher air content induced water surface movement and bubbles were smaller due to break up.

1. Introduction

One of the ways to prevent Primary Containment Vessel (PCV) failure during severe accident is to actuate containment venting [1]. The pressure inside the containment is decreased by opening flow paths to the outside. However, venting may lead to direct release of fission products (FPs) to the environment. In order to reduce the risk, benefits of pool scrubbing effect can be used in Boiling Water Reactors (BWR).

In the case of Reactor Pressure Vessel failure, increased pressure in the Drywell (D/W) is reduced by conducting Wetwell (W/W) venting by releasing mixture of steam and non-condensable gas to the Suppression Pool (S/P) via vent pipes. Big bubbles form at the outlet of the downcomer pipe, detach and rise. Large bubble starts to collapse and breaks to many small bubbles. When the steam within the bubble reaches the surface of the bubble, condensation occurs and bubble decrease in size. On the other hand, rising bubble without condensing steam expands due to the decreasing water head towards the surface of the pool. Fission products, trapped inside the bubbles, are expected to be captured in the water once it reaches the interface of the bubble surface and water. The removal of FP in the water pool is referred to as "Pool scrubbing effect".

On March 11, 2011, the Great East Japan Earthquake caused the accident in Fukushima Daiichi Nuclear Power Plant [2]. The Station Blackout resulted in the core meltdown and the explosion of reactor building at Unit 1 and 3. A great amount of radioactive materials was released to the environment. Because of loss of AC/DC power, reactor cooling systems failed. Pressure inside containment vessel started to rise and several PCV venting were conducted in Unit 1 and in Unit 3. Mixture of steam and non-condensable gases with fission products (FPs) flew into wetwell. Non-condensable gases passed through the pool and reached the surface of water. Large amount of steam increased the temperature of the pool and it was suggested that thermal stratification occurred above the exit of vent pipe. Analysis of Radiation Monitoring data shows large peaks of radiation after initiated W/W venting, which demonstrates that pool scrubbing was not as effective as expected [3]. The efficiency of scrubbing effect under depressurization and thermal stratification was remarkably reduced. Hence, more analysis is needed to fully understand the effect of venting conditions on FPs release to environment and how efficiency of pool scrubbing effect was reduced during Fukushima Daiichi accident [4].

Hydrodynamic parameters of bubble have impact on pool scrubbing effect. However, there is only few recent available data to evaluate quantitatively the bubble behavior. Many experiments on pool scrubbing effect were conducted in 80's and 90's. Hakii *et.al.* [5] analyzed influence of parameters, such as injection nozzle diameter, scrubbing depth, and steam fraction in carrier gas, for pool scrubbing. Kulhman *et.al.* [6] tested the influence of FP size and water temperature for decontamination factor. They noticed the importance of bubble diameter and its rise velocity for FP retention. Hydrodynamic behavior of S/P was analyzed in THAI facility [7], [8]. Dependency of bubble size on downcomer pipe diameter and the injected gas flow rate was analyzed. Formation and disappearance of thermal stratification in suppression pool was experimentally analyzed by Yamauchi *et.al.* [9], and Bo *et.al* [10]. In a scaled-down torus shaped suppression pool, they observer thermal stratification above the outlet. Reduced subcooling temperature did not allow steam bubbles to condensate effectively, which led to strong mixing and dissipation of stratification. Previously conducted experiments provide good results leading to better understanding of basic pool scrubbing phenomena. However, most of the studies analyze steam or steam-air injections under constant and atmospheric pressure. Also, high flow rate causes mixing in the pool. In the case of the severe accident, especially during wetwell venting, pressure is changing, and under this condition, bubble behaviour differs. The lack of experimental data for depressurization and thermal stratification conditions does not allow to accurately evaluate the influence of hydrodynamic bubble parameters on retention of FPs in the pool under these conditions.

In this paper, experimental results of temperature distribution in S/P, air content and pressure in the W/W influence on steam-air bubble behaviour in suppression pool under different conditions are presented. The main long-term goals of the study are to observe bubble behaviour, to measure bubble size change, and traveling time to reach pool surface during BWR wetwell venting under depressurization and/or thermal stratification conditions. Therefore, the effect of bubble behaviour on pool scrubbing should be clarified.

2. Test facility

2.1. Description of the test facility

The designed test facility (Figure 1) consists of two pressure vessels – drywell (D/W) and wetwell (W/W), downcomer pipe, steam generator, air compressor, heat exchanger and High speed camera (HSC) and light source. Water temperature in the wetwell can be changed by steam injection using wetwell bypass line, and circulation pump is used to get initial uniform pool water temperature.

D/W vessel is 1.2 m long, 0.345 m in diameter. The vessel has a layer of rock wool insulation. W/W is 2.5 m high and 0.355 m in diameter (design pressure 0.6 MPa). Observation windows on the W/W vessel are arranged to observe bubble behaviour at 3 different positions above the pool floor. Downcomer submergence can be changed by changing water level and/or length of the pipe. There are

3 possible downcomer diameters and two possible downcomer lengths (Table 1). Presented cases were conducted using 12.48 mm diameter downcomer.

Table 1. Possible submergence and size of downcomer pipe.

Submergence, m	Downcomer diameter ID, mm
0.7	12.48
0.95	35.08
1.4	52.48

The piping lines are divided into main line with wetwell bypass line, S/P overflow line, W/W water inlet/outlet line and circulation line, discharge/SRV line. S/P overflow line allows to keep water level in the vessel constant. Pressure reducing valve (PRV) on the discharge line allows to keep the W/W pressure constant or depressurize the vessel. Discharge line is connected with SRV lines on D/W and W/W and the heat exchanger. Circulation line with installed magnetic pump is used to obtain uniform temperature distribution in W/W for experiments without thermal stratification.

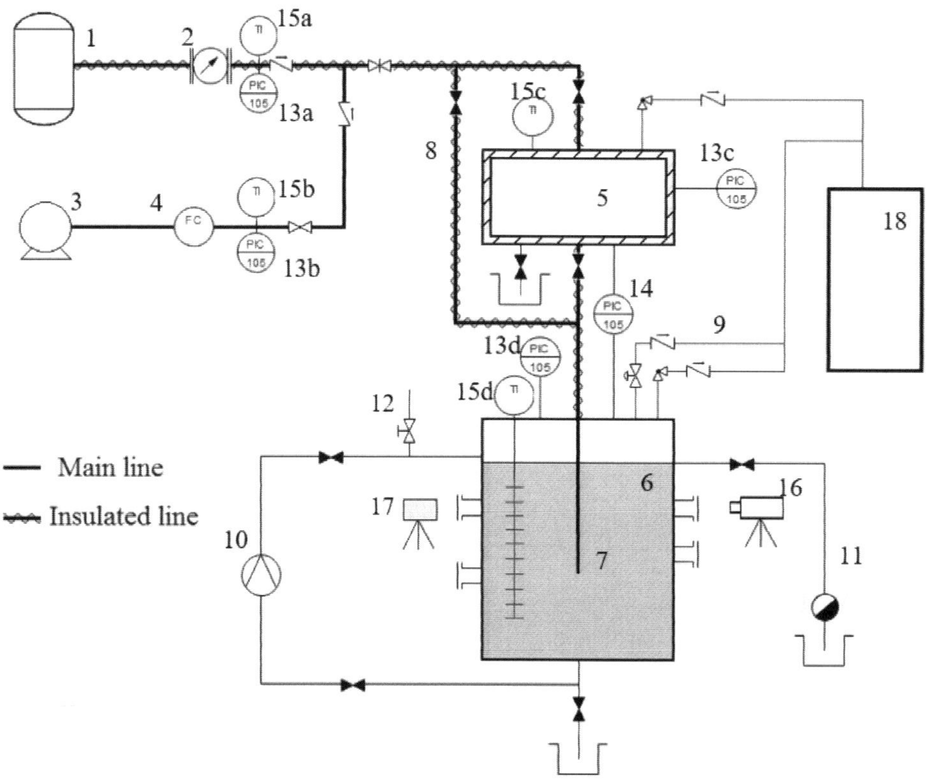

Figure 1. Schematic diagram of the test facility. 1 – steam generator; 2 – vortex flow meter; 3 – air compressor; 4 – air flow regulator; 5 – drywell; 6 – wetwell; 7 – vent line (downcomer pipe); 8 – wetwell bypass line; 9 – gas discharge line; 10 – circulation line/circulation pump; 11 – overflow line/steam trap; 12 – water supply line; 13 – pressure gauge; 14– differential pressure gauge; 15 – thermocouples; 16 – High Speed Camera; 17 – light source; 18 – heat exchanger.

2.2. Instrumentation

Pressure in D/W ($P_{D/W}$) and W/W ($P_{W/W}$) is measured by pressure transducers Panasonic DPH-L113V (measurement accuracy – 2% of the full scale). Pressure on the steam and air lines is measured by pressure transducers Panasonic DPH-L113 (measurement accuracy – 2% of the full scale). Differential pressure gauge Yokogawa DPharp EJA110 (measurement accuracy – 0.2% of the full scale) is used to measure pressure difference between D/W and W/W. Steam flow rate ($\dot{M}_s$) is measured using TLV EF73 vortex flow meter (measurement accuracy for the low flow rate – 10% of the full scale), air flow rate is measured and regulated with PFM5 flow switch (measurement accuracy – 3 % of the full scale). Temperature in D/W ($T_{D/W}$) and on the air line is measured with a K-type thermocouple (measurement error – 2.2°C), steam line temperature (T_{st}) is measured with a T-type thermocouple (measurement error – 1°C). 10 K-type thermocouples are used to measure temperature in the W/W and to evaluate thermal stratification after steam-air release. Three thermocouples are installed below downcomer outlet (T1-T3), one thermocouple is at the outlet level (T4) and six thermocouples are above the outlet, but below water surface. Temperature is measured 7 cm away from the wall of the vessel to avoid the thermal boundary layer caused by the cooling of the non-insulated W/W. Data is recorded using Yokogawa MX100 data logger.

2.3. Bubble tracking
Bubble observation was performed adopting shadowgraphy technique. Bubbles were recorded using 2 High Speed Cameras (HSCs) and, on the opposite side of the W/W, a metal halide lamp was positioned to illuminate the inside of the vessel. The camera Photron Mini AX50 (500 fps, 1/25000 shutter speed, image resolution 1024x1024) was used to observe bubble behavior at the pool surface. The other camera Photron Fast Cam 1024 PCO, model 1KC (500fps, 1/1000 shutter speed, image resolution 1024x1024) was used to observe bubble formation at the downcomer outlet. Images were analyzed using MATLAB R2017b built-in toolboxes. The size of forming bubble at the downcomer outlet was obtained manually measuring selected images. Bubbles around the pool surface and its boundaries were automatically detected and tracked in each frame. Then, bubble velocity and size were obtained.

3. Initial conditions
Initial conditions of experimental Runs are given in Table 2. Five experiments were conducted under one parameter change. For all the cases, inner diameter of downcomer was 12.48 mm. It allowed to have more uniform size of the bubbles around the pool surface for simplified case. Submergence was set to 1.4 m, which corresponds to submergence in Fukushima Daiichi Unit 3. Bubbles were observed at two positions: around downcomer exit and around pool surface. Size of initial forming bubbles and rising bubbles at the top of pool surface were compared. W/W pressure was close to atmospheric for the Runs 1, 2, 3, and 5.

Table 2. Test matrix.

Run number	1 (ref.)	2	3	4	5
Downcomer ID, mm	12.48 (3/8")	12.48 (3/8")	12.48 (3/8")	12.48 (3/8")	12.48 (3/8")
Submergence, m	1.4[a]	1.4	1.4	1.4	1.4
Position of observation	Downcomer exit/Pool surface	Downcomer exit/Pool surface	Downcomer exit/Pool surface	Downcomer exit/Pool surface	Downcomer exit/Pool surface
W/W pressure, MPag	0.04	0.04	0.04	0.30	0.04
S/P temperature, °C	38.3	36.5-76.3	65.8	44.3	30.5
S/P temperature distribution	Uniform	Thermal stratification	Uniform	Uniform	Uniform
Steam flow rate, kg/h	2.77	2.77	2.77	18.71	5.69
Air content	4.92%	4.92%	4.92%	4.79%	22.02%

[a] - corresponding to submergence in Fukushima NPP Unit 3.

For the Run 4 W/W pressure was increased by closing discharge line and increasing the mass flow rates of steam and air. S/P temperature was changing from 30.5 °C to 76.3 °C. Runs 1, 3, 4, and 5, were done

IOP Conf. Series: Materials Science and Engineering **249** (2017) 012017 doi:10.1088/1757-899X/249/1/012017

having uniform S/P temperature, and the Run 2 was carried out in thermally stratified condition above the downcomer exit. Steam mass flow rate was set observing the differential pressure between D/W and W/W and bubble formation at the downcomer exit. Target air content was 5% in the Runs 1, 2, 3, and 4, and it was 20% in the Run 5. However, due to sensitivity of the flow regulating valves only close values were achieved.

4. Experimental procedure

After the test facility was prepared for operation, piping lines were heated up by steam release. By opening steam and air lines, D/W was heated up and remaining air from the vessel was purged. At the same time S/P was heated up releasing steam-air mixture to the W/W. Pool water was recirculated to get uniform temperature in S/P for the runs with uniform temperature distribution. For the case with thermal stratification, release of steam-air mixture was carried on without recirculating the pool. Steam and air flow rates were set to obtain predetermined air content, avoiding high pressurization of D/W. Value of differential pressure between D/W and W/W was almost equal the pressure corresponding to water head in the downcomer. After the required conditions in S/P pool were set, images of forming bubbles at the downcomer outlet and bubbles reaching the pool surface were recorded with HSCs.

5. Results and discussion

5.1. Bubble behaviour

In the case of low uniform temperature (Run 1), bubbles slid up along the downcomer pipe. Number of bubbles was small (Figure 2 (a)), and big bubbles were observed frequently. Bubbles were rising close to the downcomer. Assuming that most of the steam condensed around the outlet of downcomer due to high subcooling, bubbles contained mostly air. Big bubbles reaching the surface of the pool could have been caused by bubble coalescence after detachment from the downcomer outlet, and by decreasing water head. Bubbles collided and coalesced, and become stable enough to avoid breaking up before reaching the pool surface. The case 1 was followed by case 2 and case 3. Temperature profile in the S/P and position of termocouples are shown in Figure 3.

Thermal stratification started to develop from the beginning of steam-air mixture release to the S/P (Run 2). Due to low mass flow rate, steam-air mixture did not penetrate S/P water below downcomer outlet and mixing was not induced. Two distinct temperature layers can be seen in the temperature profile (Figure 3 (a)). Temperature T4 at the downcomer outlet level showed the boundary between two layers. Bubbles travelled only through higher temperature region. Therefore, bubbles were affected by thermal stratification layer around the outlet of downcomer, where the temperature gradient was high. Going further from the outlet, temperature was increasing evenly. Bubbles were forming in a ring shape, and started to break up around the outlet. The size of bubbles tented to change further due to break up and coalescence. Bubbles dispersed away from the downcomer (Figure 2 (b)).

In the case of higher uniform temperature (Run 3), rising bubbles dispersed further to the periphery of the vessel (as much as it was possible to see through the window). With increasing pool temperature, steam condensation rate decreased and number of bubbles increased (Figure 2 (c)). Most of the bubbles were small in size. Bigger bubbles were appearing only occasionally, which implies that even after the collision and coalescence, bubbles were not stable enough to stay big and broke up. Due to decreased rising bubble velocity, bubbles were staying at the pool surface.

In the pressurized W/W (Run 4), pool temperature was kept low, therefore, as in the first case, steam mostly condensed around the outlet of the downcomer. Only few big bubbles were observed and they were smaller in size comparing with the case 1 (Figure 2 (d)). Although, bubbles were not fast enough to break the water surface. After small bubbles reached the pool surface, majority of them were stopped or even bounced back from the interface of water. Bubble behaviour around the water surface in pressurised W/W was similar to the case with higher temperature in the S/P.

IOP Conf. Series: Materials Science and Engineering **249** (2017) 012017 doi:10.1088/1757-899X/249/1/012017

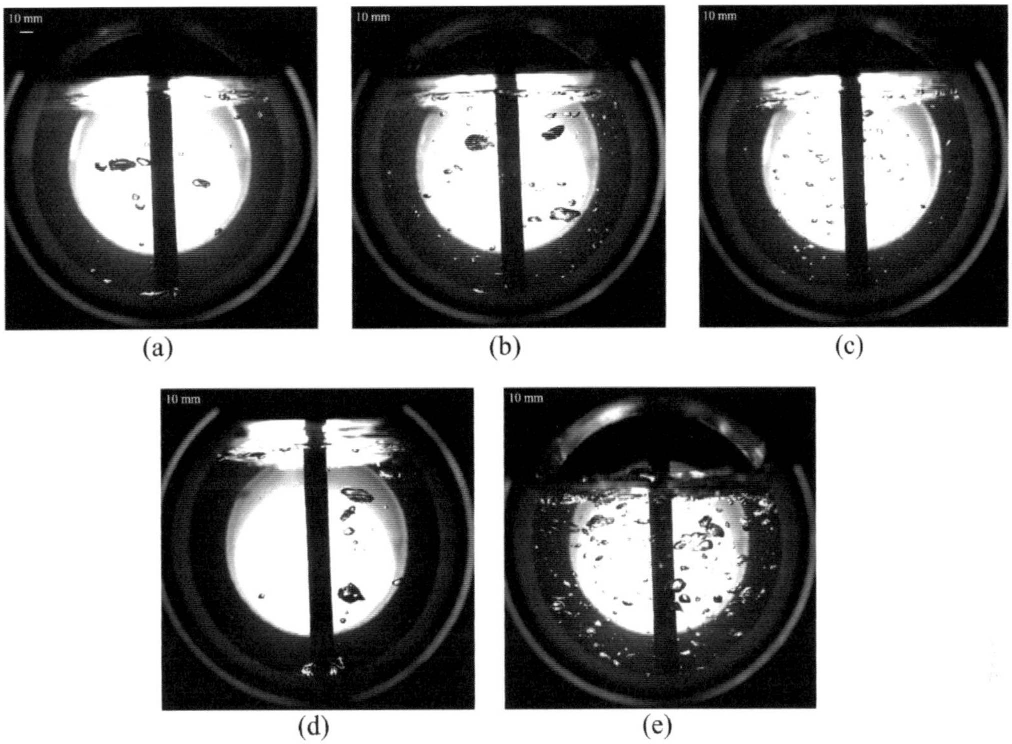

Figure 2. Bubble images close to the pool surface: (a) – low S/P temperature; (b) – stratified S/P; (c) – high S/P temperature; (d) – pressurized W/W; (e) – higher air ratio.

Having high air content (Run 5), the number of bubbles reaching the surface distinctly increased (Figure 2 (e)). Bubbles were forming faster and detaching from the downcomer outlet more frequently. Higher mass flow rate of the steam-air mixture induced movement of the pool surface. Rising bubbles strongly dispersed in the radial direction around the pool surface. Bubbles were colliding and breaking up a lot, although coalescence was observed even before the bubble reached the pool surface.

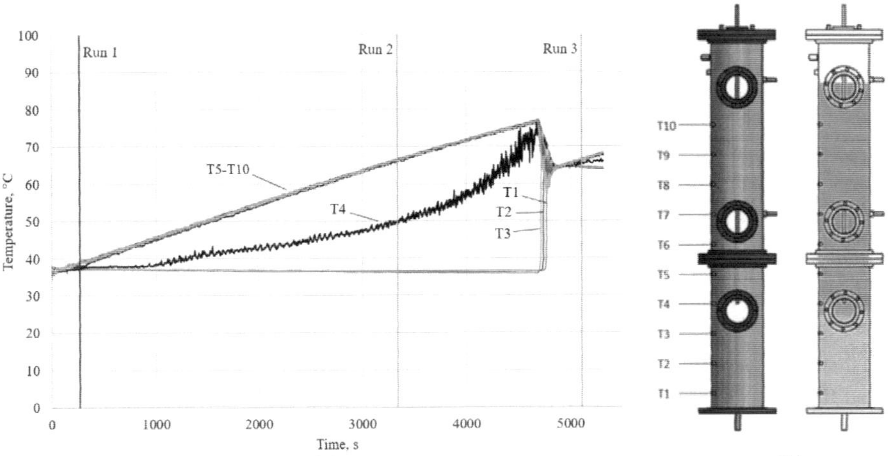

(a) (b)

Figure 3. S/P in cases 1, 2 and 3: (a) – temperature profile, (b) – position of thermocouples in the W/W with temperature distribution in the case of thermal stratification.

5.2. Bubble size and velocity

Velocity and size distribution for all cases is given in Figure 4. Average size of forming and rising bubbles, and velocity of the rising bubbles around the pool surface are given in Table 3. The smallest detectable size of the bubbles was assumed to be 2 mm. In the low temperature S/P (Run 1), the size of formed bubble at the downcomer outlet was 18.3 mm. Average size of rising bubbles was 6.7 mm. Rising bubble size around the pool surface ranged up to 33.4 mm, although majority of the bubbles were smaller than 10 mm (Figure 4 (a)). Bubble velocity mostly ranged from 0.20 m/s to 0.55 m/s. Average velocity of rising bubbles was 0.353 m/s.

In the stratified S/P (Run 2), the size of forming bubble increased to 21.7 mm. Bubble size in this case was 4.8 mm on average, with the maximum size of 23.3 mm. Majority of the bubbles were smaller than 8 mm (Figure 4 (b)). Bubbles tended to travel slower, 0.317 m/s on average. Bubble velocity changed in the same range as in the Run 1.

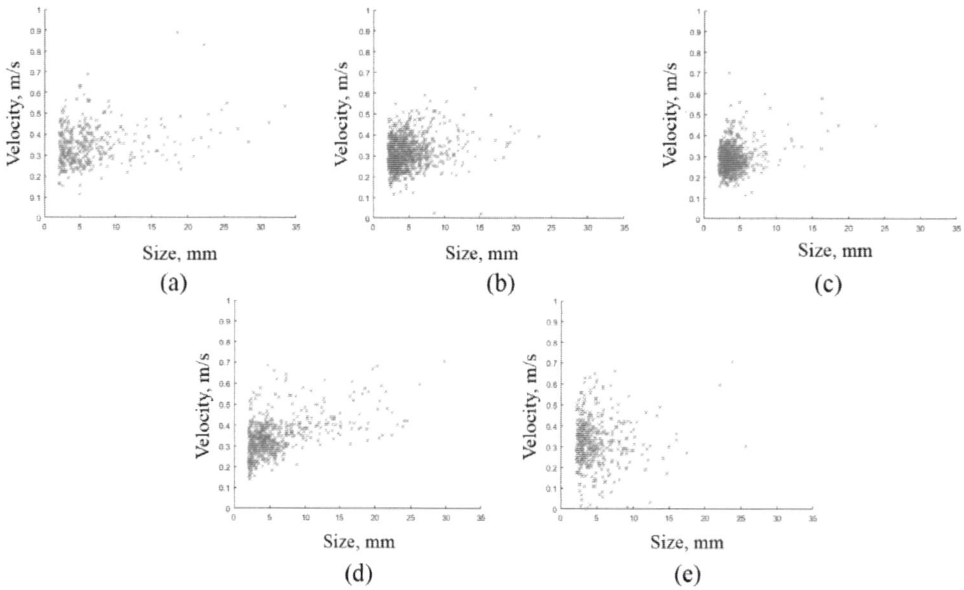

Figure 4. Size and velocity distribution around the pool surface: (a) – low S/P temperature; (b) – stratified S/P; (c) – high S/P temperature; (d) – pressurized W/W; (e) – higher air ratio.

In the high temperature S/P (Run 3), the size of formed bubble was 24.6 mm. Bubble size around the pool surface was 4.0 mm on average, with the maximum size of 23.7 mm. Bubbles around the pool surface were more uniform in size, majority of the bubbles were smaller than 6 mm (Figure 4 (c)). Comparing low and high temperature cases, the velocity of bubbles decreased, and on average it was 0.291 m/s. The velocity mostly ranged from 0.2 m/s to 0.4 m/s.

Table 3. Size and velocity.

Run number	1 (ref.)	2	3	4	5
Initial bubble size at the downcomer exit, mm	18.3	21.7	24.6	21.3	19.9
Average bubble size around pool surface, mm	6.7	4.8	4.0	6.4	5.1
Max size around pool surface, mm	33.4	23.3	23.7	29.8	25.6
Average bubble velocity around pool surface, m/s	0.353	0.317	0.291	0.352	0.320

In the pressurized W/W (Run 4), formed bubbles were 21.3 mm in size. Bubbles reaching the pool surface were 6.4 mm size on average, maximum detected size was 29.8 mm. Majority of the bubbles were smaller than 7 mm (Figure 4 (d)). Size and velocity distribution in this case is similar to the Run 1 (Figure 4 (a)), solely bigger bubbles were traveling slightly faster. Otherwise, velocity of bubbles was similar as in the first case, 0.352 m/s on average.

Lastly, in the higher air content case (Run 5), forming bubbles were 19.9 mm in size (Figure 4 (e)). In this case, bubble size reaching the surface was 5.1 mm on average, and maximum size was 25.6 mm. Bubbles were mostly smaller than 10 mm. Rising bubble velocity varied in a bigger range than in other cases. It changed from 0.1 m/s up to 0.7 m/s. Bubbles travelled at 0.320 m/s velocity on average.

6. Conclusion

Rising bubble behaviour was observed in Suppression Pool under different initial conditions. Experimental data was analyzed to evaluate the effect of S/P thermal stratification, air content in steam-air mixture, and pressure on bubble behaviour. With increasing S/P temperature, forming bubble size increased while bubbles were colliding and breaking up more easily. Therefore, the size and velocity of the bubbles decreased. Bubbles dispersed away from the downcomer pipe to the radial direction in the pool. The smallest and the slowest bubbles were observed in the case of the highest temperature S/P. In the pressurised W/W bubble parameters were similar to the low S/P temperature case. Although bubble behaviour around the water surface was resembling to the high temperature S/P. Bubbles stayed and floated at the pool surface before collapsing. High air content induced unstable movement of the pool surface by rising bubbles, although bubble size was smaller due to increased probability of break up.

References

[1] Dana L Kelly 1991 Overview of containment venting as an accident mitigation strategy in US light water reactors *Nuclear Engineering and Design* **131** 253-61

[2] AESJ, The Fukushima Daiichi Nuclear Accident Final Report of the AESJ Investigation Committee, Springer, (2014).

[3] TEPCO. The 4th Progress Report on the Investigation and Examination of Unconfirmed and Unresolved Issues on the Development Mechanism of the Fukushima Daiichi Nuclear Accident. TEPCO Progress report, (2015).

[4] Mizokami Sh, Yamada D, Honda T, Yamauchi D and Yamanaka Y 2016 Unsolved issues related to thermal-hydraulics in the suppression chamber during Fukushima Daiichi accident progression *Journal of Nuclear Science and Techology* **53** 630-8

[5] Hakii J, Kaneko I, Fukusawa M, Yamashita M and Matsumoto 1990 Experimental study on aerosol removal efficiency for pool scrubbing under high Temperature Steam Atmosphere *Proceeding of the 21st DOE/NRC nuclear air cleaning conference* **2** 21-0918

[6] Kulhman M R, Gieseke J A, Merilo M and Oehlberg R 1985 *Proceeding of American Nuclear Society meeting on fission-product behaviour and source term research* Scrubbing of fission product aerosol in LWR water pool under severe accident conditions. Experimental Results 633-48

[7] Gupta S, Schmidt E, Laufenberg B, Freitag M, Poss G, Funke F and Weber G 2015 THAI test facility for experimental research on hydrogen and fission product behavior in light water reactor containments *Nuclear Engineering and Desing* **294** 183-201

[8] Gupta S 2015 Experimental Investigation relevant for hydrogen and fission product issues raised by the Fukushima accident *Nuclear Engineering and Technology* **47** 11-25

[9] Yamauchi D, Jo B, Erkan N, Takahashi Sh, Sagawa W and Okamoto K 2016 Verification of thermal stratification characteristics using scaled-down suppression pool of the Fukushima Daiichi nuclear power plants *Mechanical Engineering Letters* **2** 16-00092

[10] Jo B, Erkan N, Takahashi Sh, Song D, Sagawa W and Okamoto K 2016 Thermal stratification in a scaled-down suppression pool of the Fukushima Daiichi nuclear power plant *Nuclear Engineering and Design* **305** 39-50

IOP Conf. Series: Materials Science and Engineering **249** (2017) 012018 doi:10.1088/1757-899X/249/1/012018

Flows due to pressure induced dissociation-formation of gas hydrates

J R Agudo[1], S Kwon[1], R Saur[1], S Loekman[2], G Luzi[1], C Rauh[3], A Wierschem[4] and A Delgado[1,4]

[1]Institute of Fluid Mechanics, FAU Busan Campus, University of Erlangen-Nuremberg,
46-742 Busan, Republic of Korea
[2]Institute of Chemical Reaction Engineering, FAU Busan Campus, University of Erlangen-Nuremberg,
46-742 Busan, Republic of Korea
[3]Department of Food Biotechnology and Food Process Engineering
Technische Universität Berlin
D-14195 Berlin, Germany
[4]Institute of Fluid Mechanics, University of Erlangen-Nuremberg,
D-91058 Erlangen, Germany

Antonio.delgado@fau.de

Abstract. During the last decade, Gas Hydrates (GH) have attracted the interest of the scientific community for engineering applications. Carbon dioxide hydrate (CO_2H), for instance, may play an important role for capture and sequestration methods in order to reduce global climate change. Despite the extensive literature, the transport phenomena involved during CO_2H formation are not yet fully understood. CO_2 transfer from gas or liquid phase to the bulk of water is expected to happen not only by molecular diffusion but also driven by natural convective currents induced by CO_2 dissolution in water. Using particle tracer methods, we experimentally characterize the flow velocity of the bulk of water during CO_2H formation. For that purpose, CO_2H is grown inside an optical cell with a volume of 12 mL at various pressures and temperatures. Due to CO_2 dissolution, convection currents are noticed prior to hydrate formation. Our experimental results point to a significant correlation between this process and the subsequent hydrate formation. Two well-differentiated hydrate growth patterns were observed depending on the hydrate induction time and the corresponding CO_2 concentration distribution inside water. For long induction times, CO_2 can be provided from the water phase resulting in rapid growth. Short induction times resulted in slow growth at the interface creating a solid barrier accompanied by a significant drop in the flow velocity. In some cases, the hydrate layer appeared to be unstable and convection could restart.

1. Introduction

Gas Hydrates (GHs) are crystalline solids consisting of gas molecules, also called guest or former molecules, enclosed in water cavities composed of hydrogen-bonded water molecules, also referred as host molecules [1]. Methane, ethane, propane or CO_2 are typical sorts of gas hydrate former. There is no bonding between frozen water and former gas molecules which are free to rotate inside the frozen water cage. The stabilization resulting from gas molecule is supposed to be due to van der Waals forces (attraction forces between molecules) and hydrogen bonds [2]. Given the enormous amounts of methane enclosed in form of hydrate that is encountered in deep ocean floors [3], Methane Hydrates (MHs) are considered as one of the most important future source of hydrocarbon fuel [4].

Unlike MH, the existence of CO_2 Hydrates (CO_2Hs) as permanent disposal in Earth seems to be unrealistic as experimentally observed by [5]. However, CO_2H is assumed to be abundant in the solar system playing a potential role in shaping the Martian landscape [6], or as a source of potable water for future human habitation [7]. Coming back to earthling applications, CO_2H formation-dissociation processes have recently attracted the interest of the scientific community for bioenergy applications. They are, for instance, of vital importance for CO_2 capture and sequestration methods, known as CCS. CCS are considered as one of the potential ways to reduce global climate change by preventing CO_2 gas emissions to the atmosphere. The first step of this process consists of separating CO_2 gas from exhaust emissions, coming for instance from fossil fuel power plants [8, 9]. Here, hydrated-base CO_2 separation points to a very promising alternative to current methods, e.g. absorption, adsorption, membrane separation or cryogenic fractionation [9, 10, 11, 12]. The last step of CCS process is the CO_2 storage into a safe and long-term system. Here, sequestering a large amount of CO_2 into deep ocean set-up has been recently contemplated in literature. The replacement of CO_2 in marine MH deposits would permit CO_2 storage and release of methane as energetic input at the same time [9, 13]. Other novel industrial applications include CO_2H-based water desalinization [14, 15, 16] or CO_2H-based liquid concentration process for the food industry [17, 18]. To find a reliable, reproducible and efficient experimental method for GH production-dissociation process, is crucial for studying its viability in the aforementioned applications.

Despite the extensive literature, the transport phenomena involved during CO_2H formation are not yet fully understood. CO_2 transfer from gas or liquid phase to the bulk of water, for instance, is expected to happen not only by molecular diffusion but also driven by natural convective currents induced by CO_2 dissolution in water [19, 20, 21]. These currents, that might play a key role in the CO_2 transport prior to hydrate formation, have not considered yet in classical mass transport models for CO_2H formation. Using particle tracer methods, we experimentally characterize the flow velocity of the bulk of water during CO_2H formation. For that purpose, CO_2H is grown inside an optical cell with a volume of 12 mL at various pressures and temperatures. Due to CO_2 dissolution, convection currents are noticed prior to hydrate formation. Our experimental results point to a significant correlation between this process and the subsequent hydrate formation.

2. Experimental set-up

CO_2H is formed in a cylindrical optical cell with a total volume of 12 mL, and dimensions of 25 mm in height and diameter, designed for pressures up to 350 bars. The pressure cell contains four sapphire glasses for optical access and a cooling jacket to keep the temperature constant at 273.5 ± 0.5 K with the help of a circulating bath. The temperature is controlled with a K-type thermocouple coupled to the center of the cylindrical cell. The pressure is continually recorded outside the reactor by a pressure transducer within a relative uncertainty lower than 1%. Figure 1 illustrates the optical cell used for the experiments.

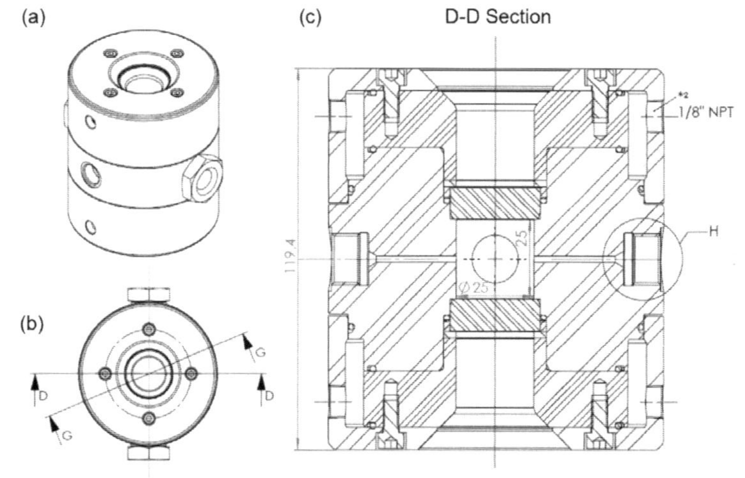

Figure 1. Optical cell with a volume of 12 mL and a total of four sapphire glasses used for the experiments (a). Top view (b) and vertical sectional view (c). The measurements are shown in mm.

To form CO_2H, 5.5 or 7 mL of deionized water (18.2 MΩ X cm of resistivity) are added first to the optical cell. After flushing the system, high purity CO_2 (99.95%) is injected into the vessel to achieve the proper pressure to form hydrate. Two different pressures were selected to form hydrate, 36 and 56 bars. The former is the equilibrium pressure between gaseous and liquid CO_2 at the operating temperature. Under this pressure, there are three phases in the reactor prior to hydrate formation: liquid water, liquid CO_2, and gaseous CO_2 from the bottom to the top in the reactor. The latter pressure corresponds to the maximum provided by the CO_2 bottle at the same temperature. In this case, only liquid water and liquid CO_2 are encountered in the reactor before hydrate formation.

To characterize the flow in the bulk of water, the water is seeded with Polyamide Seeding Particles (PSP-5) of 5 microns diameter with a density of 1030 kg/m^3. At operating conditions, the settling velocity of PSP remains about 3 order of magnitude smaller than typical velocities due to buoyancy driven convection flows at onset of experiment. Given the long-term duration of the experiments, up to 48 hours for hydrate formation, a continuous cold LED source of 55 W with a light guide of 15 mm in diameter is selected for the volume illumination of the cell. The use of this low power light source showed a marginal impact on the bulk temperature and allows us to continue the particle image velocimetry (PIV) measurements in the bulk of water once the hydrate growths in the interface. Two different imaging systems are simultaneously used for conducting the measurements. A 1280 x 1024 CMOS camera coupled to a 180 mm, f/3.5 Macro lens is used for recording from the front at a working distance of 730 mm from the center of the cylindrical cell. The field of view, in this case, is a circular window of 15 mm in diameter which corresponds to the dimensions of the sapphire glass integrated into the side of the reactor. A 1920 x 1080 CMOS camera equipped with a modular zoom lens system that incorporates a tilted mirror is used for recording from the back. The working distance is 53 mm. The magnification of the lens system ranges between M = 1.1 and M = 7.8, and the numerical aperture between NA = 0.036 and NA = 0.12, respectively. The field of view ranges between 4.2 x 5.6 and 0.61 x 0.81 mm, while the depth of field ranges between 0.42 and 0.038 mm, respectively. Figure 2 illustrates the experimental set-up. While the front camera is used to measure the flow velocity in the entire bulk of water (see Figure 3), the back camera permits us to focus on the interface, where hydrate formation is likely to happen (see Figure 4). For the front and the back system, the images are recorded at a video rate of 2 and 5 Hz, respectively. The particle image fields are analyzed by classical cross-correlation implemented by fast Fourier transform using Insight4G®. The size of the interrogation area is 32 x 32 pixels in both systems. This corresponds to a spatial resolution of approximately 0.4 mm in the front system. Defining a peak to noise peak ratio of 1.47, the analyzed particle-image fields reveal typical ratios of good vectors above 80%. The PIV analysis, however, produces a few obvious false positive velocities vectors which are several orders of magnitudes larger than the average. Additionally, all points outside of the processing mask are zero. Thus, to remove the false positive values every point with a velocity above the 95%-tile of non-zero velocities was cut. When calculating the average velocity at a specific point in time, only non-zero velocities were considered. In order to improve readability, the velocities were averaged over 30 seconds.

The accuracy of the PIV system installed in the back was checked with the known solution of a Couette flow velocity profile induced by an MCR302 rotational rheometer with a parallel-disk configuration. A circular container with transparent Plexiglas sidewalls and a chamfer in the front was concentrically coupled to the rheometer. For more details in this experimental set-up, we refer to [22, 23]. The distance from the rotating plate to the surface is fixed at 2 mm and the angular frequency of the rotating plate was set to obtain velocities of the order than those observed during GH formation experiments. The PIV measurements revealed a deviation lower than 5% with respect to the expected flow velocity confined between the plates. A comparison in velocity between the front and the back PIV system showed typical deviations lower than 10%.

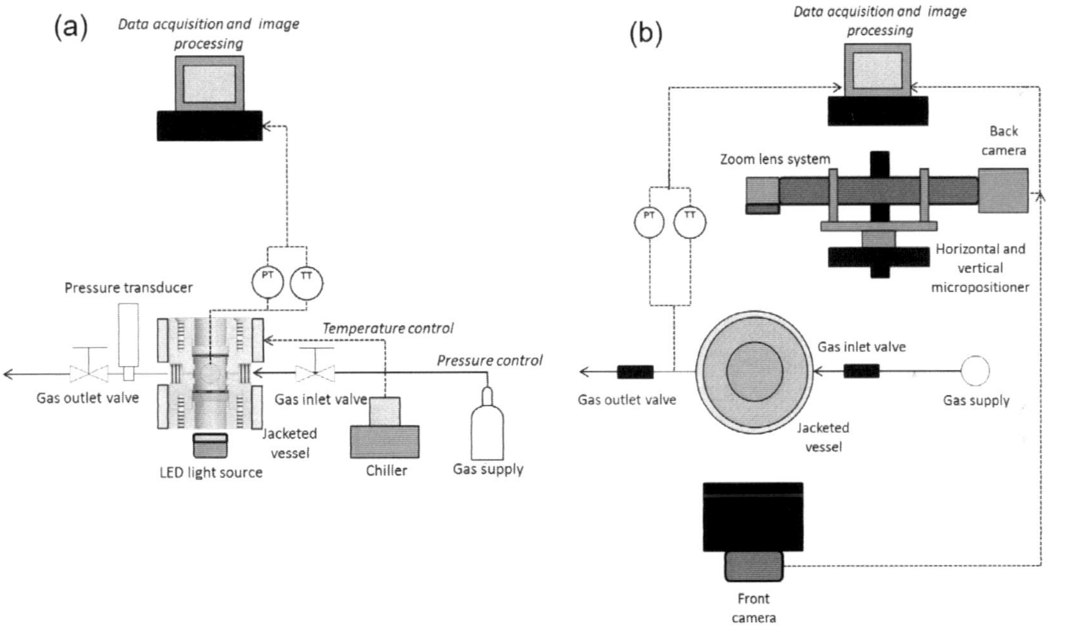

Figure 2. Sketch of the experimental set-up for PIV measurements during CO_2 hydrate formation. Front view (a) and top view (b).

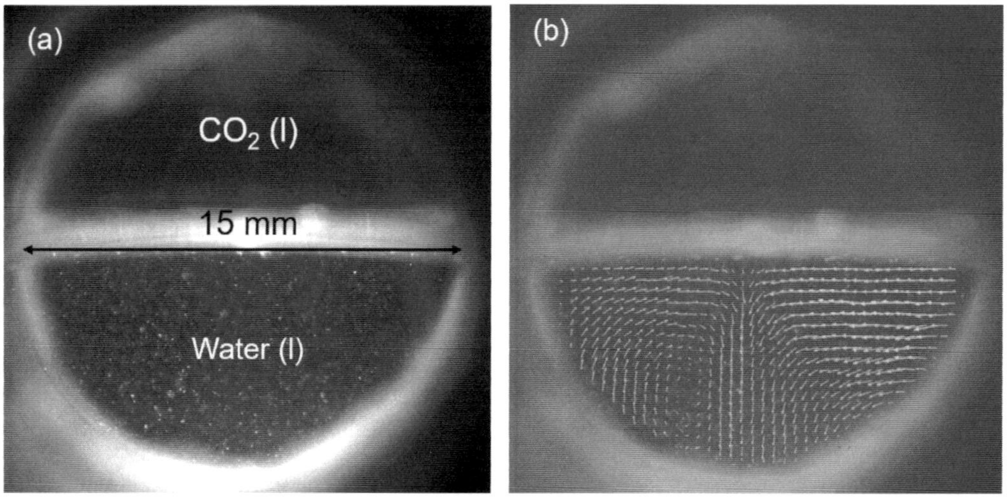

Figure 3. Digital image of the seeded flow inside the optical cell recorded from the front imaging system at 56 bar and 273.5 K (a). Velocity vector map obtained by the aforementioned system prior to hydrate formation (b).

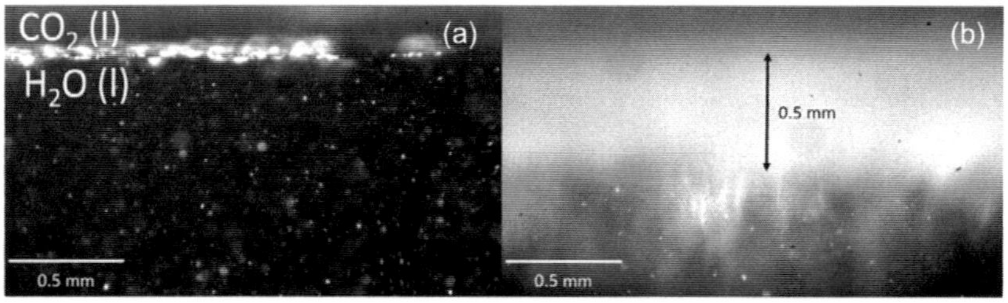

Figure 4. Digital image of the seeded flow inside the optical cell recorded from the back imaging system at 56 bars and 273.5 K prior to hydrate formation (a). Frame representing the onset of hydrate formation in the experiment (b).

3. Results and discussion

In a system as shown in Figure 3, CO_2 is first expected to dissolve into the water phase by molecular diffusion through the interface [20]. Since the dissolution of CO_2 increases the density of water [24], a gradient of density is expected to induce the instability of the boundary layer saturated with CO_2 [19, 20]. This instabilities yields in recirculation flow as shown in Figure 3(b). The criteria to predict the existence of this natural convection is usually characterized by the Rayleigh number, which depends on the fluid properties and geometry dimensions. The Rayleigh number can be defined as [20, 21]:

$$Ra = \frac{\Delta\rho g R^3}{\rho_i \nu D} \tag{1}$$

where $\Delta\rho$ is the density difference of the fluid, g is the gravity acceleration, R is the radius of the cylindrical cell, ρ_i is the initial density of water, ν is the kinematic viscosity of water and D is the diffusion coefficient of water. Natural convection typically occurs when Ra>2100 [21]. The other independent dimensionless number that governs the flow is the Re numbers defined as:

$$\mathrm{Re} = \frac{uR}{\nu} \tag{2}$$

where u is the maximum in-plane velocity registered by the PIV system. At similar pressure conditions than in our experiments, gradients of density driven by CO_2 dissolution up to 8 kg/m^3 has been reported in literature [21]. Accordingly, Rayleigh numbers of the order of 10^7 are expected in our system, and buoyancy-driven convection is likely to happen prior to hydrate formation in the reactor. On the other hand, with maximum measured velocities of about 0.3 mm/s, the Reynolds number remains about 2 in our experiments indicating the existence of laminar flows.

3.1. Effect of seeding tracers on the Gas Hydrate formation

Particle image velocimetry relies on light scattering particles being suspended in the water phase. The flow velocity measurement inadvertently introduces small particles that may cause hydrate to nucleate faster as observed in other recent experiments with suspended particles [25, 26]. To investigate this possibility, experiments without seeding and with two different kinds of tracers, Silver Coated Hollow Glass Spheres 10 μm (S-HGS-10) and Polyamide Seeding Particles 5μm (PSP-5), are conducted. The experiments are compared in terms of induction time and growth pattern. At saturated conditions, we performed a total of 12 experiments with 5.5 mL of water; 3 experiments with non-seeded water, 4 experiments with PSP-5 tracers and 5 experiments with S-HGS-10 tracers. The stochastic nature of the crystallization phenomenon yielded to a broad range of induction times. In the case of trials carried out without tracers or with PSP-5 tracers, the induction time ranged between 2 and 10 hours with exception of a single with PSP-5 tracers event that took 27 hours to nucleate. In the case of experiments performed with S-HGS-10 tracers, however, the nucleation occurred in less than 15 min in all of the cases. While the induction time of experiments with PSP-5 tracers scattered similarly to experiments without tracers, experiments done with S-HGS-10 tracer showed almost immediate growth. This fact suggests that S-HGS-10 particles act as nucleation spots.

This is further corroborated, by the way hydrate nucleated inside the optical cell. In all the cases, experiments with PSP-5 seeding particles showed the typical growth from the water-former interface, as one would expect in the absence of any contaminants (see Figure 5(a)-(d)). The same nucleation pattern happening at the interface was indeed observed in experiments with non-seeded water. On the other hand, in all the experiments conducted, hydrate grew from all over the reactor when 10 micrometer-sized particles were used (see Figure 5(e)-(g)). Nucleation seems to happen in the bulk of water but also in the reactor walls as can be observed in Figure 5(e).

Given that the two types of particles are different in size and material it is not clear what caused the difference in behavior. Literature indicates that the larger size of suspended particles has a shortening effect on the induction time [26]. However, in that case, the pertinent size is that of the agglomerates, and it is not clear whether the size itself or some underlying property of the particles that caused the larger size is to blame for the lower induction lag. Furthermore, was it suggested by [26] that the functional oxide group was essential and [25] reported that suspended silver nanoparticles had no effect on the induction time. Literature seems to be lacking studies, investigating the effect of suspended particles in the micrometer range on hydrate formation. This task, however, remains out of the scope of the present study that relies on the characterization of the flow velocity prior to hydrate formation. Any further experiments were done using the Polyamide particles, where no significant difference with non-seeded experiments was observed in the induction time, nor in the growth pattern.

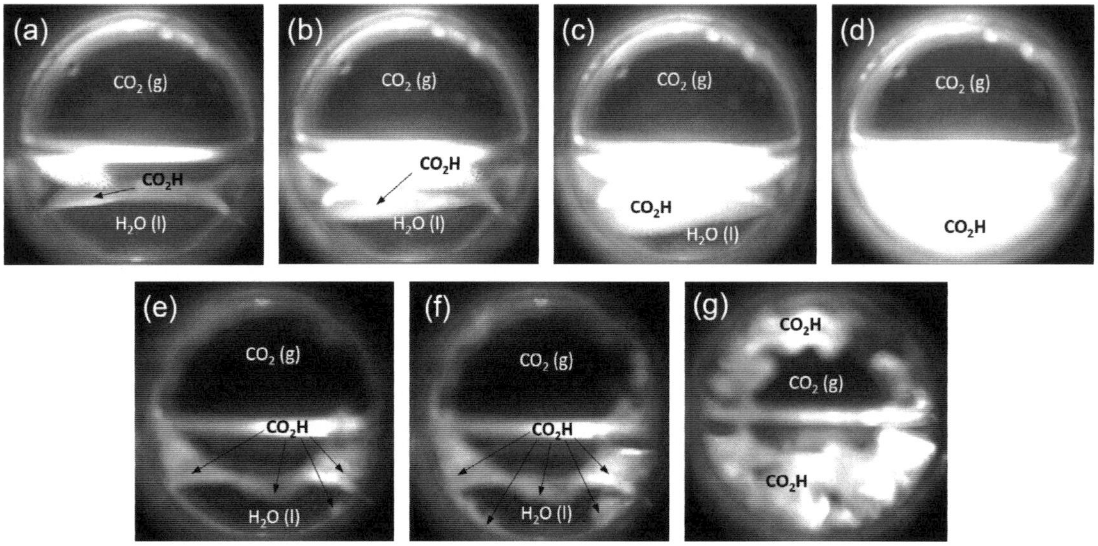

Figure 5. Snapshots illustrating the CO_2H formation with PSP-5 tracers at the interface (a)-(d) and CO_2H formation with S-HGS in the bulk of water and at the walls of the reactor (e)-(g). Experiments performed with 5.5 mL of water and CO_2 at 36 bars and 273.5 K. The induction time of hydrate formation was 2 h and 23 minutes (a) and 3 minutes (e), respectively. The time elapsed between (a) and (d) is about 5 seconds. The time elapsed between (e) and (g) is about 10 minutes.

3.2. Buoyancy-driven convection inducing flow currents prior to Gas Hydrate formation

Two preliminary experiments were conducted to clarify that buoyancy-driven convection due to CO_2 dissolution is the main cause of induced-flow currents before hydrate is formed in the optical cell. In a first experiment, once the operating temperature is achieved, CO_2 is injected into the reactor to an initial pressure of 32 bars. Pressure, temperature and averaged velocities in the bulk of water are recorded simultaneously. Figure 6 depicts the results. The data are plotted up to 6 hours. No hydrate formation was noticed during that time in this experiment. Pressure and average velocity experience a gradually decrease with time until a plateau at about 5 hours is reached. The temperature increases by approximately 1 K as a consequence of the augment in pressure. Unlike the long-term decrease of pressure with time, the measured temperature reaches a constant plateau after approximately 1 min, and keep constant during the rest of the experiment (see inset of Figure 6(a)). Since the temperature remains constant, the pressure drop in the system is mainly due to CO_2 dissolution in water [27].

Comparison between pressure and velocity trends show a remarkable similarity between each other. This indicates that CO_2 dissolving in the bulk of water but not fluctuations in temperature is the main source of the density-driven natural convection phenomenon. Note that a constant pressure in the system indicates that CO_2 is fully distributed in the bulk of water at the corresponding pressure and temperature. No gradient of CO_2 concentration exists in the bulk of water, and convection currents disappear as shown in Figure 6(b).

To check that flow velocities in the bulk of water are not driven by mechanical agitation after an increase of pressure, a second experiment is conducted by injecting water non-soluble gas helium (He) up to 56 bars at operating temperature. Open symbols in Figure 7 represent the averaged flow velocity as a function of time. Solid symbols depict the averaged velocity for an experiment performed with CO_2 at same conditions. The experiment performed with CO_2 reveals again a gradual decrease of velocity until a plateau at a marginal velocity is achieved after about 5 hours. The averaged velocity in the experiment conducted with He drops to the same value but immediately after the pressure is increased in the optical cell. The pressurization and the corresponding velocity measurements for the experiment performed with He can be seen in Figure 7(b) and (c), respectively. The red line indicates the instant in which pressure is increased in the system. From Figure 7(c) it is clear that although increasing the pressure induces some initial movement, the effect dies down after a few seconds. This corroborates that without the dissolution-driven Rayleigh convection the measured velocities are significantly lower. Note that convection induced by a temperature gradient should still be visible when using a non-soluble gas.

The trend of velocity in the experiment performed with CO_2 at 56 bars is very similar to that observed at 32 bars (see Figure 6(b)). This can be explained by the fact that solubility isopleths of CO_2 hardly increases with pressure at operating temperature (see for instance P-T diagram for CO_2 solubility by [28]). This indicates that similar gradients of density are expected at saturated and supersaturated conditions.

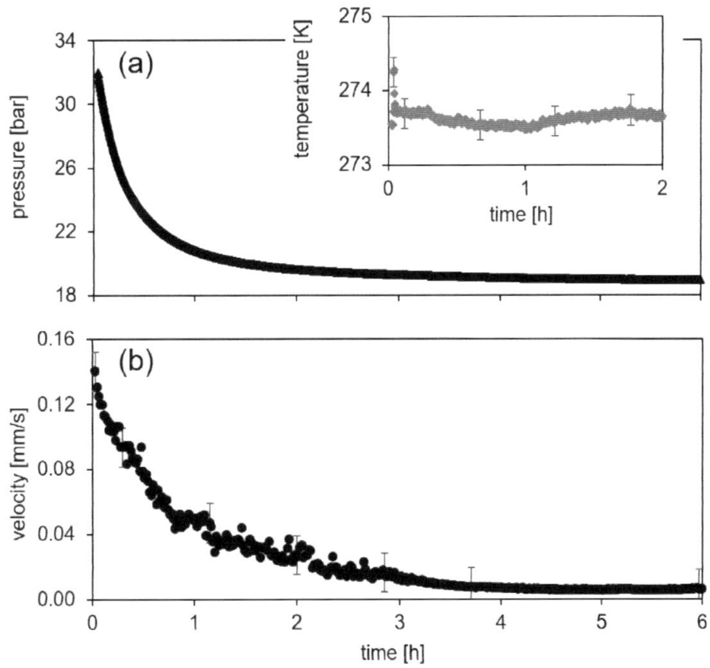

Figure 6. Pressure (a) and averaged velocity in the bulk of water (b) as a function of time. The inset depicts the temperature as a function of time. Experiment performed with 7 mL of water and CO_2 at 36 bars and 273.5 K.

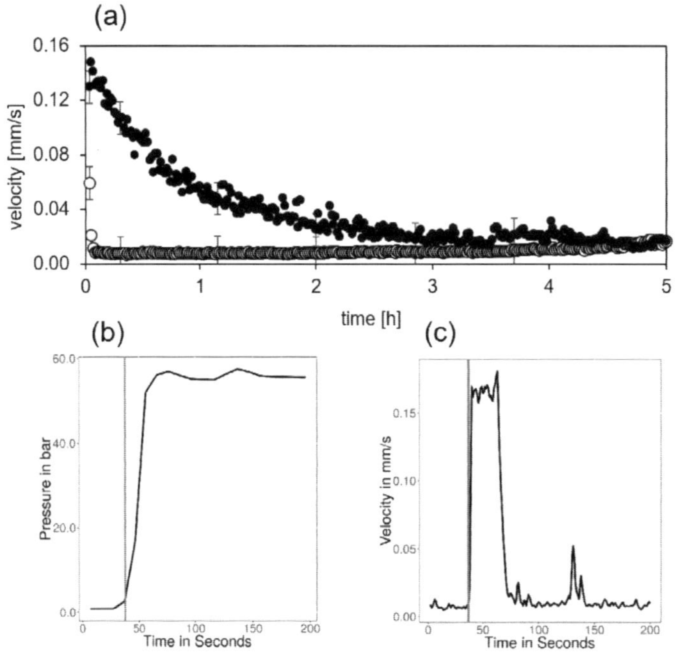

Figure 7. Averaged velocity in the bulk of water as a function of time (a). Solid and open symbols represent experiments performed with CO_2 and He, respectively. Experiments performed with 7 mL of water at 56 bar and 273.5 K. Pressure (b) and averaged velocity (c) right after increase of pressure for the experiment conducted with He.

3.3. Flow velocity measurements during GH formation

We performed a total of 45 experiments using SPS-5 as tracers for measuring the flow velocity in the bulk of water before and during the hydrate formation. From all 45 experiments, in 26 of them, the hydrate grew in the optical cell in less than 48 h. In what follows, we refer to those experiments as successful. For sake of feasibility, we stopped those experiments in which hydrate did not grow after 48 hours. In all of the successful experiments, independently of the volume of water used, i.e. 5.5 or 7 mL, and independently of the pressure selected, i.e. 36 or 56 bar, the initial nucleation occurred at the interface with inductions times ranging between 15 min and 45 hours. Under our experimental conditions, therefore, we could not find any significant effect of the pressure, or of the volume of water selected on the induction time.

Figure 8 depicts the snapshots for two different successful experiments performed with 7 mL of water at saturated conditions. Under the same conditions, the induction time was approximately 15 min (a)-(d) and 2 hours and 31 min (e)-(h), respectively. In both cases, a thin hydrate layer of about 0.5 mm (see Figure 4(d)) is formed first at the interface and we observe gas bubbles rising from the interface (see Figure 8(b) and (f)). This release of bubbles was only noticed when there is both a gas and a liquid CO_2 phase present. We hypothesize this phenomenon is driven by the heat supplied by the exothermic GH formation or due to the decrease of equilibrium CO_2 concentration in water with hydrate formation [29]. In experiments under saturated conditions, the sudden release of bubbles was used as a clear indication for the onset of hydrate formation.

Interestingly, the growth rate velocity between both experiments is completely different. In the experiment with a long induction time, we observed a rapid hydrate growth toward the bulk of water. After 2 seconds, the hydrate layer has covered a large portion of the water (Figure 8(g)) and after 6 seconds most the hydrate has grown throughout the reactor and there seems to be no water left (Figure 8(h)). On the other hand, in the experiment with a short induction time, the hydrate growth rate is much slower. After 2 seconds, the hydrate layer has barely increased (Figure 8(c)) and it requires about 23 min to cover completely the bulk of water (Figure 8(d)). This pattern in growth rate was observed in all experiments with an induction time lower than 30 minutes. Those experiments that took longer resulted in a rapid growth. This fact is directly related to how fast CO_2 is dissolved into water, which in turns is strongly influenced by the Rayleigh convection.

To clarify this point, Figure 9(e) illustrates the averaged velocity in the bulk of water as a function of time for the experiment depicted in Figure 5(a)-(d), blue triangles, and for the experiment depicted in Figure 8(a)-(d), red squares, respectively. The former corresponds to a long induction time experiment, 2 hours and 23 minutes, with the corresponding rapid growth pattern while the latter is an example of short induction time experiment with a much slower growth rate velocity. For sake of comparison, the averaged velocity of the experiment depicted in Figure 6 is shown in the figure, see black circles. In that case, no hydrate was formed during the first 6 hours. All the experiments were performed at saturated conditions, i.e. 56 bar and 237.5 K.

As shown in Figure 9(e), the long induction time example, blue triangles, presents a steeper slope on the initial stage of the averaged velocity as compared to the other two cases. This may be explained by the smaller volume of water used in this case, 5.5 mL instead of 7 mL. A faster distribution of CO_2 is therefore expected in the bulk of water. The final averaged velocity, however, achieves a similar final value than the experiment performed with 7 mL. The hydrate nucleation occurred after 140 minutes when the flow velocity is strongly diminished, indicating that convection currents due to CO_2 concentration gradients have been dropped and CO_2 is well distributed in the bulk of water. This fact explains the immediate growth of hydrate after nucleation which covered the entire bulk of water within a couple of seconds. Figures 9(c) and (d) depict the contour of velocity in the bulk of water 30 seconds before nucleation, and at onset of nucleation, respectively. The latter corresponds to the velocity profile of the snapshot depicted in Figure 5(a). The velocity seems to be higher in the left half of the water volume, just below the left side of the interface where hydrate grew (see Figure 5(a)). This may indicate that this increase of velocity is driven by the rapid hydrate crystallization.

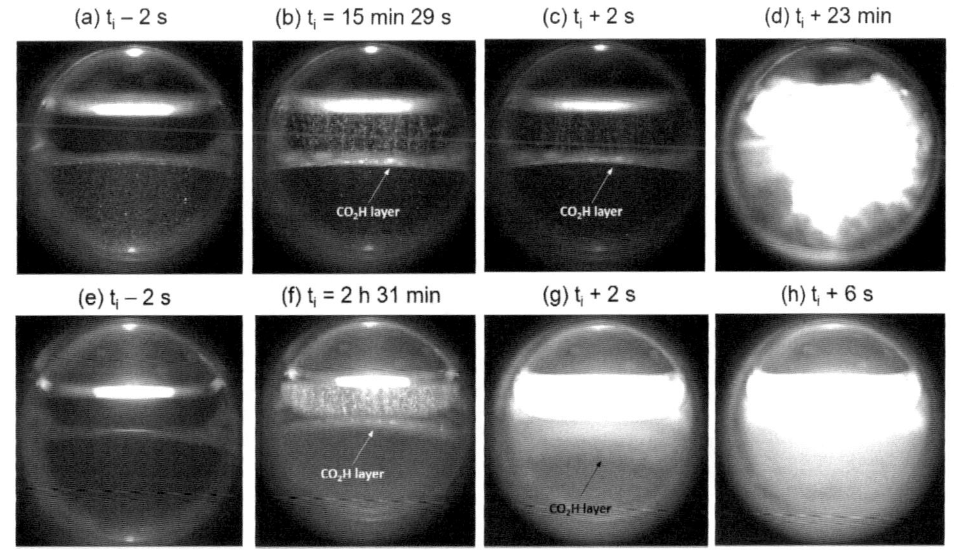

Figure 8. Snapshots illustrating the CO_2H formation for short (a)-(d) and long induction time experiments (e)-(g). Experiments performed with 7 mL of water and CO_2 at 36 bars and 273.5 K.

As expected, the averaged velocity in the short induction time experiment, red squares, presents a similar initial trend that the trial where non-hydrate was growth during the first six hours, black circles. The nucleation happened after about 15 min when the flow velocity is still very high indicating that convection currents are still present due to the gradient of CO_2 concentration. After this short period of time, there simply is not enough CO_2 dissolved in the whole water to sustain growth beyond a thin layer at the interface. The hydrate growth apparently becomes mass-transfer limited with a much lower growth rate velocity. We hypothesize that there should be a specific time threshold at which the concentration quickly changes to a sufficient level of CO_2 concentration. Before this point has been reached only the thin diffusion layer contains enough CO_2 to form hydrate, while shortly afterward the whole of the water can be transformed. The characteristic time for that threshold is dependent on how fast the CO_2 is dissolved in water and should be therefore dependent on the Rayleigh number. In our experiments, the threshold time was found to be 30 min.

Interestingly, the curve of averaged velocity shows a significant drop right after hydrate nucleates. For this kind of low induction lag experiments, PIV can be done continuously throughout nucleation and hydrate growth. Similar drops of velocity were observed right after nucleation in all the experiments with a short induction time. The dramatic drop in velocity can be also clearly discerned from the contour of velocity 30 second before nucleation, Figure 9(a), and at onset of nucleation, Figure 9(b). We consider two explanation for this phenomenon so far. One the one hand, the hydrate layer formed at the water-former interface is impermeable to gas transport as stated by [30], resulting in a sudden stop of Rayleigh convection. Secondly, the freezing in velocity could be due to the initial formation and agglomeration of labile cluster resulting into an increase of macroscopic viscosity. At this stage of research, more experiments are necessary to corroborate this hypothesis.

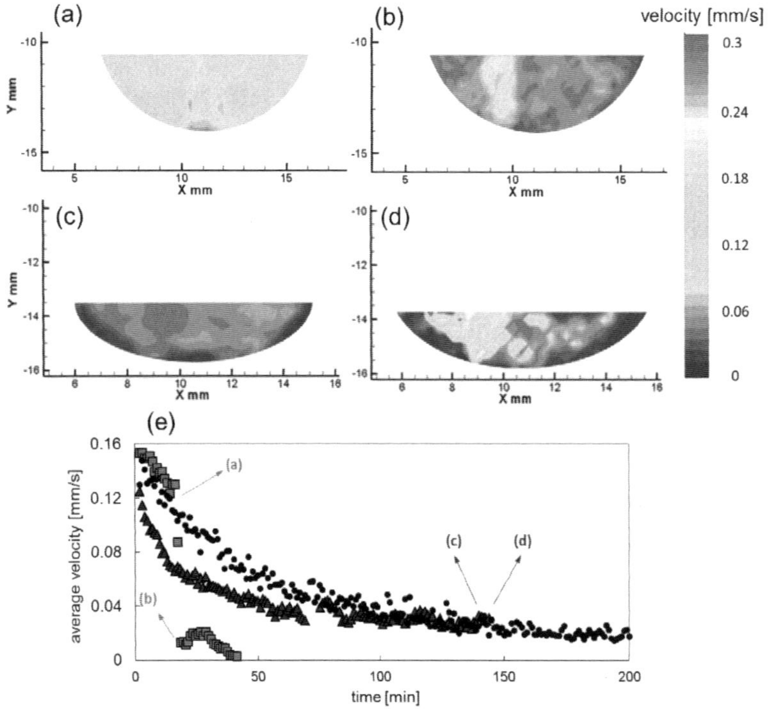

Figure 9. Contour of instantaneous flow velocity for a short induction time experiment 30 seconds before nucleation (a), at onset of nucleation (b) and for a long induction time experiment 30 seconds before nucleation (c) and at onset of nucleation (d). Averaged velocity over time (e). Red squares and blue triangles represent the short and long induction time experiment, respectively. Black circles represent an experiment in which hydrate did not grow over the first six hours.

IOP Conf. Series: Materials Science and Engineering **249** (2017) 012018 doi:10.1088/1757-899X/249/1/012018

4. Conclusions

We characterize the flow velocity field of the bulk of water during CO_2 hydrate formation in an optical cell with a volume of 12 mL filled with 5.5 or 7 mL of deionized water. For that purpose, two different imaging systems are simultaneously used for recording the particle-image fields from the front and the back, respectively. Classical particle image velocimetry (PIV) analysis is performed to obtain the map of velocity vectors. The system on the back, with a higher spatial resolution, is used to focus on the interface where hydrate is expected to form. The system from the front is used to analyze the velocity on the entire bulk of water. We first study the impact of tracers on the CO_2 hydrate formation. Glass spheres of 10 μm seem to act as nucleation spots resulting into a much faster hydrate formation in the bulk of water. Polyamide seeding particles of 5μm, however, displayed little effect on the hydrate growth process and are therefore deemed acceptable as tracer particles. Recirculation flows are noticed as a consequence of density-driven natural convection. We observe a gradual decrease of the averaged velocity with time as a consequence of this phenomenon. We performed a total of 45 experiments varying the volume of water and the pressure of the system, 36 and 56 bar, while the temperature was kept at 273.5 ± 0.5 K. Under those conditions, 26 of the experiments resulted in hydrate formation in less than 48 hours. The induction time varied widely between 15 min and 45 hours in those experiments. Hydrate was nucleate always at the interface. We could not recognize any significant impact of pressure or volume of water on the induction time. However, we identified two well-differentiated hydrate growth patterns depending on the hydrate induction time and the corresponding flow velocity on the bulk of water at that moment. The flow velocity depends on the CO_2 concentration distribution inside water due to Rayleigh convection phenomenon. For induction times above 30 min, CO_2 can be provided from the water phase resulting in rapid growth. Induction times below 30 min resulted in slow growth at the interface creating a solid barrier accompanied by a significant drop in the flow velocity.

References

[1] Sloan E D and Koh A C 2008 *Clathrate Hydrates of Natural Gases* (New York: CRC Press)

[2] Khan M R 2011 *Advances in clean hydrocarbon fuel processing: Science and technology* (Cambridge: Woodhead Publishing Limited)

[3] Matsumoto R, Ryu B J, Lee S R, Lin S, Wu S, Sain K, Pecher I, Riedel M 2011 *Mar. Petrol. Geol.* **28** 1751

[4] Max N D, Johnson A H and Dillon W P 2006 *Economic Geology of Natural Gas Hydrate* (Dordrecht: Springer)

[5] Brewer P G, Friederich G, Peltzer E T and Orr Jr F M 1999 *Science* **284** 943

[6] Genov G, Falenty A and Kuhs *The possible role of CO_2 hydrates for terra-forming processes on Mars (European Geosciences Union 2005)* **7** 05532

[7] Pellenbarg R E, Max M D and Clifford S M 2003 *J. Geophys. Res.* **108** E4 2156

[8] Yang M, Song Y, Ruan X, Liu L, Zhao J and Li Q 2012 *Energies* **5** 925

[9] Castellani B, Filipponi M, Nicolini A, Cotana F and Rossi F 2013 *Journal of Energy and Power Engineering* **7** 883

[10] Kikkinides E S, Yang R T and Cho S H 1993 *Ind. Eng. Chem.* **32** 2714

[11] Scondo A and Sinquin A 2011 *Effect of additives on CO_2 capture from simulated flue gas by hydrates formation in emulsion 7th International Conference on Gas Hydrates*, Edinburgh, Scotland, UK

[12] Babu, P, Linga, P, Kumar, R, Englezos P 2015 *Energy* **85** 261-279

[13] Sivaraman, R 2003 *Gas TIPS News Letter* Fall 2003 Gas Technology Inst

[14] Park, K, Hong, S Y, Lee, J W, Kang, K C, Lee, Y C, Ha M G, Lee, J D 2011 *Desalination* **274** 91-96

[15] Babu, P, Kumar, R, Linga, P 2014 *Desalination* **274** 91-96

[16] Kang, K C, Linga, P, Park, K, Choi, S J, Lee, J D 2014 *Desalination* **353** 84-90

IOP Conf. Series: Materials Science and Engineering **249** (2017) 012018 doi:10.1088/1757-899X/249/1/012018

[17] Li, S, Shen, Y, Dongbing, L, Fan, L, Zhang, Z, Li, W 2014 *Advance Journal of Food Science and Technology* Bd **6 6** 780-783
[18] Li, S, Shen, Y, Dongbing, L, Fan, L, Tan, Z 2015 *Chemical Engineering Research and Design* **93** 773-778
[19] Farajzadeh, R, Hamidreza, S, Zitha, P L J, Bruining, H 2007 *Int. J. Heat Mass Transfer* **50** 5054-5064
[20] Farajzadeh, S, Zitha, P L J, Bruining, H 2009 *Ind. Eng. Chem. Res.* **48** 6423-6431
[21] Khosrokhavar, R, Elsinga, G, Farajzadeh, R, Bruining H 2014 *J. Pet. Sci. Eng.* **122** 230-329
[22] Agudo J R and Wierschem A 2012 *Phys. Fluids.* **093302** 24
[23] Agudo J R, Dasilva S and Wierschem A 2014 *Phys. Fluids.* **053303** 26
[24] Gmelin, L 1973 **8** Auflage Kohlenstoff Teil C3 Verbindungen
[25] Mohammadi A, Manteghian M, Haghtalab A, Mohammadi A H, Rahmati-Abkenar M 2014 *Chemical Engineering Journal* **237** 387395
[26] Nesterov A, Reshetnikov A, Manakov A Y, Rodionova T, Pauk-shtis E, Asanov I, Bardakhanov S, Bulavchenko A 2015 *Journal of Molecular Liquids* **204** 118125
[27] Lirio C. F. S, Pessoa F L P and Uller A M C 2013 *Chem. Eng. Sci.* **96** 118123
[28] Diamond, W L, Akinfiev, N N 2003 *Fluid Phase Equilibria* **208** 265-290
[29] Ohmura R, Shimada W, Uchida T, Mori Y H, Takeya S, Nagao J, Minagawa H, Ebinuma T and Narita H 2014 *Philosophical Magazine* **84** (1) 116
[30] Sun C, Peng B, Dandekar A, Ma Q and Chen G 2010 *Annual Reports Section "C" (Physical Chemistry)* **106** 77100

IOP Conf. Series: Materials Science and Engineering **249** (2017) 012019 doi:10.1088/1757-899X/249/1/012019

Pressure Propagation of Impinging Jet with Cavitation by Numerical Analysis

Daisei KANAMORI[1], Fumihiro INOUE[2], and Yutaka OHTA[3]

[1] Graduate Student, Waseda University, Ookubo Shinjuku Tokyo, Japan, d.kanamori@ruri.waseda.jp

[2] Department of Mechanical Engineering, Shonan Institute of Technology, Tsujidou, Fujisawa, Kanagawa, Japan

[3] Department of Applied Mechanics and Aerospace Engineering, Waseda University, Ookubo Shinjuku Tokyo, Japan

Abstract. In recent years, cavitating jet has attracted attention as an application of water jet technology. In its application, it is important to clarify the jet flow structure and the behavior of bubble cloud collapse. Therefore, in order to visualize the cavitating jet flow structure and elucidate the behavior of collapsing of cavitation bubble clouds, we conducted numerical simulations with gas-liquid two-phase media model. We validated the numerical model by comparing the numerical results with the theoretical and experimental results and had a good agreement. In the case of gas-liquid two-phase free jet, cavitation bubble clouds emit periodically and transfer at a regular speed. And some bubble clouds merge with a preceding bubble clouds. Comparing with liquid single-phase jet, the core region is maintained to the further downstream and we show the usefulness of the cavitating jet. In the case of gas-liquid two-phase impinging jet, after a cavitation bubble cloud collides with wall, it is broken by applying pressure and generates a shock wave. At this time, the impact pressure becomes maximum. Thereafter, the shock wave affects other cavitation bubble clouds and break these. The collapsed cavitation bubble cloud rebounds and collapses again near the collision wall surface.

1. Introduction

Water jet in air or water is widely used in various situations such as cleaning, excavation, cutting, and destruction due to collision effect on the object [1, 2, 3, 4, 5, 6]. In recent years, cavitating jet in water has attracted attention as an application of water jet technology. Cavitation is capable of generating a very high impact pressure in the collapse of cavitation bubbles and has been applied to new fields such as decomposition of organic compounds and sterilization of sewage. In the case of cavitating jet, shear layer is generated near the nozzle outlet and velocity gradient changes abruptly between a high flow velocity region due to the high-pressure water and a surrounding fluid. And an unsteady vortex is generated in the shear layer. When the pressure inside the vortex decreases to the saturated vapor pressure of the fluid, vortex cavitation occurs. Vortex cavitation develops as it moves downstream, or it coalesces with other vortex cavitations to form a cavitation bubble cloud. It is known that the cavitation bubble cloud is periodically released and exhibits unsteady behavior.

In application of cavitating jet, it is important to clarify jet structure of jet and behavior of bubble cloud collapse. Until now many experimental researches have been conducted concerning the shape and size of the nozzle, the standoff distance, and the cavitation number etc. [7]. Other Studies have also been made to elucidate the jet structure and breakdown of the cavitation bubble cloud using the Schlieren method, high-speed video camera, and PIV measurement [8, 9]. However, due to the influence of the cavitation bubble cloud, it is very difficult to observe its internal structure.

In this study, we conducted a numerical simulation with introducing a gas-liquid two-phase local homogeneous medium model which is effective for unsteady numerical analysis to visualize the jet structure of the cavitating jet and clarify the behavior of the bubble cloud collapse and to evaluate the impact pressure by the collision jet on the wall.

2. Computational Procedure

2.1. gas-liquid two-phase local homogeneous medium model

In this study, the gas-liquid two-phase local homogeneous medium model developed by Okuda et al. was used [10]. This model treats the gas-liquid two-phase medium as a homogeneous pseudo-single-phase medium with finite voids locally. That is, it is assumed that innumerable and infinitely fine bubble particles or droplet particles are homogeneously distributed in the microvolume element in the gas-liquid two-phase medium. With the assumption of local homogeneity, the mixed density ρ of the two-phase medium is calculated as follows by using the gas phase density ρ_g, the liquid phase density ρ_l and the local void fraction (volume fraction) α.

$$\rho = (1-\alpha)\rho_l + \alpha\rho_g \tag{1}$$

We use an equation for the state of an ideal gas for the gas phase and Tammann's state equation for the liquid phase.

$$\begin{aligned} p_g &= \rho_g R_g T_g \\ p_l + p_c &= \rho_l K_l (T_l + T_0) \end{aligned} \tag{2}$$

where p and T are pressure and temperature. Also, R_g = 287.0 J / (kg / K), K_l = 472.2 J / (kg / K), p_c = 1944.6 MPa, and T_0 = 3836.93 K are the gas constant, the liquid constant, the liquid pressure constant, and temperature constant. Suppose that the pressure and temperature of the gas and liquid phases are in equilibrium, the equation of state of the homogeneous medium is written as follows.

$$\rho = (1-\alpha)\frac{p + p_c}{K_l (T + T_0)} + \alpha \frac{p}{R_g T} \tag{3}$$

Here, the following relational expressions hold between the local void fraction α and the quality (mass fraction of gas phase) Y.

$$\rho(1-Y) = (1-\alpha)\rho_l \qquad \rho Y = \alpha\rho_g \tag{4}$$

From these expressions, the state equation of the gas-liquid two-phase medium is given as follows using Y.

$$\rho = \frac{p(p + p_c)}{K_l (1-Y) p (T + T_0) + R_g Y (p + p_c) T} \tag{5}$$

2.2. Governing Equations

The governing equations are an equation of continuity for a gas-liquid two-phase medium, a 2-D compressible gas-liquid two-phase Navier-Stokes equation assuming an isothermal process, and an equation of continuity for a gas phase. Equations are discretized by a finite volume method. When the above three equations are summarized in vector form, these can be written as follow.

$$\frac{\partial \boldsymbol{Q}}{\partial t}+\frac{\partial \boldsymbol{E}}{\partial x}+\frac{\partial \boldsymbol{F}}{\partial y}=\frac{\partial \boldsymbol{E}_{\mathrm{v}}}{\partial x}+\frac{\partial \boldsymbol{F}_{\mathrm{v}}}{\partial y} \tag{6}$$

$\boldsymbol{Q}$ is a vector of conserved quantity, $\boldsymbol{E}$ and $\boldsymbol{F}$ are flux vectors representing convection terms, and $\boldsymbol{E}_{\mathrm{v}}$ and $\boldsymbol{F}_{\mathrm{v}}$ are flux vectors representing viscous terms, which are given as follows.

$$\boldsymbol{Q}=\begin{bmatrix}\rho\\ \rho u\\ \rho v\\ \rho Y\end{bmatrix}\quad \boldsymbol{E}=\begin{bmatrix}\rho u\\ \rho u^2+p\\ \rho uv\\ \rho uY\end{bmatrix}\quad \boldsymbol{F}=\begin{bmatrix}\rho v\\ \rho uv\\ \rho v^2+p\\ \rho vY\end{bmatrix}\quad \boldsymbol{E}_v=\begin{bmatrix}0\\ \tau_{xx}\\ \tau_{yx}\\ 0\end{bmatrix}\quad \boldsymbol{F}_v=\begin{bmatrix}0\\ \tau_{xy}\\ \tau_{yy}\\ 0\end{bmatrix}$$

$$\tau_{xx}=\frac{2}{3}\mu\left(2\frac{\partial u}{\partial x}-\frac{\partial v}{\partial y}\right)\quad \tau_{yy}=\frac{2}{3}\mu\left(2\frac{\partial v}{\partial y}-\frac{\partial u}{\partial x}\right)\quad \tau_{xy}=\tau_{yx}=\mu\left(\frac{\partial u}{\partial y}+\frac{\partial v}{\partial x}\right) \tag{7}$$

Where u and v are velocity components in the X and Y directions, respectively. And τ is a viscous stress. It should be noted that the viscosity coefficient μ of the gas-liquid two-phase local homogeneous medium is evaluated by the following equation to calculate the viscous stress.

$$\mu=(1-\alpha)(1+2.5\alpha)\mu_{\mathrm{l}}+\alpha\mu_{\mathrm{g}} \tag{8}$$

The convective flux was evaluated using flux difference splitting (FDS), and extended to a third order using the monotonic upwind scheme for conservative laws (MUSCL) interpolation. The viscous flux was discretized as the second-order central difference with Gauss' theorem. For the time integration second-order Runge–Kutta method was used.

2.3. Computational Domains and Boundary Conditions

In this study, 2-D free jet field and impinging jet field are analysed. And axial symmetric flow is assumed. The nozzle orifice diameter D is 1.0 mm and the minimum grid width is 0.1 mm at the nozzle orifice portion. Overviews of the computational domains are shown in figure 1, and the grids are thinned for easy viewing.

An overview of the computational domain used for free jet field analysis is shown in figure 1 (a). The computational domain is taken from 15D upstream to 60D downstream with respect to the nozzle exit and the width is 20D in the radial direction. A multiblock grid is adopted, and the computational domain is divided into three regions, a nozzle front region, a nozzle orifice region, and a jet flow region. The number of grid points in the axial direction and radial direction in each region are 55 × 50, 55 × 15, and 355 × 305, respectively. Therefore, the total number of grid points is about 11 million points. For free jet analysis, analysis was conducted under two conditions. In the case of the liquid single-phase state, the nozzle inlet pressure p_1 = 0.2 MPa and the outlet pressure p_2 = 101.325 kPa. The cavitation number σ (= (p_2 - p_v) / (p_1 - p_2)) is calculated to be 1.0. In the case of the gas-liquid two-phase state, the nozzle inlet pressure p_1 = 1.0 MPa and the outlet pressure p_2 = 101.325 kPa, cavitation number σ is calculated to be 0.11. At the inflow boundary, total pressure was fixed and free inflow conditions was set. At the outflow boundary, static pressure was fixed and free flow condition was set. At the wall boundary, adhesive condition was applied to velocity and von Neumann condition was imposed on density and pressure.

An overview of the computational domain used for impinging jet flow analysis is shown in figure 1 (b). Unlike the free jet analysis, a collision wall with a width of 10D was set at the position of the standoff distance X/D = 10D which is the first peak of the erosion curve. The computational domain is divided into four regions, a nozzle front region, a nozzle orifice region, a collision wall front region, and a collision wall upper region. The number of grid points in the axial direction and radial direction in each region are 55 × 50, 55 × 15, 105 × 305, and 255 × 105, respectively. Therefore, the total number of grid points is about 6.2 million points. In the impinging jet analysis, analysis was carried out only in the gas-liquid two-phase state with the cavitation number σ = 0.11. The boundary conditions were the same as free jet analysis.

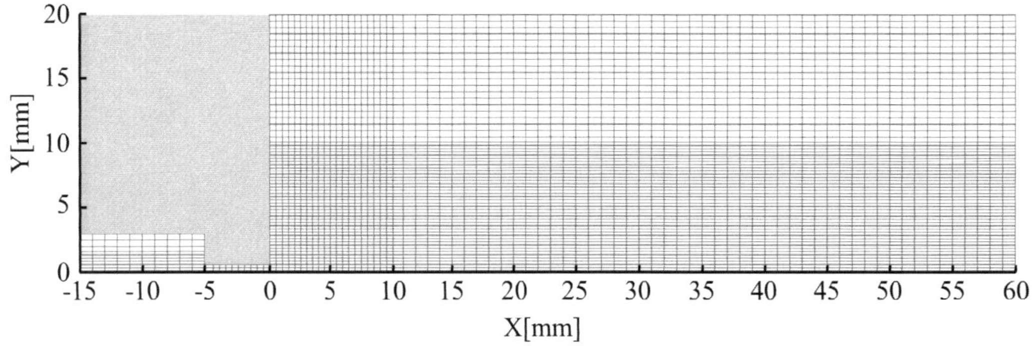

(a) Free Jet

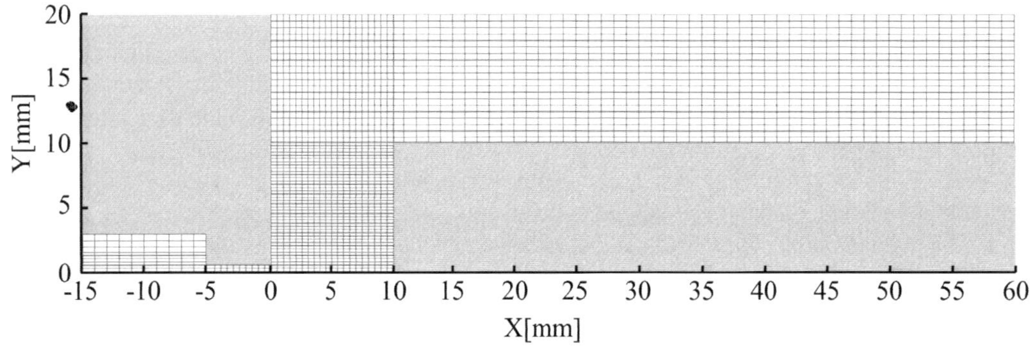

(b) Impinging Jet

Figure 1. Overview of the computational domain

3. Results and Discussion

3.1. Free Jet Analysis

First we confirm the validity of the analysis model by comparing with Geortler's and Tollmien's theoretical curves [11, 12]. The velocity distribution in the cross section at each standoff distance X/D is shown in figure 2. In this graph, the distance Y from the jet axis and the axial flow velocity are nondimensionalized by the half-width of the jet b and the maximum axial flow velocity u_m. b is the distance from the jet axis where the axial flow velocity u is equal to $0.5u_m$. In the case of the liquid single-phase state (figure 2 (a)), the velocity distribution for each cross section shows a similar distribution and is in good agreement with Geortler's and Tollmien's theoretical curves. On the other hand, in the case of gas-liquid two-phase state (figure 2 (b)), velocity distributions have quadratic curve-like distribution, but it cannot be said that they are similar to each other. Differences from the theoretical curves are found both in the jet flow inner and outer regions, and the difference is larger in the outer region ($1.0 < b$). This is thought that flow becomes complicated due to a vortex cavitation.

Figure 3 shows the axial flow velocity distribution of the jet flow field. Jet flows from the left to the right of the page. According to figure 3 (a), it can be said that the flow diffuses stable. In addition, the initial length of the initial region *Xc* is about 7.0 mm. On the other hand, in the case of gas-liquid two phase state (figure 3 (b)), the jet does not show a stable diffusion appearance and the jet width is suppressed to about 2.0 mm. It is considered that shear vortices are formed at the outer edge of the main jet. Further, the initial length *Xc* gets longer to about 20.0 mm and the core region is maintained further downstream. In other words, it can be said that the usefulness of the cavitating jet was demonstrated such that jet energy can be propagated to the downstream by using the cavitating jet and the standoff distance *X/D* can be lengthened.

With respect to the half-width of the jet *b*, we can write $b = C_2X$ from the theoretical equation of the jet. The slopes C_2 obtained from Geortler's and Tollmien's theoretical solutions are 0.114 and 0.097. In addition, Rajaratnam set $C_2 = 0.10$ as an experience value. In the liquid single-phase analysis, the slope C_2 was calculated to be 0.103, and the validity of this analysis model could be confirmed quantitatively.

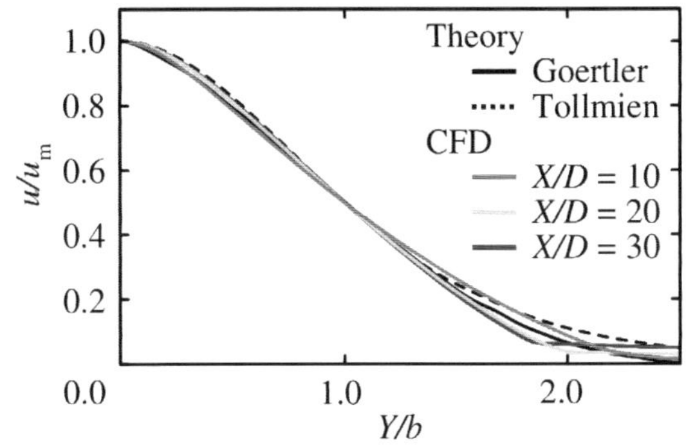

(a) Non-cavitating Jet ($\sigma = 1.0$)

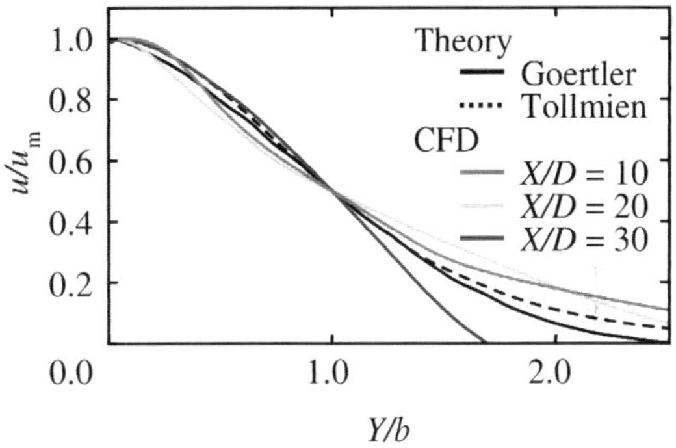

(b) Cavitating Jet ($\sigma = 0.11$)

Figure 2. Velocity distribution in the cross section at each standoff distance *X/D*

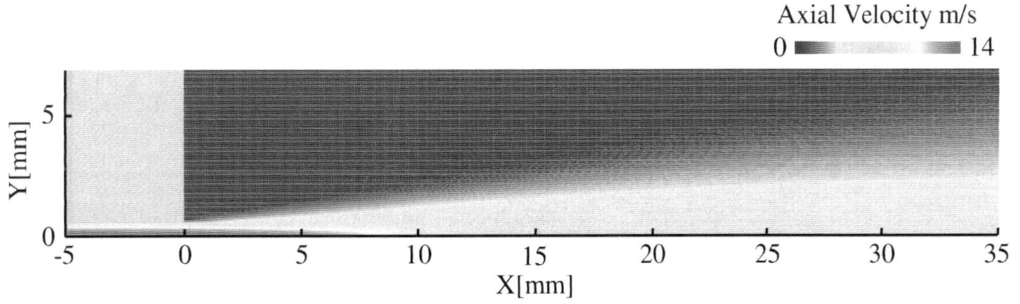

(a) Non-cavitating Jet ($\sigma = 1.0$)

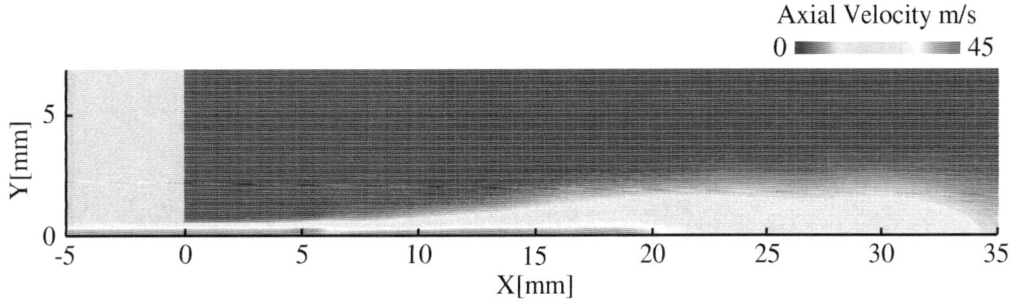

(b) Cavitating Jet ($\sigma = 0.11$)

Figure 3. Velocity distribution of Non-cavitating and cavitating jet

IOP Conf. Series: Materials Science and Engineering **249** (2017) 012019 doi:10.1088/1757-899X/249/1/012019

Figure 4 shows the time course of the pressure distribution in the range up to the standoff distance *X/D* = 30 in order to clarify the periodic behavior of the cavitation bubble cloud released from the nozzle exit. The pressure at the deep blue part reaches saturated vapor pressure of water, 2332 Pa, which indicates that cavitation occurs. Looking at the pressure distribution, we can see the cavitation bubble clouds occur periodically. The generated cavitation bubble clouds advect at a substantially constant speed like the path A, C, and E. The advection velocity is the same as the main stream velocity at that point. It has also been observed that some cavitation bubble clouds catch up with the bubble clouds generated earlier like the path D and F and coalesce.

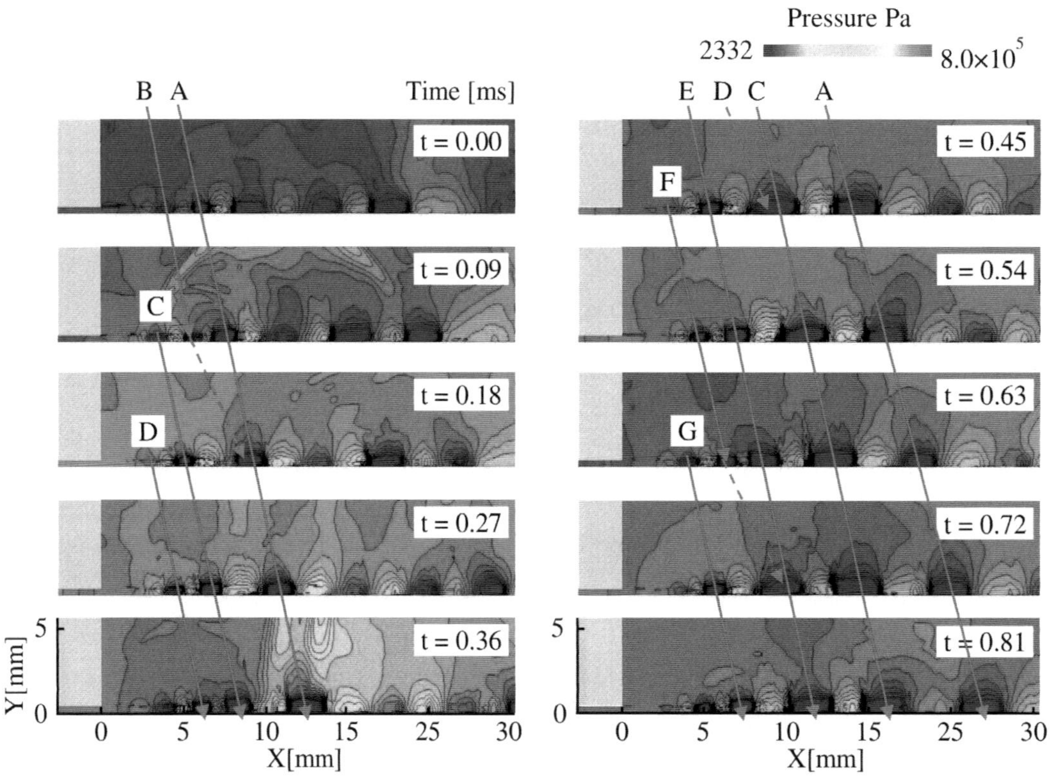

Figure 4. Time course of the pressure distribution

3.2. Impinging Jet Analysis

Figure 5 shows absolute velocity distribution and pressure distribution. Please be aware that the length of the flow velocity vector on figure 5 (a) is corrected to be a uniform length, not shows the actual magnitude of the velocity. From figure 5 (a), the jet collides with the wall surface at the standoff distance *X/D* = 10 and flows along the wall as a wall jet. Poreh and Cermak [13], Bradshow and Love [14], Tani and Komatsu [15], Poreh et al. [16], and Beltaos and Rajaratnam [17] made the wall impinging jet experience and reported that the wall impinging jet radially spreads along the wall surface as a radial wall jet after hitting the wall. However, in this analysis, the jet didn't diffuse radially. This is thought to be that the turbulence model is not used in this analysis, that is, mixing and diffusion action by turbulence are not taken into consideration and cavitation. Also, looking at the pressure distribution (figure 5 (b)), cavitation bubble clouds are periodically generated like the free jet flow field. Cavitation bubble clouds also exist after the jet collides with the wall surface.

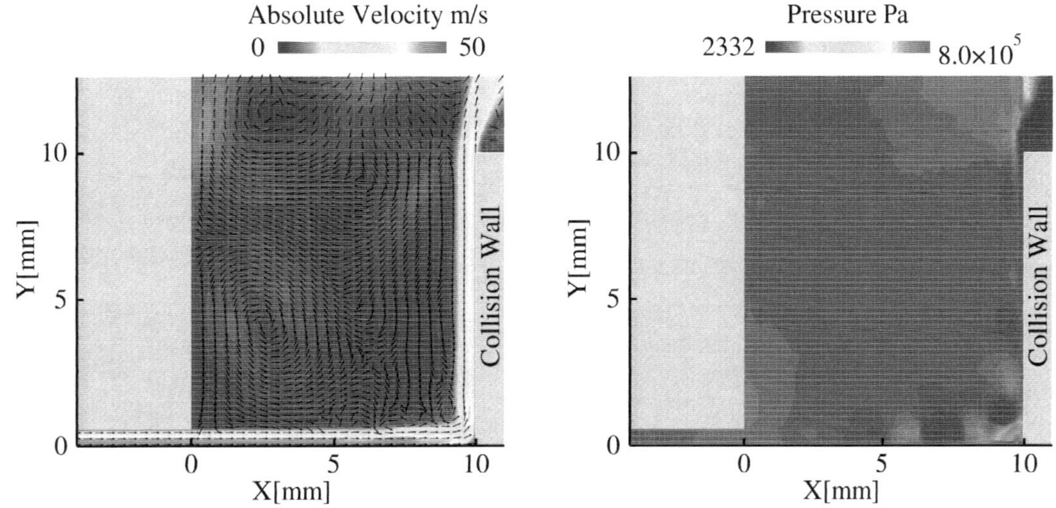

(a) Absolute velocity distribution with vector (b) Pressure distribution

Figure 5. Flow pattern of impinging jet

Next, we investigated the propagation of the shock wave and pressure wave caused by the cavitation bubble cloud collapse when the cavitation jet collides and to evaluate the impact pressure on the wall surface. Figure 6 shows the time course of the pressure distribution, figure 7 shows the pressure distribution along the Y axis on the collision wall. Unlike the free jet flow field, cavitation bubble clouds are adhering to the wall without coalescence (figure 6). At the time t = 0, the cavitation bubble cloud collides, the pressure rises, and the void fraction decreases. A decrease in the void fraction means an increase in sound velocity, and the region transits from the supersonic region to the subsonic region. And then a shock wave is generated and the bubble cloud collapses (t = 2). At this time, the impact pressure increases to the maximum of 1.4 MPa (figure 7). The generated shock waves propagate with collapsing other bubble clouds and generate new shock waves and pressure waves. However, shock waves and pressure waves are strong in the direction along the wall surface and can be seen to be propagating with avoiding bubble clouds existing on the upstream side (t = 4). They propagate further upstream (t = 6, 8) and extend to the interior of the nozzle (t = 10, 12). Since new cavitation bubble clouds are released after the shock waves and pressure waves arrive at the nozzle part, it is suggested the relationship between the arrival of these waves and the emission of new cavitation bubble clouds. After that, the pressure fluctuates gets smaller, the crushed cavitation bubble clouds raise the void fraction again due to the rebound phenomenon, and become cavitation bubble cloud (t = 14 ~ 20). Ultimately, it collapses again on the collision wall surface and generates a shock wave (t = 22 ~ 26). In general, it is known from the erosion test results that erosion marks made by cavitation jet have ring-shaped, suggesting that this cavitation bubble cloud collapse pattern contributes to the formation of ring-shaped erosion marks. After that, periodic unsteady behavior is shown as well.

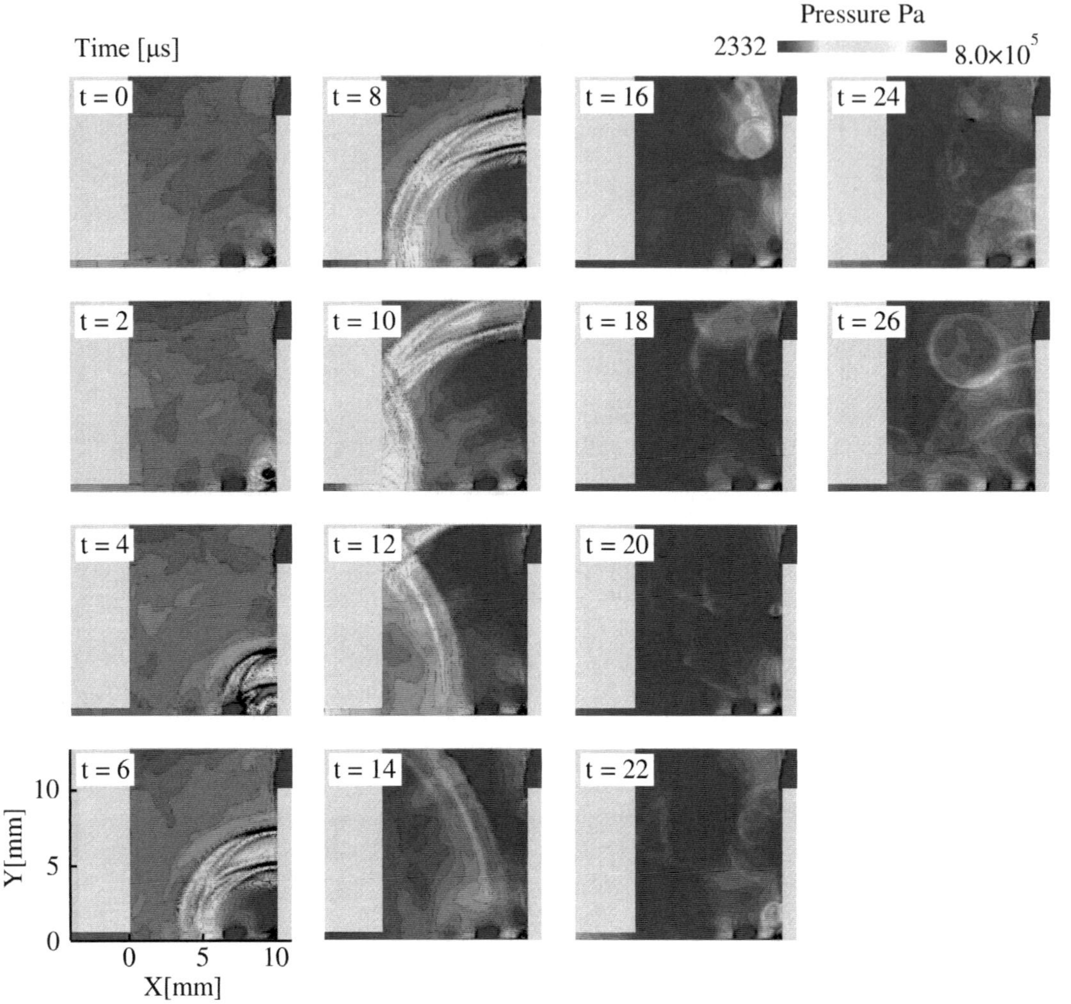

Figure 6. Time course of pressure distribution

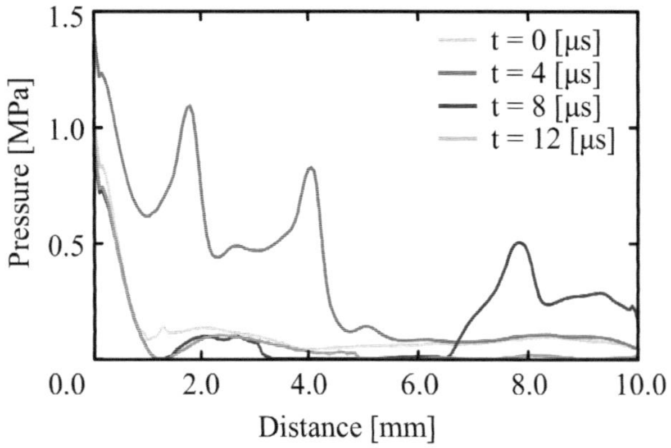

Figure 7. Pressure distribution on the collision wall

4. Conclusions

We introduced the gas-liquid two-phase local homogeneous medium model and conducted a numerical simulation to visualize the jet structure of the cavitating jet and clarify the behavior of the bubble cloud collapse and to evaluate the impact pressure by the collision jet on the wall.

The results can be summarized as follows:

(1) The numerical simulation results of the free jet flow field in the liquid single-phase state and the gas-liquid two phase state using the gas-liquid two-phase local homogeneous medium model showed good agreement with the theoretical solution and experimental results. The effectiveness of the simulation model could be shown.
(2) Cavitating jet had smaller jet width than water jet in liquid single-phase state, and it was possible to propagate jet energy further downstream. Cavitation bubble clouds were periodically emitted and advected at constant speed. Some bubble clouds coalesced with the preceding bubble clouds.
(3) Cavitation bubble clouds also showed cyclic behavior even in collision jet field, and they collapsed at wall collision and generated shock waves and pressure waves. It is suggested that these waves correlate with cavitation bubble cloud generation.
(4) After bubble clouds collapsed at the wall, generated shock waves again near the collision wall due to the rebound phenomenon. It is thought that the characteristic ring-shaped erosion marks in the erosion test is formed by these impact pressures.

References

[1] Yamauchi A and Shimizu S 1987 Erosion Due to Impingement of Cavitating Jet *J.Fluids Engineering* **109** 442-447
[2] Soyama H, Yamaguchi Y, Adachi Y, Sato K, Shinoda T and Oba R 1993 High-Speed Observation of the Cavitation Cloud around a High-Speed Submerged Water-Jet *Transactions of the Japan Society of Mechanical Engineers Series B* **59** (562) 1919-1924
[3] Kato H 2003 Environmental Protection Using Cavitation *Technology Turbomachinery* **31** (11) 648-654
[4] Tsujimura 1997 *J. Jet flow engineering* **14** (2) 21-25
[5] Yokota 1999 *J. Jet flow engineering* **16** (1) 17-22
[6] Soyama 1996 *J. Jet flow engineering* **13** (1) 25-32
[7] Shimizu S 1996 Effects of nozzle shape on structure and drilling capability of premixed abrasive water jets *Japan Society of Mechanical Engineers Series B* **62** (603) 84-91
[8] K K Ooi 1985 *J. Fluid Mech.* **151** 467-390
[9] Oguma Y, Peng G and Shimizu S 2017 *J. Jet flow engineering* **33** (1) 10-18
[10] Okuda K and Ikohagi T 1996 Numerical Simulation of Collapsing Behavior of Bubble Clouds *Japan Society of Mechanical Engineers Series B* **62** (303) 38-43
[11] Nomura Y 1981 *Jet flow* (Tokyo: Morikita Publishing Co.)
[12] Sakouchi T 2004 *Jet flow* (Tokyo: Morikita Publishing Co.)
[13] Poreh M and Cermak J E 1959 Flow characteristics of a circular submerged jet impinging normally a smooth boundary *Porc. 6th Mid-Western Conf. on Fluid Mechanics* 198-212
[14] Bradshaw P and Love EM 1961 The normal impingement of circular air jet on a flat surface (London: Aeronautical Research Council R and M) **3205** 1-8
[15] Tani I and Komatsu Y 1964 Impingement of around jet on a flat surface *Proc. 11th Conge. Appl. Mech. Munich* 672-676
[16] Poreh M, Tsuei, Y G & Cermak and J E 1967 Investigation of a turbulent radial wall jet *Trans. ASME J. Appl. Mech.* **34** 457-463
[17] Beltaos S and Rajaratnam N 1973 Impinging circular turbulent jets *Proc. ASCE J. Hydraul Div.* **100** 1313-1328
[18] Fukushi A, Shin B R, Sato K and Ikohagi T 2000 Numerical Simulation of High-Speed

Submerged Water Jet with Cavitation *J. Jet flow engineering* **17** (3) 16-21
[19] Fukushi A, Nakamori I and Ikohagi T 2002 Numerical Simulation of High-Speed Submerged Water Jet with Cavitation *Japan Society of Mechanical Engineers Series B* **68** (673) 46-53
[20] Inoue F, Outa E, Tajima K and Machiyama T 1993 An experimental Study on Control Valve Cavitation (3rd Report, Vortex Shedding and Cavitation Inception on a Contourde Plug) *Japan Society of Mechanical Engineers Series B* **59** (562) 131-138
[21] Sato K, Ohjimi S and Sugimoto Y 2009 Collapsing and Impulsive of Cavitation Clouds on Cavitating Water-Jet Impinging on Solid Wall *Japan Society of Mechanical Engineers Series B* **75** (750) 47-56

IOP Conf. Series: Materials Science and Engineering **249** (2017) 012020 doi:10.1088/1757-899X/249/1/012020

Aerodynamic characteristics and heat radiation performance of sportswear fabrics

H Koga[1], M Hiratsuka[2], S Ito[2] and A Konno[2]

[1],Graduate School of Kogakuin University, 1-24-2 Nishi-Shinjuku, Shinjuku-ku, Tokyo, 163-8677 Japan.
[2] Kogakuin University, 1-24-2 Nishi-shinjuku, Shinjuku-ku, Tokyo, 163-8677,Japan,;

ito@cc.kogakuin.ac.jp

Abstract. Sports such as swimming, speed skating, and marathon are sports competing for time. In recent years, reduction of the fluid drag of sportswear is required for these competitions in order to improve the record. In addition, sweating and discomfort due to body temperature rise during competition are thought to affect competitor performance, and heat radiation performance is also an important factor for sportswear. The authors have measured fluid force drag by wrapping cloth around a cylinder and have confirmed their differences due to the roughness of the fabric surface, differences in sewing. The authors could be verified the drag can be reduced by the position of the wear stitch. This time, we measured the heat radiation performance of 14 types of cloths whose aero dynamic properties are known using cylinders which are regarded as human fuselages, and found elements of cloth with heat radiation performance. It was found to be important for raising the heat radiation performance of sportswear that the fabric is thin and flat surface processing.

1. Introduction

With the advent of a new type of swim suit represented by a laser racer in recent years, the field of sportswear development attracts attention. Although the swim suit was banned, it proved that the time record was shortened by sportswear. It is important to pursue improvement of sportswear performance in time competing games. Due to its hydrodynamic performance, the swim suit mentioned above was also applied to various kinds of sporting events due to its functionality [1]. Improvement in the performance of sportswear that suits within the regulations is pursued in sports competing for recording not only for swimming events. Since sports motion involves a reciprocating action, Luth et al. [2] measures the hysteresis of fabric drag using a cylinder, and the rough material is said to have less drag reduction and less hysteresis. Chowdhury et al. [3][4] predicted the drag of the human body by measuring ski jumping and bicycle attitude aerodynamic position mainly and treating the human body as a combination of cylinders and measuring the drag of the inclined cylinder that is composed.

In order to find a sportswear with good aerodynamic characteristics and heat radiation performance, we measured the fluid force of several kinds of cloths using cylinder regard as a human torso, and also checked the drag at actual wearing using a mannequin did. The influence on arm torso was also measured by PIV. Furthermore, perspiration and discomfort due to body temperature rise during competition are thought to affect competitor performance, and heat radiation performance is also an important factor for sportswear, so we confirmed the heat radiation performance as well.

IOP Conf. Series: Materials Science and Engineering **249** (2017) 012020 doi:10.1088/1757-899X/249/1/012020

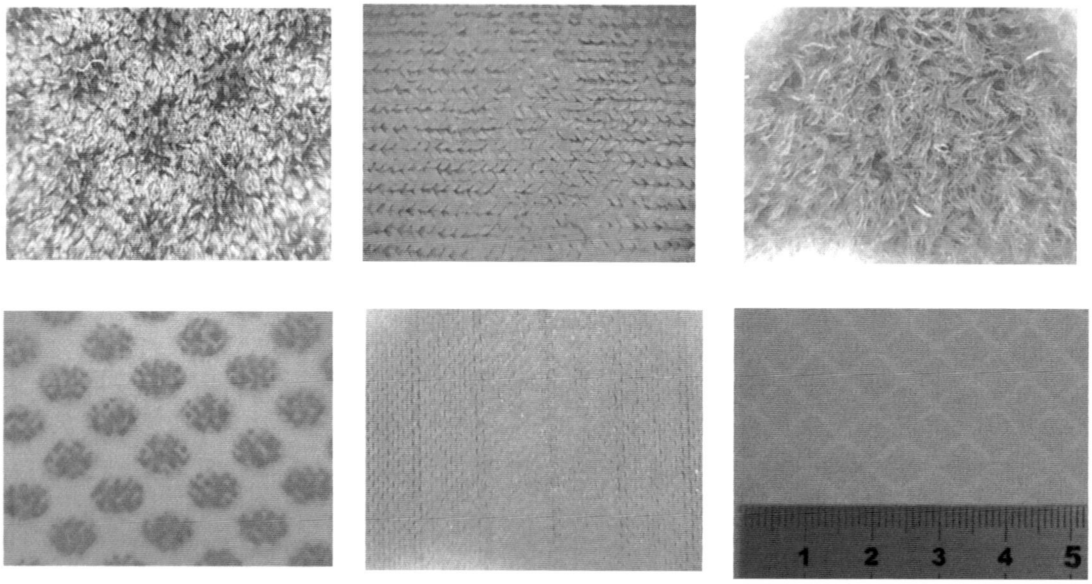

Figure 1. Examples of Fabrics investigated on aerodynamic properties

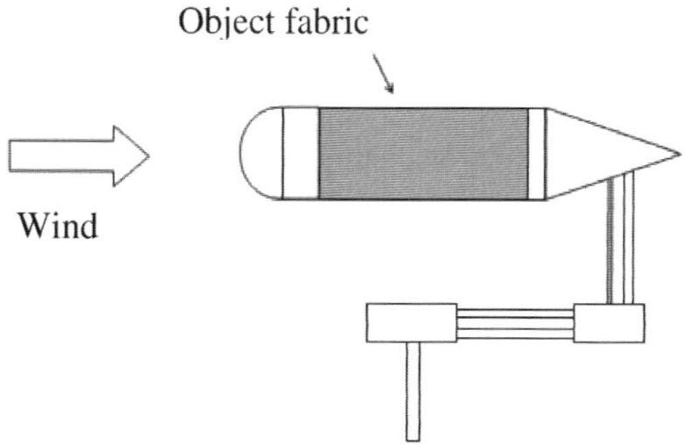

Figure 2. Pressure drag measurement system **Figure 3.** Friction drag measurement system

2. Experiment

A part of the sample cloth materials is shown in Figure 1. They were composed of different fabrics (woven fabric, knitted fabric), different surface roughness, fabrics with different surface finish.

2.1 Fluid-Force measurement

There are two types of drag: pressure drag by shape and friction drag by surface friction. Regarding the pressure drag which varies according to the flow pattern, a phenomenon called drag crisis in which the drag coefficient decreases abruptly as the Reynolds number, *Re*, increases in a uniform flow past a sphere or a cylinder is known. Achenbach [5], [6] conducted wind tunnel experiments on spherical balls, *Re* at which drag crisis occurs (Critical *Re* number) was about 3×10^5. In addition, it was found that the critical *Re* decreases by giving roughness to the sphere [5]. By using this mechanism, it is possible to grasp minute changes in fabrics as drag change. This pressure drag is proportional to the square of velocity. On the other hand, the friction drag is also related to the roughness of the surface and is proportional to the magnitude of the velocity. Using a low turbulence wind tunnel with a rectangular cross section of 380 mm × 380 mm, we changed the wind velocity from 6 to 30 m/s every 1 m/s in order to change the Reynolds number, and we measured the fluid drag of poly vinyl chloride, PVC, pipes with ϕ114mm, 240mm tall. Figure 2 shows the state of a vertical support type experimental apparatus for

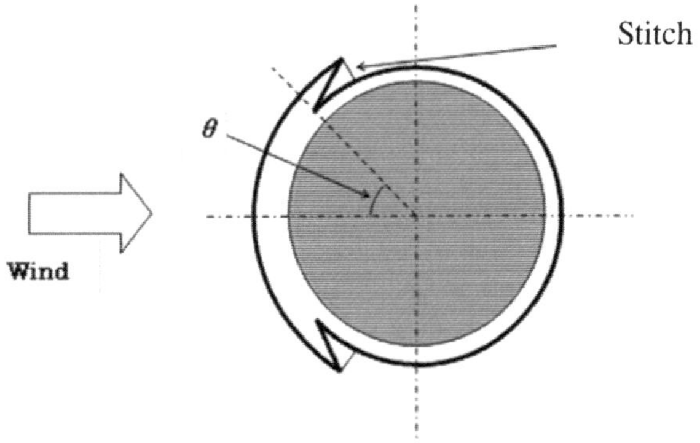

Figure 4. Outline of stitch position

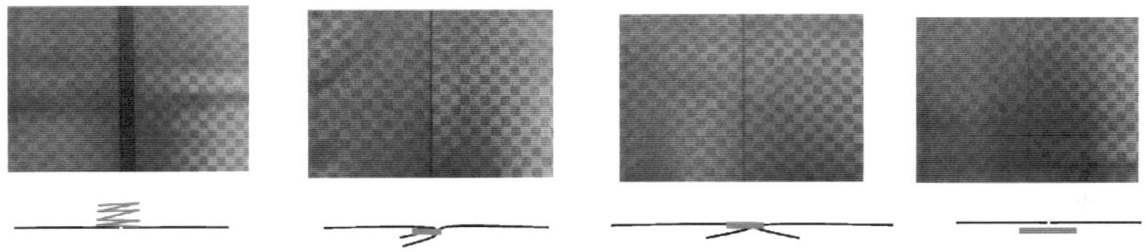

Figure 5. Seam/Stich of cloth

pressure drag, and Figure 3 shows a cobra type experimental apparatus for friction drag. In the friction drag measurement, the friction drag of the cloth was measured by removing the tare drag of the setting jig in order to measure the friction of only the cylinder to which the cloth was attached.

2.2 Stich position and its shape
In the pressure drag, the drag by different sewing positions in the same fabric wrapped around a cylinder as shown in Figure 4 was measured, and the drag which changed according to the difference of the stich shape as shown in Figure 5 was confirmed. Furthermore, the pressure drag of the cylinder with false stitch attached by a thick thread as shown in Figure 6 was also measured.

2.3 Mannequin Experiment
As shown in Figure 7, using a Goettingen type large wind tunnel facility with a size of 1800 mm × 1800 mm, 0.1% maximum wind speed of 56 m/s, the pressure drag of a full-size mannequin was measured, dressed in whole-body tights with false stitchs as 2, 5 thread of ϕ0.7mm thick flatly or stereoscopically, with masked tape at the curved side surface.

2.4 Experiments on cylinders with arms
Fluid force measurement and PIV measurement were carried out on a cylindrical model with arm on which cloth was attached, reflecting the result of using a human body model. By connecting two small cylinders of ϕ30 mm to PVC cylinder of ϕ114 mm shown in Figure 8, the cylinder with arms model simulates the body trunk with arm. The cloth was attached on them and the fabric having the lowest

IOP Conf. Series: Materials Science and Engineering **249** (2017) 012020 doi:10.1088/1757-899X/249/1/012020

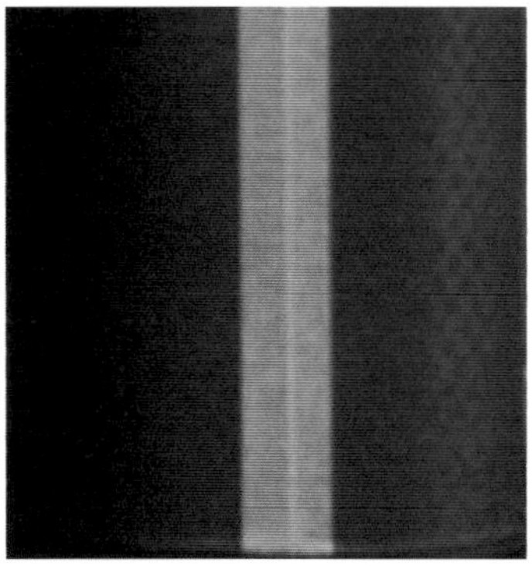

Figure 6. A thread with adhesive tape simulating a stitch

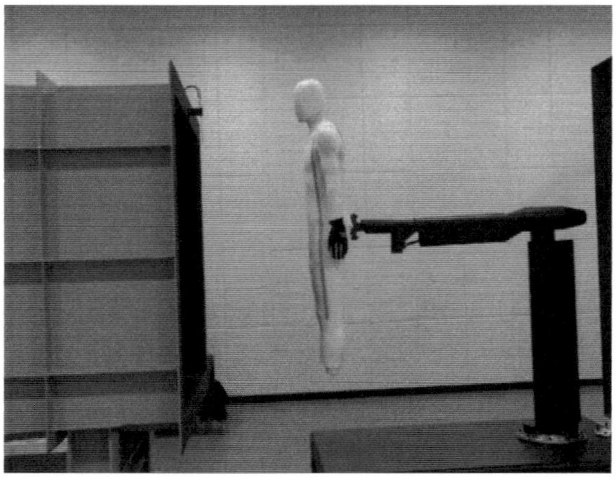

Figure 7. Full-size mannequin with false stitches at wind tunnel facility

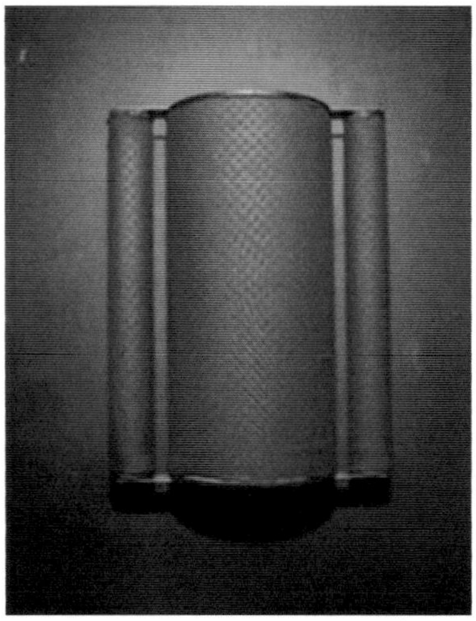

Figure 8. Test cylinder with arms

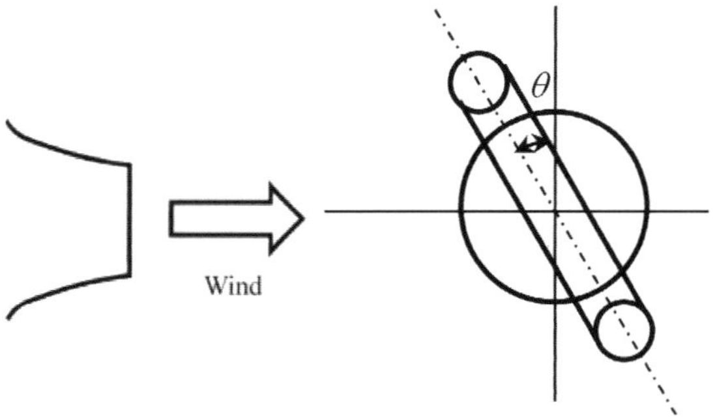

Figure 9. Definition of slant angle θ

drag was used, and the stitch position with the lowest pressure drag was made at an tilt angle θ of 60° as shown in Figure 9 with respect to the wind direction. PIV measurements were made at 10 m/s and 27 m/s before and after drug crisis based on the drag test results. In the PIV measurement, an average image obtained by translating the light source and the camera in order to cover the narrowness of the area was combined.

2.5 *Heat radiation experiment*

14 types of cloths were attached around the cylindrical model partly shown in Figure 1 and the ground plate was placed at the lower end. Heat radiation performance measurement was carried out using infrared thermography camera (R 500 EX) thermography of Nippon Avionics Co., Ltd. In the cylindrical model, a silicone rubber heater was wrapped around a vinyl chloride cylinder of ϕ 114 mm, and cloth was stuck thereon. The heat radiation performance test was conducted in a state where the room

Figure 10. Thermal Experimental setup

temperature was controlled to 22 °C by air conditioning. By setting the voltage to 20.4 Volt using the direct current stabilized power supply, the average surface temperature of the heater was set to 40 °C, at first. A cover was placed on the test cylinder and the energization was cut off in order to keep its temperature. After driving the wind tunnel, when reaching the prescribed speed, i.e. 3, 6, 9, 12, 15 m/s, remove the cover and energize the heater. The ventilation time for each wind speed was shown in Table 1. And the surface temperature was photographed by thermography. Regarding the ventilation time, it was set the time when the heater without cloth became below 35 °C as shown in Table 1. The ventilation time was different for each wind speed.

Table 1. Wind velocity and ventilation time

Wind velocity [m/s]	3.0	6.0	9.0	12.0	15.0
Ventilation time [s]	90	75	60	45	30

3. Result and Discusion

The drag coefficient C_D, the friction drag coefficient C_f and the Reynolds number R_e are represented in Eqs. (1), (2) and (3). A: section area of the cylinder [m^2], d: diameter of the cylinder [m], ν: kinematic viscosity of air [m^2/s], D: drag force [N], D_f: drag force by friction[N], ρ: air density[kg/m^3], U: velocity of uniform flow[m/s].

$$C_D = \frac{D}{\frac{1}{2}\rho U^2 A} \tag{1}$$

$$C_f = \frac{D_f}{\frac{1}{2}\rho U^2 A} \tag{2}$$

$$Re = \frac{Ud}{\nu} \tag{3}$$

3.1 Fluid-Force measurement

3.1.1 Pressure drag coefficient C_D

Figure 11 shows some results of pressure drag. The drag crisis could not be seen on the base material of the PVC pipe which has a smooth surface shown in the curve of Tare. In some fabrics, drag crisis has occurred from the area of $Re = 0.4 \times 10^5$,i.e. curve "a", depending on the roughness of the fabric. As the *Re* number increases, that is, as the speed increases, the laminar boundary layer transitions to the turbulent boundary layer, and the separation point shifts backward along with the cylinder surface, and the drag reduction occurs in the region with low *Re* in this region. The value of C_D after the criticality gradually increased with the roughness of the surface was seen.

3.1.2 Friction drag coefficient C_f

The surface of PVC pipe which has smooth surface showed the lowest value in Figure12. Fabrics except Fabric "a" have to gradually rise as the speed rises. The "a" has a fabric with relatively high roughness, but due to the unique roughness shape C_f was going down as the velocity rose. The friction drag increases as transitioning from the laminar boundary layer to the turbulent boundary layer. The C_f of the fabric similar to the smooth surface is low, and the C_f of the texture transitioning to the coarse turbulent

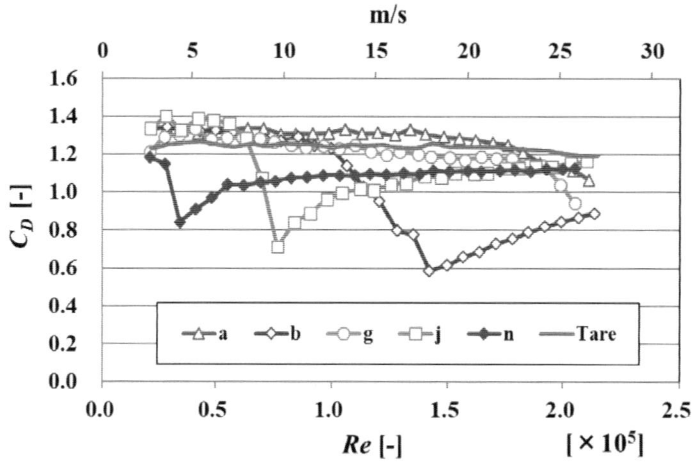

Figure 11. Pressure drag coefficient

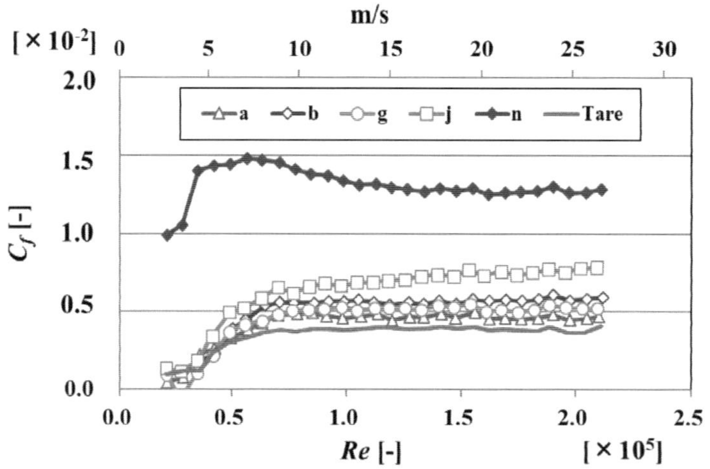

Figure 12. Skin friction drag coefficient

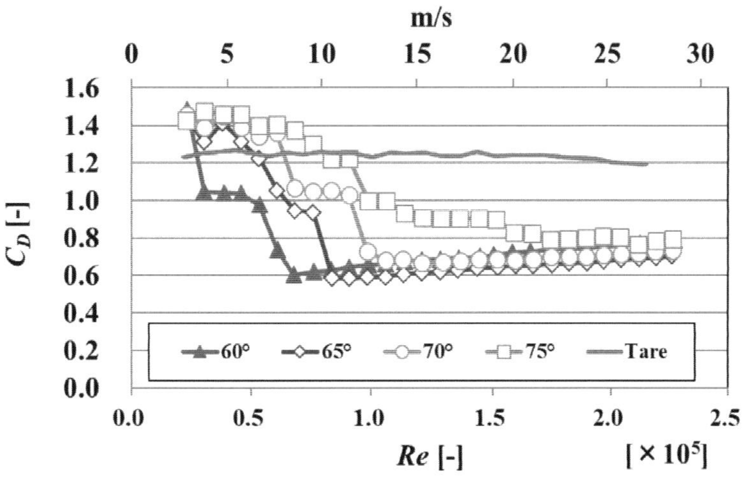

Figure 13. Differences of pressure drag by the positions of stiches

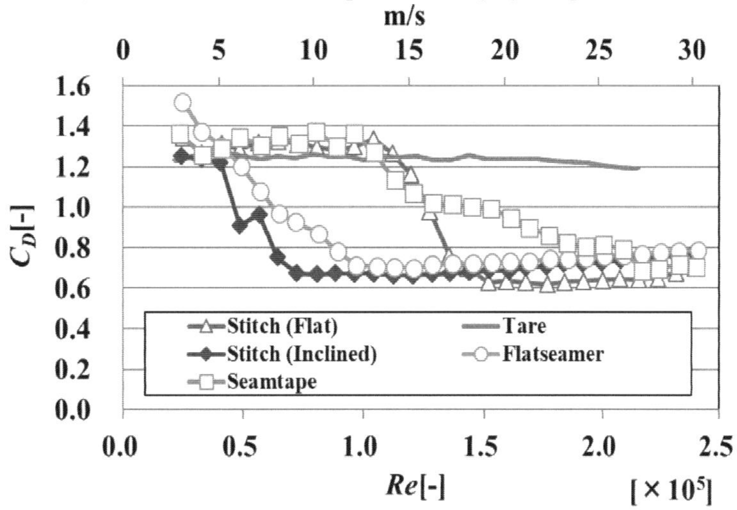

Figure 14. Differences of pressure drag among the different stitch types

boundary layer tends to be turbulent high. It was also found that the influence of *Re* number on friction drag was not large.

3.2 Stich position and its shape

3.2.1 Difference in stitch position

The pressure drag varies greatly due to the difference in turbulent transition position. Figure 13 shows the change in the pressure drag caused by the difference in the position of the cylinder symmetrical stitch using the same fabric. The data of the surface of PVC pipe is a line without mark indicated by Tare, but in this Reynolds number region it does not show drag crisis. However, due to the difference in positions of stitches, The "60°" reaches the supercritical region from the Reynolds number around 0.6×10^5, shows $C_D = 0.6$ and half the drag coefficient of the $C_D = 1.2$ of the smooth surface. The Reynolds number becomes high, it is understood that the values of C_D s of every position is maintained to the same extent.

3.2.2 Difference in stitch shape

As mentioned above, the pressure drag varies depending on the difference in the separation point, an attempt was made to forcibly generate a turbulent flow transition by changing the stitch shape as shown in Figure 5 and to reduce the drag. The stitch position is constant, and the drag force depending on the stitch shape is shown in Figure 14. It can be seen that drag crisis also occurs from around 0.6×10^5 shown in the figure and keeps the low drag as it is. It can be seen that there is a stitch shape that has the effect of fixing the separation point.

3.3 Mannequin Experiment

Figure 15 shows the C_D of the mannequin by the false stitch on the mannequin side surface by various combination of thick threads. The drag crisis seen at the stitch position of the cylinder did not appear in the mannequin and whole the C_D curves are shown as a nearly flat. The wake, related to the pressure drag, started from the flow separation point of the body trunk by the stitch was thought to have absorbed into the wake of the whole mannequin by the influence of the arm.

Figure 16 shows a mannequin CD dressed in various different roughness sportswear, Tights, BA, RA, RB, FA, on a mannequin. In $Re < 2.3 \times 10^5$, there is a difference depending on wears, but in practical competition speed or Reynolds number, all the wears have nearly same value of C_D. In this regard, it is the same result as the effect of C_D of trunk stitches, and it is necessary to consider the influence of the arm.

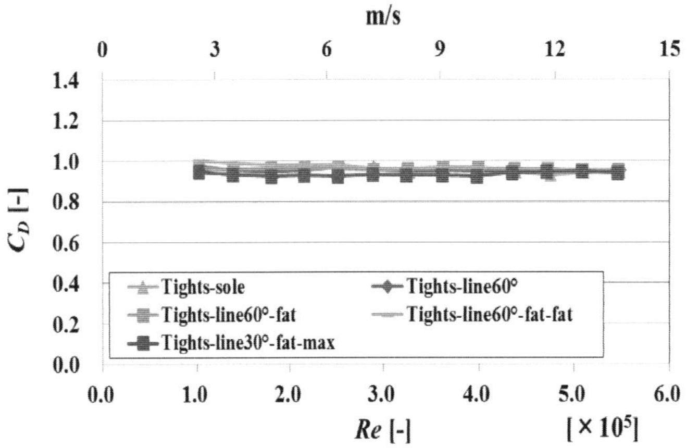

Figure 15. Effect of drag coefficient by the position of seams

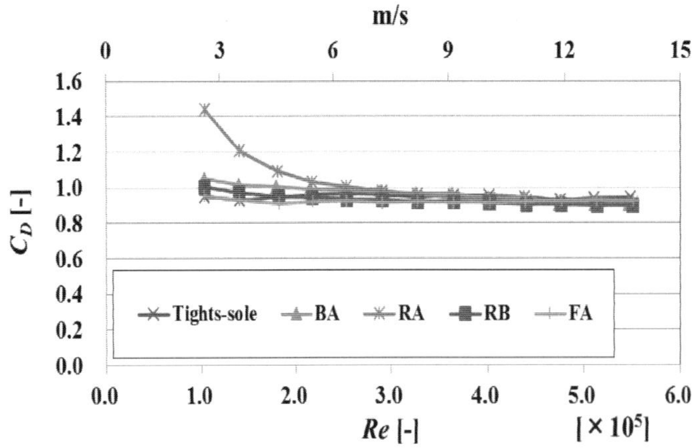

Figure 16. Effect of drag coefficient by different sports wears.

3.4 Experiments on cylinders with arms

3.4.1 Fluid force experiment

Fig.17 shows the result of fluid force measurement with a false stitch placed in the position with the smallest drag of the cylinder. Regarding the drag coefficient C_D, in order to make it easy to compare the magnitude of the drag, the dimensionless processing is performed with the same frontal projected area of only the central representative cylinder. By attaching the arm, the C_D value increased dramatically. Drag crisis induced by stitches occurred in a single cylinder model did not arise either. This is thought to be because the wake flow generated from the stitch as drag crisis was absorbed in the wake flow generated by the arm, so that the decrease of the band width of the wake flow as drag crisis did not occur. Figure 17 also shows the C_D change of cylinders with arms of different arm tilt angles with red circles and blue triangles. However, there was little difference in the C_D value. In the case of tilt angle θ, since separation proceeds by the arm protruding forward as compared with the case of no tilt, it seems that there is no big difference in the wake width as a total.

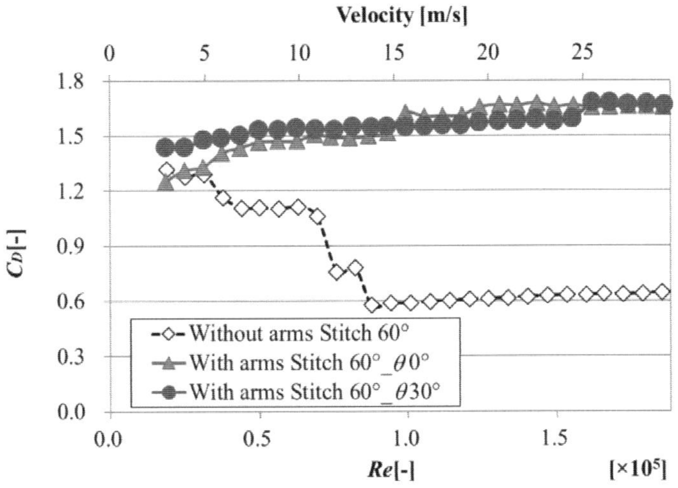

Figure 17. Effect of drag coefficient C_D with/without arms.

3.4.2 PIV measurement

The results of the PIV measurement of the cylinder wake flow are shown in Figures. 18, 19, and 20. Each image represents a flow field of a large area of about 400 mm × 200 mm combining the average flow images of the front and rear of the wake flow. Each shows a velocity vector field of "a single cylinder", "a cylinder with an arm", "a cylinder with an arm tilted by 30°." In each figure, "a" and "b" indicate wind speeds of 10 m/s and 27 m/s, respectively. The color bars in each figure show the range of the wind speed, "a" and "b" are 0 to 10 m/s and 0 to 27 m/s, respectively. In Figures 18, the cylinder has a stitch on the fabric surface. Figure 18a and 18b are the appearance of the flow field before and after the drag crisis. In Figure 18a, the flow separates from the cylindrical surface at 90° ahead, but in Figure 18b, the separation point is displaced backward due to the influence of the stitch. As a result, it is understood that the wake width of Figure 18a is narrower than Figure 18b. This causes drag reduction and drag crisis. However, in Figures 19 and 20 showing the wake velocity field of the cylinder with arms, the width of the wake band is wider than the single cylinder in either case. Moreover, there is a strong ejection flow to the outside from the gap between the arm and the center cylinder as the same speed as the main flow velocity in each velocity condition. It can be said that the witdh is correlated with the result of fluid force measurement. Their C_D curves do not show drag crisis. This is the reason why the turbulent separation from a single cylinder is canceled and the drag crisis does not exist within this experimental range in the arm cylinder, and also the pressure drag curve is flat in measuring the fluid force of the mannequin. Changing the tilt angle of the cylinder with arms did not affect the C_D value, but also in the PIV measurement, as shown in Figures 19 and 20, no significant difference was found in each downstream flow field as well.

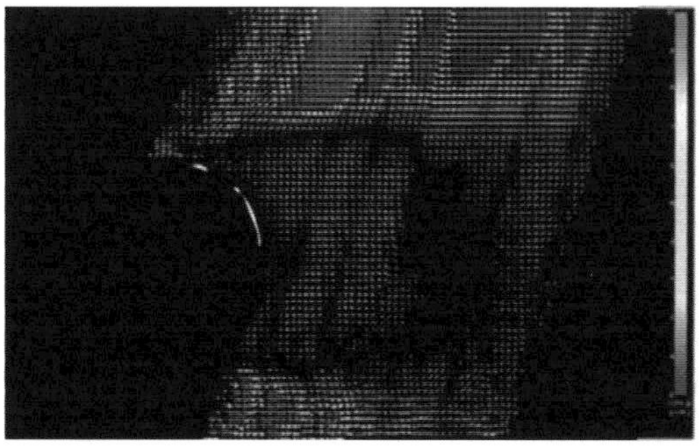

Figure 18a. Wake velocity field of single cylinder at 10 m/s, before drag crisis.

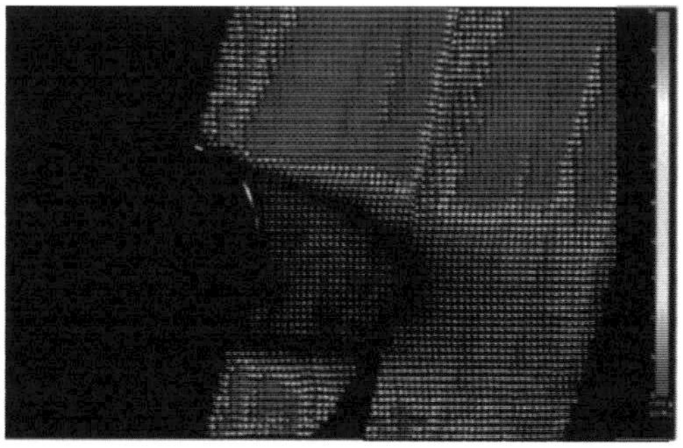

Figure 18b. Wake velocity field of single cylinder at 27 m/s, after drag crisis.

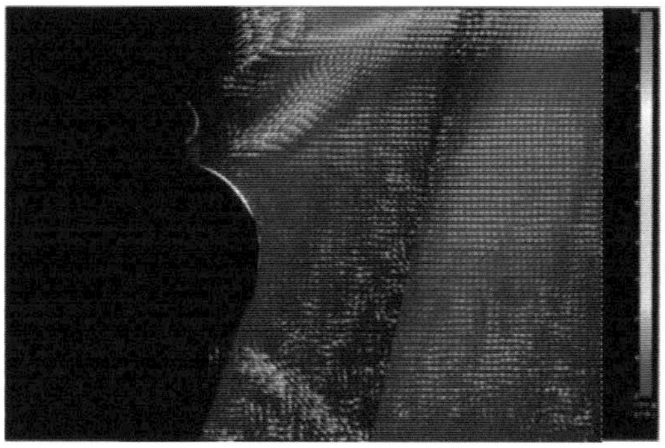

Figure 19a. Wake velocity field of cylinder with arms, no tilt at 10 m/s.

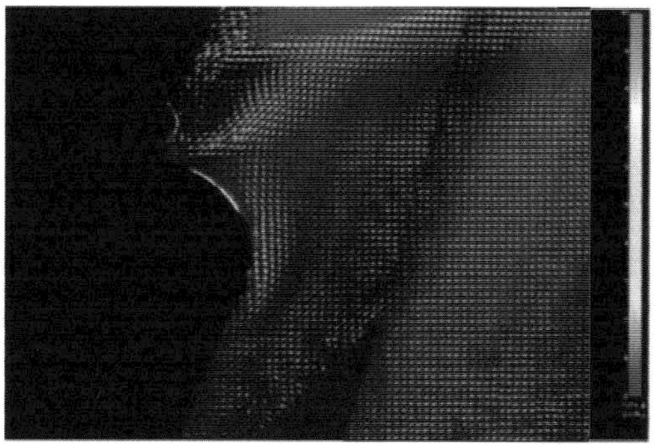

Figure 19b. Wake velocity field of cylinder with arms, no tilt at 27 m/s.

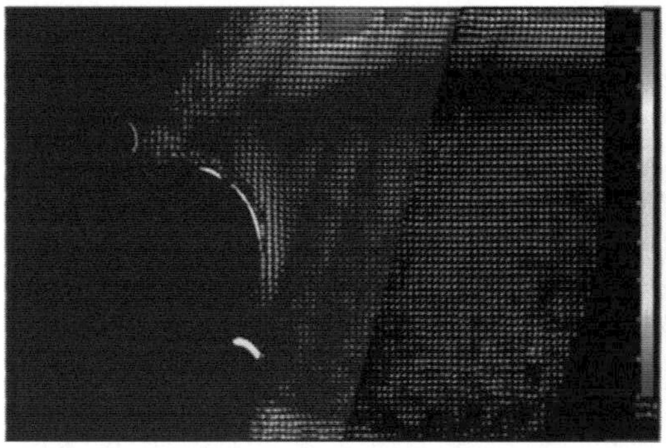

Figure 20a. Wake velocity field of cylinder with arms, tilt angle 30°, at 10 m/s.

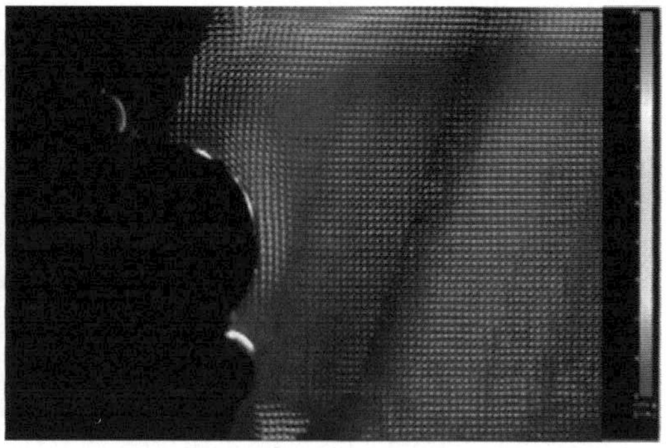

Figure 20b. Wake velocity field of cylinder with arms, tilt angle 30°, at 27 m/s.

3.5 Heat radiation experiment

Figure 21 shows a surface temperature image of a heater with a thermographic camera. In consideration of the flow in the vicinity of the end portion of the cylinder, the average temperature of the square area shown in the center portion in the figure was taken as the measurement temperature of the cylinder. Figures 22a to 22e show the initial temperature and the temperature difference at the end of measurement. for each cloth by setting wind speed. That is, Figures 22 show the heat radiation performance. In each figure, excellent performance indicate as blue hatch, bad one shows as red stripe. At the velocity $u = 3$ m/s, $Re = 0.2 \times 10^5$, shown in Figures 22a, C_Ds of # 4 and # 5 are about 1 and it is lower than that of other fabrics ($C_D > 1.2$). However in terms of heat radiation performance, # 3 is outstanding and superior. In this Reynolds number region, no drag crisis occurred in the flow field, and the surface should be laminar flow, which indicates that the heat-transfer performance is good due to thin, surface-flattened fabric with high thermal conductivity.

Figure 21. Image of thermo graphy camera and the area of average tempaerature

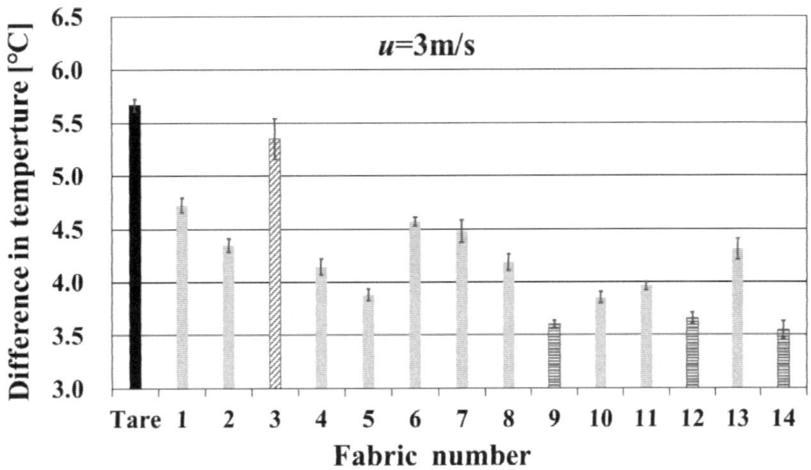

Figure 22a. Heat radiation performance at 3m/s

The heat radiation performance at the velocity u = 6 m/s, Re = 0.4 × 10^5, and u = 9 m/s, Re = 0.6 × 10^5, is shown in Figures 22b and 22c respectively. According to the drag test results, the C_D values of # 4, # 5, # 11, and # 12 are about C_D = 1. Concerning the C_D value of #8, in the recovery mode after drag crisis in these Re, the C_D value=0.9. However, the heat radiation performance was excellent for the fabrics such as # 1, # 3 and # 6. They were thin and the surface was in flat or fine processing. In the laminar flow area, the material with high heat transfer coefficient was found that it is thin and that the surface is in flat processing.

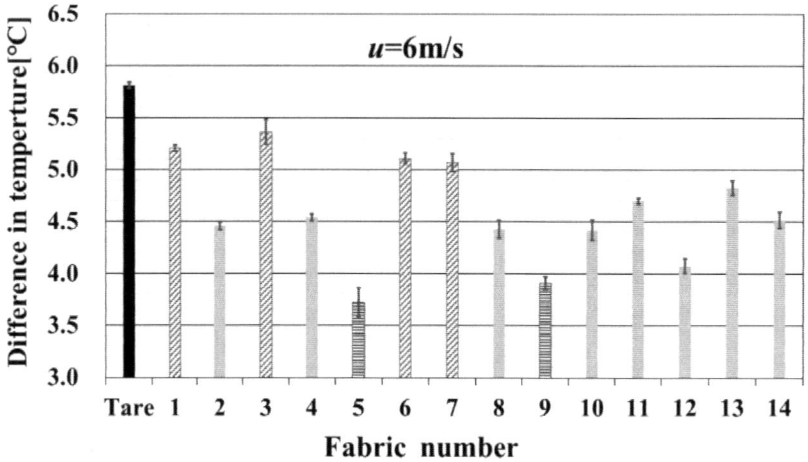

Figure 22b. Heat radiation performance at 6m/s

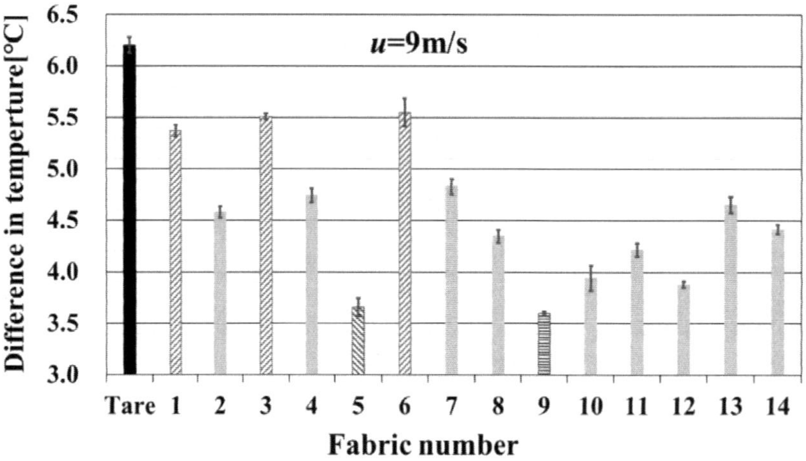

Figure 22c. Heat radiation performance at 9m/s

The results at the velocity u = 12 m/s, Re = 0.8 × 10^5, and u = 15 m/s, Re = 1.0 × 10^5, are shown in Figures 22d and 22e respectively. According to the drag test results, # 9 and # 10 are in drag crisis recovery and their C_D = 0.9. Concerning # 4 and # 5, their flow conditions are during drag crisis and their C_D = 0.7. # 3 and # 12 were also in C_D crisis while C_D = 0.8.

In heat radiation performance, those of # 1, #3 and # 6 are high because their surfaces are on turbulent condition. This seems to be due to turbulent on the fabric surface and this condition increased heat transfer efficiency.

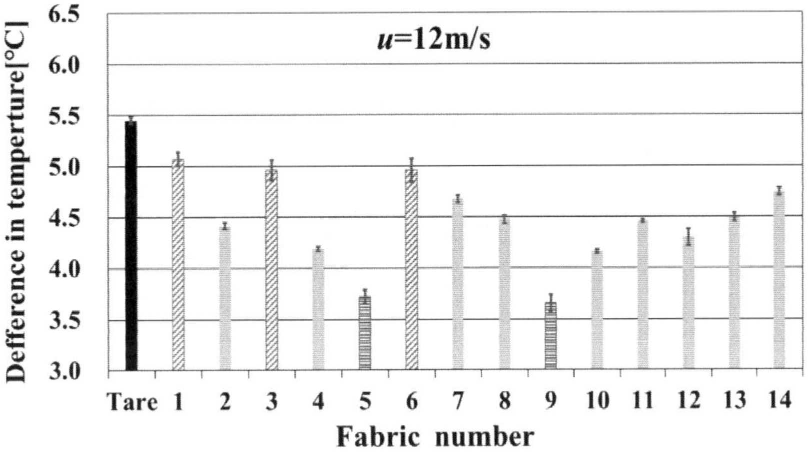

Figure 22d. Heat radiation performance at 12m/s

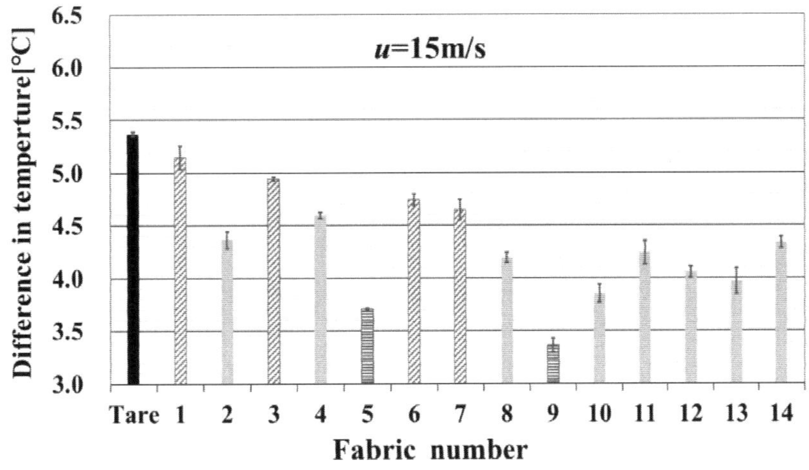

Figure 22e. Heat radiation performance at 15m/s

4. Conclusion

In search of sportswear with low drag and high heat radiation performance, we experimentally investigated changes in fluid drag and heat radiation performance due to various fabrics.
As a result, the following knowledge was obtained:

1. Due to the fabric properties, a sudden decrease in pressure drag, drag crisis, occurs similarly to spheres and cylinders.
2. Drag crisis occurrence range of *Re* varies depending on the roughness of the fabric.
3. In the friction drag measurement, C_f changes due to the difference in surface roughness.
4. It is possible to control the *Re* of occurrence of drag crisis due to the difference in position of the stitch.
5. It is possible to control the *Re* of occurrence of drag crisis due to the difference in stitch shape.
6. As a result of the Mannequin experiment in the standing position, the influence of the stitch is cancelled by the influence of the arm and does not result in the C_D reduction.
   (1) The same phenomenon was confirmed also in a cylinder with arm, and it was confirmed that drag crisis due to turbulent transition did not occur.
   (2) In the visualization of the wake flow by the PIV measurement, correlation was found between the C_D value and the wake flow appearance.
7. In order to mix aerodynamic elements and heat radiation elements, it is important to increase the heat radiation performance by using turbulent flow transitions using thin texture.
8. It is important for improving the heat radiation performance of sportswear that the texture is thin and flat surface processing.

5. References

[1] Gerard H. Kuper, Elmer Sterken, Do skin suits affect the average skating speed?, 2008 *Sports Technology*, Volume 1 Issue 4-5, pp.189-195, John Wiley and Sons Asia Pte Ltd..
[2] S.Luth L. Oggiano, L. M. Bardal and C. Saeter, L. Saetran, Dynamic measurements and drag crisis hysteresis in garment aerodynamics, 2013 *Procedia Engineering* Vol.60, pp.99-105.
[3] Harun Chowdhury, Aerodynamic Design of Sports Garments, 2012 *Applied Aerodynamics*, Dr. Jorge Colman Lerner (Ed.), ISBN: 978-953-51-0611-1, InTech,
[4] Chowdhury, H.; Alam, F.; Mainwaring, D.; Subic, A.; Tate, M.; Forster D. & Beneyto-Ferre, J., Design and Methodology for Evaluating Aerodynamic Characteristics of Sports Textiles, 2009 *Sports Technology*, Vol.2, No.3-4, pp. 81-86, John Wiley and Sons Asia Pte Ltd.
[5] E.Achenbach: Experiments on the flow past spheres at very high Reynolds numbers, 1972 *J. Fluid Mech.* 54, , pp.565-575.
[6] E.Achenbach: The effects of surface roughness and tunnel blockage on the flow past spheres, 1974 *J. Fluid Mech.* 65, pp.113-125.

IOP Conf. Series: Materials Science and Engineering **249** (2017) 012021 doi:10.1088/1757-899X/249/1/012021

Aerodynamic characteristics and PIV analyses concerning tennis balls

T Nakajima[1], M Hiratsuka[2], S Ito[2] and A Konno[2]

[1] Graduate School of Kogakuin University, 1-24-2 Nishi-Shinjuku, Shinjuku-ku, Tokyo, 163-8677 Japan.
[2] Kogakuin University, 1-24-2 Nishi-shinjuku, Shinjuku-ku, Tokyo, 163-8677,Japan.

ito@cc.kogakuin.ac.jp

Abstract. The structure of the current tennis ball is covered with a uniform felt fabric on the outer surface made of hard rubber. As the ball continues to hit, the surface felt wears. Also, air leaks out through the structure inside. These factors cause a difference in the aerodynamic characteristics of new and used balls. In this research, the aerodynamic characteristics of the old and new tennis balls rotating were investigated and the PIV results of the flow around the ball due to the wear of the felt producing these characteristics were showed. For the four types of tennis balls tested, used balls showed higher in drag coefficient than new balls, and new balls tended to be higher than used ball in lift coefficient. It is believed that deformation due to deterioration of ball rubber brings about a change in drag coefficient, which causes a change in lift coefficient due to wear of the surface felt. In the new ball, felt fluff stands, which indicates that the wake flow is large and inclined diagonally backward compared to the used ball, and the air around the ball is found to be caught by the felt fluff.

1. Introduction

The origin of tennis was born in France around the 12th century and was called Jeu de paume. At that time a ball was struck with the palm of the hand [1]. Various shapes and sizes of rackets were introduced since the 16th century to the mid 18th century. The origin of the modern tennis was invented by Major Walter Wingfield, an army officer of Wales in 1873. The structure of the current tennis ball is covered with a uniform felt fabric on the outer surface made of rubber and the inside of the ball is filled with gas above 1.8 atm. This is called as pressured ball. The rotational speed of the tennis ball is up to 5,500 rpm in serve and 4,700 rpm in stroke. In order to achieve such a high rotational speed, balls are struck strongly by gut and this repeatedly strike on ball cause the wear of the outer felt fabric and pressure decreasing of the ball. The surface texture of the felt fabric and internal pressure are the most significant factors related with the aerodynamics drag and lift force of the tennis ball which is closely linked with the trajectory of the ball [2, 3]. In addition, the pressured ball is usually enclosed in a pressured can before use. After the can opened, the air in the ball escapes through the ball's core structure and the pressure decrease slowly. These factors make differences of the aerodynamic properties and trajectory of the ball between the new and the used ball. In the professional matches, the balls are replaced every nine games after the seven games from the beginning. However, in the ordinary practice, such a frequent replacement of balls is difficult because of the cost. The difference of the trajectory between new ball and old ball are known by players sensuously, but from the scientific point of view it is not well understood. The understandings of the aerodynamics characteristics of new and old used ball are significant for the accurate ball controlling for the players and the development of new tennis ball. Therefore, in this research, we compared new

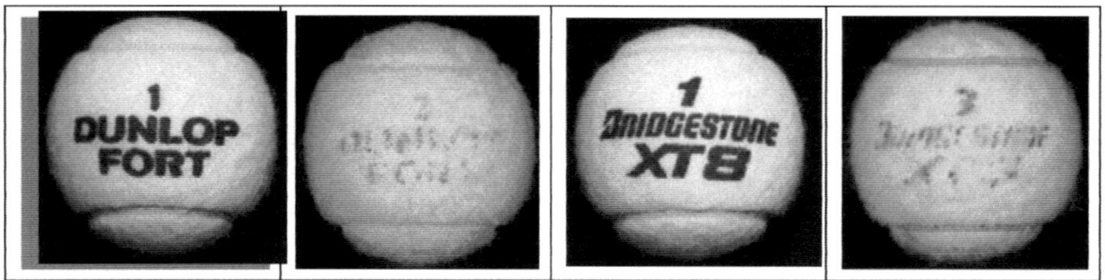

DUNLOP FORT	Used DUNLOP FORT	BRIDGE STONE XT8	Used BRIDGE STONE XT8

Figure1 Tennis balls used in experiment

balls and old used balls from a scientific / hydrodynamic point of view by fluid force and PIV measurement.

2. Experimental setup

To understand the effect of the felt wear of the ball on the fluid force and flow, we performed force measurement and PIV measurement in the wind tunnel on the four types of balls. Figure 1 shows the four types of balls measured in this study, new and old used DUNLOP FORT ball, new and old used BRIDGESTON XT8. The used balls were actually used balls about 50 hours in competitions. The balls were rotated by the mortar in the rage of 3700 ~ 4000 rpm. The rotation axis passes through the center of the ball. To remove the effect of the shape change of the ball on the measured fluid force caused by the pressure decreasing and ball rotation, we also measured the additional four types of the balls filled with urethane inside.

2.1. Fluid force measurement

In order to measure the fluid force on the rotating ball, measurements were made using a wind tunnel experimental device and a three components force detector. For each experimental condition, the wind speed is 14, 18, 20, 22, 24, 26, 28, 29, 30 m/s, the ball rotational speed was 4000 rpm. The drag coefficient and the lift coefficient were calculated from the averaged value of the three times measurement for each balls.

2.2. 2D-PIV measurement

PIV measurement was performed to visualize the flow around each ball as shown in Figure 2. The wind speed was 29 m/s and the ball rotation speed was 3700 rpm. The rotation direction was both of the forward rotation and the reverse rotation. The frequency of the high speed camera was 2000 Hz.

3. Results and discussion

3.1. Fluid force measurement

The results of the drag coefficient and the lift coefficient of the fluid force measurement in this experiment are shown in Figure 3 to Figure 6. In the both FORT and XT8 balls, the drag coefficients in the used balls were larger than that in the new balls. On the other hand the lift force in the used balls tended to lower than that in new balls. The possible causes of these differences of the aerodynamic characteristics are changes of the internal pressure and fluffing of the felt fabric. In case of the FORT, the differences of the drag coefficient and the lift coefficient in each ball were smaller than these of XT8. The felt fabric of the FORT is prepared by woven felt and the condition of the felt did not change so much between new and old one. Woven felt is made by a special method called satin weave, which is considered to have the effect of improving durability and reducing felt fluff. On the other had, since the felt of XT 8 stands long upright, both the drag and lift are considered to have increased.

From this result indicates that the felt fluff stand affected aerodynamic characteristics of the ball significantly.

3.2. 2D-PIV measurement

The images of the averaged flow velocity of PIV measurement are shown in Figure 7, Figure 8. There is a difference in the wake flow between the new balls and the used balls, which is related with the difference of the lift force and the drag force shown in Figure 3 to Figure 6. In the new ball, felt fluff stands, which indicates that the wake flow is large and inclined diagonally backward compared to the

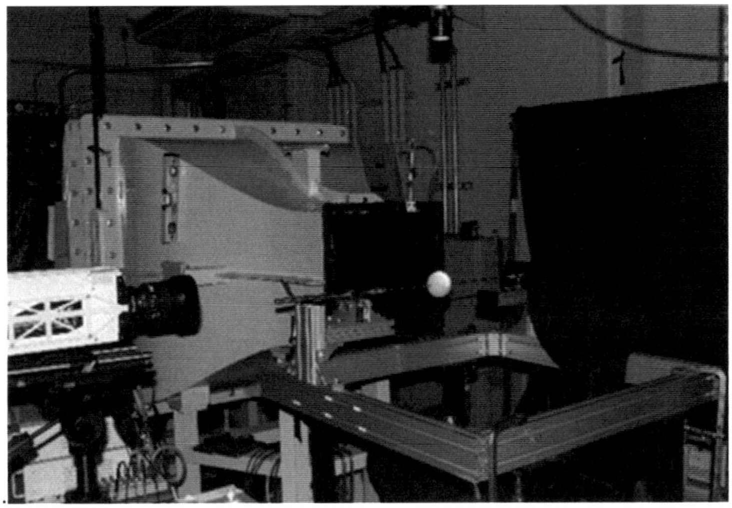

Figure 2 2D-PIV measurement instruments

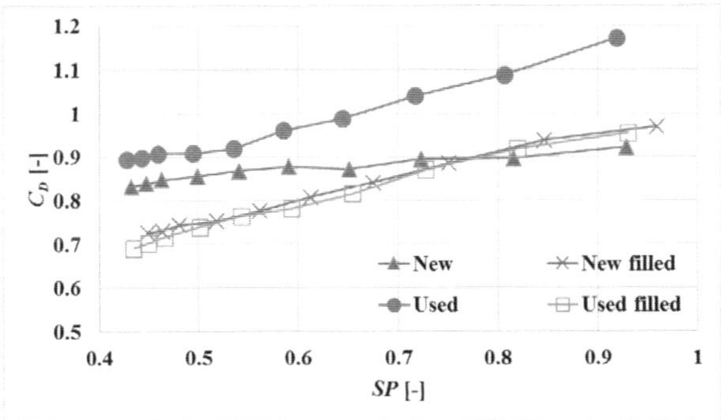

Figure 3 C_D value of FORT

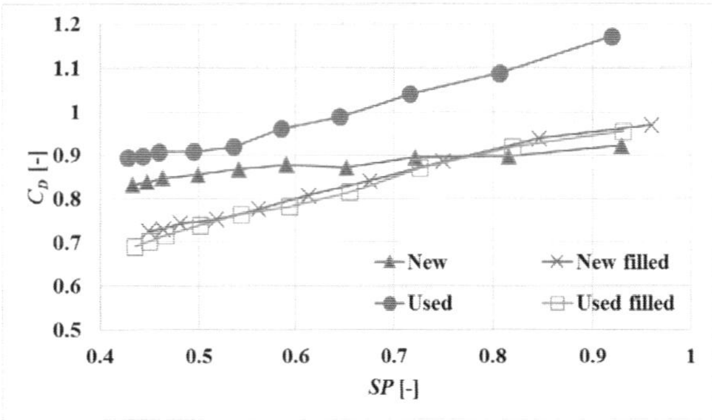

Figure 4 C_D value of XT8

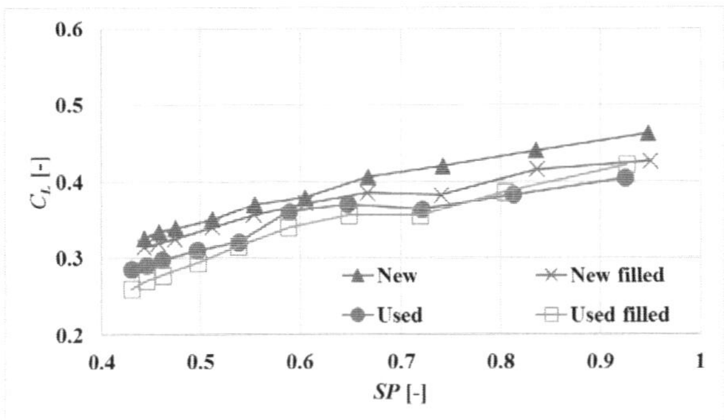

Figure 5 C_L value of XT8

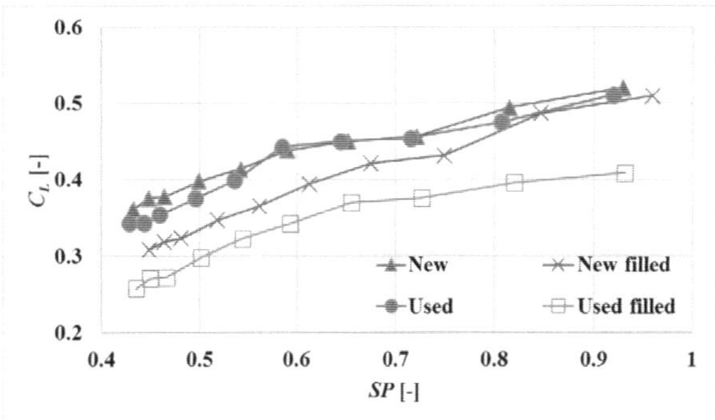

Figure 6 C_L value of XT8

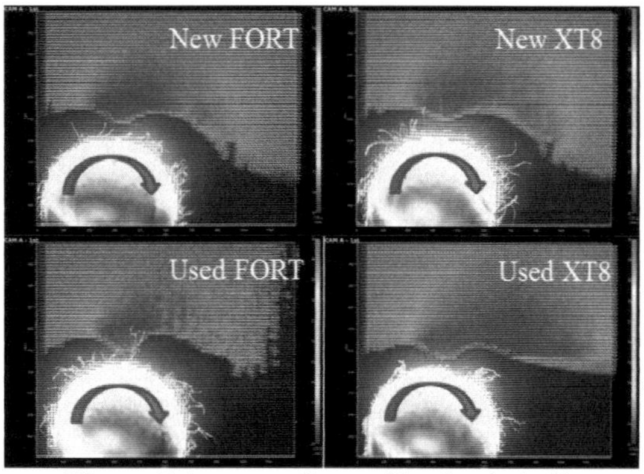

Figure 7 Average flow of backspin balls

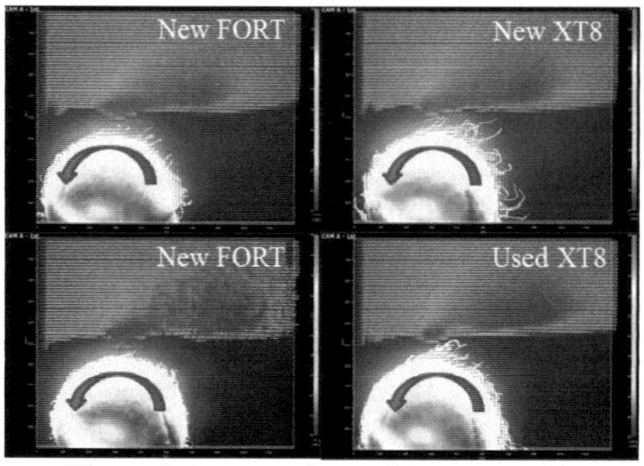

Figure 8 Average flow of Topspin balls

used ball, and the air around the ball is found to be caught by the felt fluff. Although the flow around the balls and the weak flow were visualized, the flow on the ball surface was not measured clearly. This difficulty of the observation of the flow was caused by the the felt fluffing. In the PIV measurement, if the surface of the object is smooth, the reflected laser light is easily removed and can be measured accurately. However, the surface of the tennis ball is covered with felt and it scatted the reflected laser. In addition, the color of the surface of the tennis ball is yellow and the reflected light showed halation. These characteristics of the tennis ball made the accurate measurement of the flow on the surface of the tennis ball difficult. The PIV measurement with the color changed tennis ball is future work for more accurate measurement of the surface velocity distribution of the tennis ball.

4. Conclusion

We performed fluid force and PIV measurement of the rotating new and used tennis balls to reveal the effect of the surface texture on the aerodynamics characters. The following results were obtained.

1. The effect of wear of the surface felt on the flow field and lift / drag force was measured by comparing the new and used balls. The angle change of the flow filed caused by the ball rotation and wake flow in case of the used balls were smaller than that of the new ball. This difference of the flow was reflected in the larger drag and smaller lift forces of the used ball.
2. The impact of wear on the aerodynamic characteristics was depending on the surface felt types covering the ball, especially for the fluff. In case of the ball type FORT, which is covered by the woven felt, the changes of the aerodynamics performance by the wear is smaller than the other types. The felt type of the ball is significant on the changes of the aerodynamic characters during the wearing process.

References

[1] Gillmeister H 1988 *Tennis : a cultural history* (Washington Square, New York University Press)
[2] Mehta R, Alam F and Subic A 2008 *Sport Technol.* **1** 7-16
[3] Djamovski V, Pateras J, Chowdhury H, Alam F and Steiner T 2012 *Procedia Eng.* **34** 140-145

IOP Conf. Series: Materials Science and Engineering **249** (2017) 012022 doi:10.1088/1757-899X/249/1/012022

Advanced Ignition in Supersonic Airflow by Tunable Plasma System

A A Firsov[1], E V Dolgov[1], S B Leonov[1,2], D A Yarantsev[1]

[1] Joint Institute for High Temperatures RAS, Moscow, 125412, Russia
[2] University of Notre Dame, IN 46556, USA

af@jiht.org

Abstract. The plasma-based technique was studied for ignition and flameholding in a supersonic airflow in different laboratories for a long time. It was shown that flameholding of gaseous and liquid hydrocarbon fuel is feasible by means of surface DC discharge without employing mechanical flameholders in a supersonic combustion chamber. However, a high power consumption may limit application of this method in a real apparatus. This experimental and computational work explores a distributed plasma system, which allows reducing the total energy consumption and extending the life cycle of the electrode system. Due to the circuit flexibility, this approach may be potentially enriched with feedbacks for design of a close loop control system.

1. Introduction

For the purpose of a multi-regime scramjet development, many studies were conducted to explore a plasma-based ignition system, alternative to mechanical one, that could consistently and reliably ignite non-stoichiometric mixtures at low temperature and high speed of atmospheric flight. The most technologies of ignition rely on high energy density electrical discharges to heat and ionize the mixture and initiate combustion due to a mechanism of flame propagation or chemical activation [1-3]. In the case of fuel ignition in a high-speed flow and plasma-based flameholding, there are two main approaches: (1) ignition in separation zone, which then works as a chemical reactor (pilot flame), and (2) application of high power electrical discharges in a free stream or fuel jet. In the model #1, several research groups work under combustion in supersonic flow using different geometrical configurations [4-6]. They utilize plasma only as an igniter of combustible mixture. An example of the second case is the study employed a microwave discharge and its combinations with DC discharge [7]. We advocate a viewpoint that the most successful way for the combustion promotion in high-speed flow includes not only mixture ignition, but also the mixing intensification at non-premixed conditions [8], and control of the flow structure to improve the conditions for a flame stabilization (flameholding) [9].

High power consumption of plasma assisted ignition and flameholding technology may limit its application in a real apparatus. For resolving this problem a distributed tunable plasma system is considered in this work. Two rows of electrodes were flush mounted on the ceramic wall of supersonic test cell downstream of wall-mounted injectors of gas fuel (ethylene in this work). The first electrode row was located closer to the injectors and second electrode series was installed downstream of the first row on distance which prevent the electrical breakdown from the first row to the second one. Both electrode systems were powered independently via two fast high voltage solid-state switches. Main idea of described system is to organize two low-power electrical discharges which can't ignite the fuel

IOP Conf. Series: Materials Science and Engineering **249** (2017) 012022 doi:10.1088/1757-899X/249/1/012022

separately but, working together, capable to ignite the fuel. The next step may be in decrease the power of one of plasma actuator or even turning it off, because it is expected that flameholding needs a lower discharge power than the fuel ignition.

In general, this approach is based on concept of two-stage mechanism of fuel ignition in supersonic flow described in Ref. [10]. During the first stage, the plasma induces a fuel reforming, which may be simplified as production of H_2, CH_2O, and CO. Despite the bright luminescence, this zone does not experience significant temperature and pressure increase. This so-called cold flame [11] appears as a source of active chemical species that initiates (under favorable conditions) the second stage of normal "hot" combustion, characterized by high temperature and pressure rise.

2. Description of experimental facility

The experimental facility PWT-50 of the JIHT RAS was designed as an alternative to a full-scale test rig in accordance with the following requirements: flow parameters are close to those typical for the combustion chamber of scramjet; operating time is much longer than the characteristic gasdynamic time; dimensions of the test section are much greater than the thickness of the boundary layer; gas-dynamic duct has flow expansion zones; amount of fuel injected is small in order to avoid dangerous operation, but is sufficient to observe physical and chemical effects; the number of experiments is 10-50 per working day; usability for the installation of high-voltage equipment and diagnostics. The experimental setup is equipped with a closed test section with a cross-section $Y\times Z=72\times 60$mm. The flow parameters are as follows: initial Mach number M = 2.0 or 2.3, static pressure P_{st} = 100-250 Torr, Reynolds number Re = (4-10) × 10^6 × L, thickness of the boundary layer does not exceed 1mm in the test region, typical gas flow through the channel G_{air} = 0.6-0.9 kg/s, duration of the steady-state phase t = 0.2-0.5 s, the total enthalpy of the flow is $H_{tot} \leq 300$ kW. Flow parameters vary from run to run insignificantly, which ensures high repeatability of the test results. A detailed description can be found in publication [12].

3. Details of numerical simulation

Numerical simulation (CFD) of the flowfield provides extended information on flow parameters and structure, which is challenging to acquire experimentally due to a limited set of diagnostic tools. In particular, an essentially three-dimensional distribution of fuel cross test section and along the combustion test cell is needed for a proper analysis of the chemical processes. To resolve this issue, as well as to evaluate the influence of the discharge on the supersonic flow, a three-dimensional nonstationary simulation of the supersonic flow in the test cell was performed, taking into account the direct injection of the fuel through the wall-mounted nozzles. The calculations were performed using software package FlowVision 3.10. Flow simulation was carried out by solving a three-dimensional time-dependent system of Navier-Stokes equations using a modified k-ε model of turbulence (URANS approach).

From experiments, there is known the value of static pressure in the subsonic part of the fuel injector, as well as the diameter of orifices and the static pressure in the test section of the channel. The fuel injector consists of 10 orifices with a diameter of ø1.5mm and a distance of 5 mm between the centers, located on the same line across the channel. Under the conditions corresponding to the experimental ones, the modeling of fuel injection into core flow for a range of flow rates G_{fuel}=1-7 g/s was performed. As a result of simulation the calibration curve was obtained for the fuel flow rate versus the pressure in the injector. The simulation visualizes the distribution of fuel near the wall of supersonic duct. The calculation grid was adapted to obtain good spatial resolution for fuel jets in the area of the test section wall, where the injector was installed. Number of cells was 4.2 million. Calculations were performed for half of the channel using the "symmetry" boundary condition to reduce the number of calculated cells and accelerate the calculation. Simulation results for cases with ethylene injection of G_{fuel}=1g/s and G_{fuel}=3g/s are presented in Figure 1. It is clearly seen that the discrete structure of the air-fuel mixture remains in the area of placement of the first plasma generator, and disappears to the place of location of the second plasma generator. Thus, relating to the fuel jets, the location of electrodes may be actually important for the first discharge module only. For the first discharge module, the composition of the mixture has a periodic structure and varies significantly from the poor to the rich condition over the cross section. Based on results of simulation, the composition of the mixture allows to expect fuel ignition starting from G_{fuel}=1.5 g/s. It should also be noted that the air-fuel mixture in the region of the second plasma module can be considered suitable for plasma-stimulated combustion only starting from G_{fuel}=2 g/s, because at lower flow rates the mixture is lean yet. These data should be considered as a lower estimate because, in a real test cell, oblique shocks, caused by joints of sections and optical windows, and its interaction with boundary layer increase the level of turbulence in a near-wall region, which leads to a faster mixing.

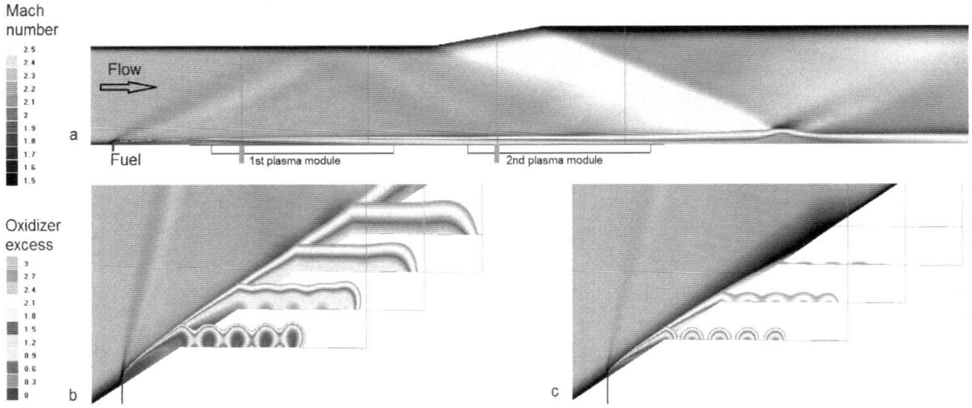

Figure 1. Result of CFD simulation for fuel injection to supersonic flow. a, b – G_{fuel}=3g/s; c – G_{fuel}=1g/s. First and third cross-sections corresponds to plasma module locations.

4. Single row plasma system

Distributed tunable plasma system considered in this work consists of two rows of electrodes which were flush mounted onto ceramic wall of supersonic duct downstream of the injectors of gaseous fuel (ethylene). First electrode line was located closer to the injectors (70 mm downstream) and second electrode series was located downstream of the first row on distance 140mm which prevent the electrical breakdown from first row to the second one. Each row of electrodes consists of 4 pairs with 5 mm gap between electrodes in pair and 7 mm gap between the electrodes of neighboring pairs. It was expected that such difference of gaps will protect plasma filaments from interaction with neighbors, and configuration presented in Figure 2(a) should be realized. But actually it was found that in this configuration the plasma filaments share the grounded electrodes, as it is shown in Figure 2(c). Such switching of discharge leads to increase the discharge length and power release (up to 50%) in comparison with expected mode.

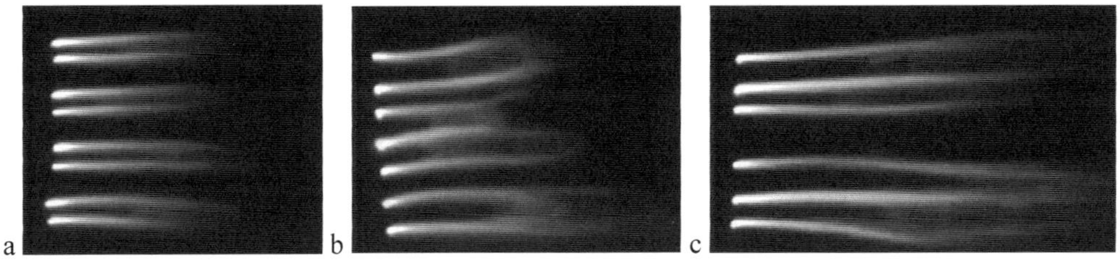

Figure 2. Modes of discharge operation. a – expected mode is realized at discharge initiation; b and c – modes with shared ground electrodes. Flow is from left to right.

Both electrode systems were powered independently via two fast high voltage solid-state switches. Discharge rows were powered by high voltage source with 5kV voltage using ballast resistors for discharge current limitation. Typical current, voltage and power time series for one discharge gap are presented in Figure 3. The length of discharge in flow is not a constant, it is significantly changes in time, and, as a result, it leads to significant fluctuations of plasma resistance which produce voltage and power oscillations. The rise of voltage corresponds to increase of plasma length and sudden voltage drop corresponds to a re-breakdown. Because of such a behavior, this type of discharge is called a Quasi Direct Current (Q-DC) discharge [13, 14].

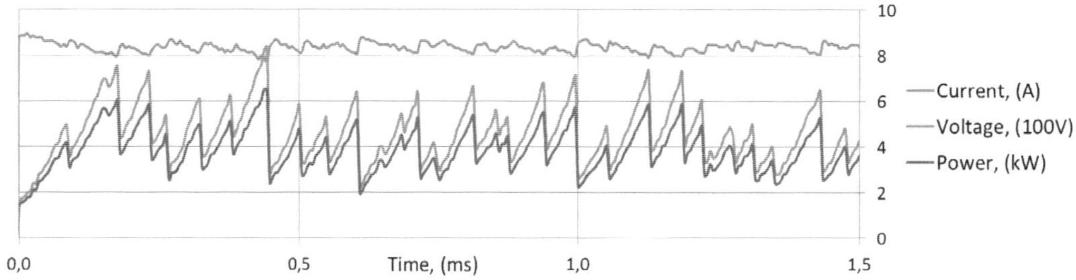

Figure 3.Typical electrical characteristics of Q-DC-discharge in supersonic flow for one gap. Increase of discharge filament length leads to increase the power release.

Influence of discharge operation mode on ignition of lean mixture was tested using second plasma module. In first case two non-interacting plasma loops (4 filaments case in Figure 4a) were used for ignition. In the second case electrical configuration with shared ground electrode were used (3 filaments case in Figure 4a). In this case a weak combustion was realized, as it is demonstrated in pressure distribution in Figure 4b. No combustion was observed in first case. This situation could be explained by increase of total power in the second case and significant local power increase caused by double current in the area of middle discharge filament. Thus, a special configuration with a shared ground could be preferable comparing to the standard electrode circuit shown in Figure 2a.

a

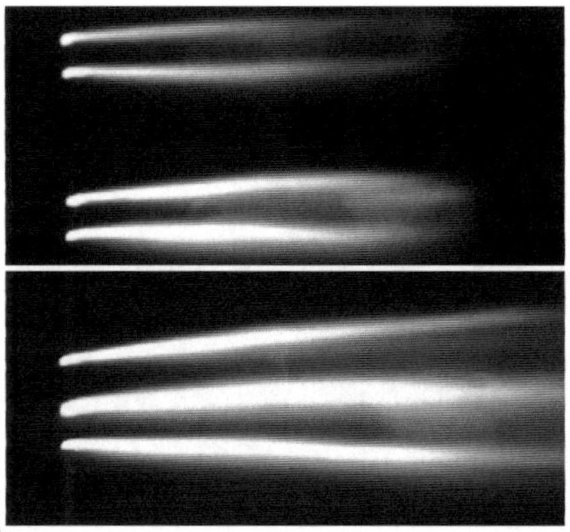

b

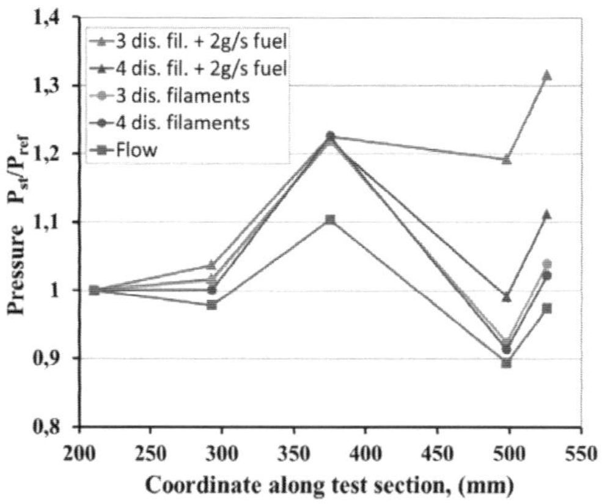

Figure 4. Influence of discharge operation mode on ignition: a – photos of discharge in case of non-interacting plasma loops (top) and shared ground electrode (bottom), b – pressure distribution;

5. Distributed plasma pattern

The test was performed with a distributed plasma pattern in a supersonic flow to confirm the concept of two-stage mechanism of fuel ignition. Fuel-air mixture, G_{fuel}=2.5-3 g/s, was affected by each actuator separately and the results were compared to the ignition data by distributed plasma system which contains two plasma modules working together. In these experiments the plasma power of the first module was about 8kW, and the plasma power of the second module was about 14kW.

All experiments were supported by schlieren visualization. Typical schlieren images are presented in Figure 5 for three cases: second discharge row only, second discharge plus fuel injection, and two discharge rows plus fuel injection. In all images upper wall demonstrates an undisturbed boundary layer. Flow disturbances near the bottom wall were caused by discharge in images (a) and (b), and by combustion in image (c). In the latter case, the wedge of the perturbations has a much larger size due to the expansion caused by flow pressure increase at combustion.

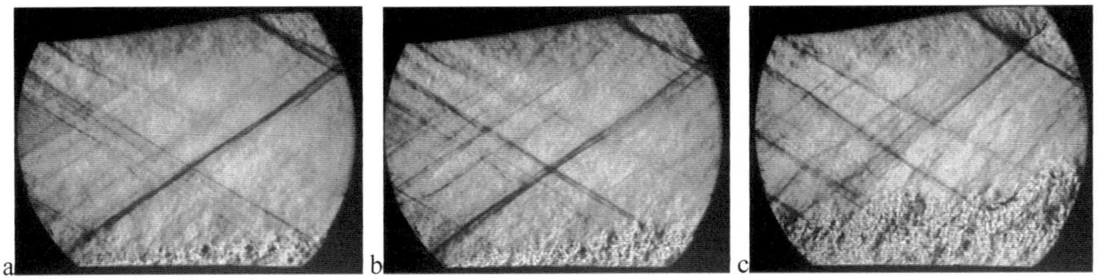

Figure 5. Schlieren visualisation: a - Dis02 only, b - Dis02+fuel 3 g/s, c - Dis01+Dis02+fuel 3 g/s. Flow is from left to right.

The fuel combustion leads to modification of flow structure: in the case of airflow without fuel injection an elongated discharge located along streamlines is realized, see Figure 6(a), and in the case of fuel combustion the discharge still follow the flow and the plasma filaments visualize the vortex structures which takes place in the combustion area. The discharge retreats from the wall, and its length is decreased, see Figure 6(b). In some cases, the resulting reverse motion led to a movement of the discharge opposite the direction of a core airflow.

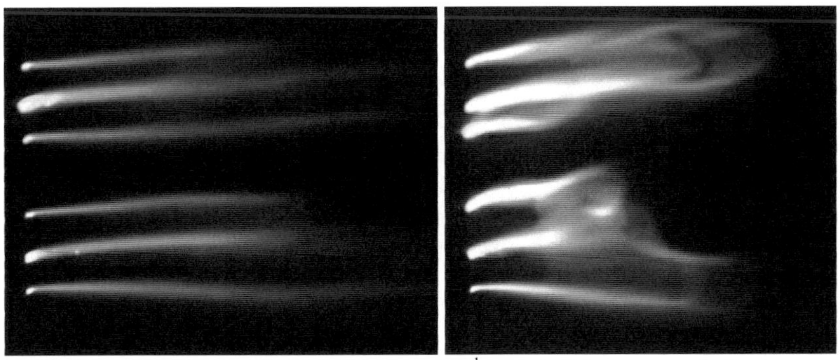

Figure 6. Photo of QDC-discharge (2nd actuator) in supersonic flow without fuel (a) and with 3g/s fuel (b).

Comparison of static pressure distribution for the described cases is presented in Figure 7. It was found that plasma modules (1st and 2nd in Figure 1(a)), working separately, cannot ignite such lean mixture at fuel mass flow rate G_{fuel}=2.5-3 g/s. Operation of the first plasma module results in an insignificant pressure increase at location near x=200 mm in the test section, but there is no active combustion downstream. Such behavior could be explained by means of occurrence of a partial fuel oxidation that did not develop to the "hot" combustion. Second plasma module results in weak combustion. It looks like that there is not enough length of the test section for a further increase of the combustion intensity and pressure. A fixed length of combustion chamber is one of limiting factors for development of scramjet: from this point of view the result at using two plasma modules at the same time looks to be the most interesting. In this case, a low pressure increase corresponds to the area between two plasma modules, whilst a rapid increase of pressure is observed in the region of the second plasma generator. The second plasma generator does not allow the flame to be blown out, and the preliminary processing performed by the first generator helps to accelerate the combustion process in the area of the second plasma generator. This test may be an implicit verification of a two-stage ignition mechanism. A more subtle use of this effect could allow reducing the power required for the plasma-assisted combustion in a high-speed flow.

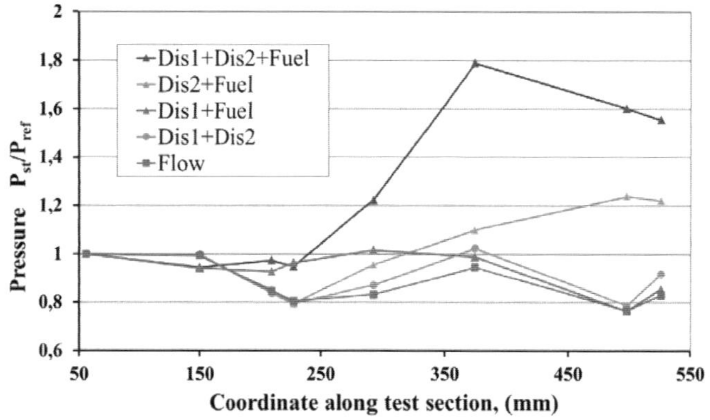

Figure 7. Pressure distribution: comparison of independent discharge systems cases with distributed plasma system which consist of two actuators working simultaneously.

Conclusions

Experimental study of distributed plasma system for ignition and flameholding for lean mixtures in supersonic flow was performed. The numerical simulation helped to improve a geometrical configuration of the test cell. It was found that the tested electrical scheme with shared ground electrode has some advantages in comparison with non-interacted plasma loops. Also it was shown that distributed plasma system, consisting of two plasma modules, accelerates the combustion process comparing to a single plasma module. These test series, at some extent, verify the two-stage mechanism of plasma-assisted ignition/flameholding.

References

[1] Starikovskaia S M 2006 *J. Phys.D: Applied Physics* **39**, R265-R299
[2] Adamovich I V, Choi I, Jiang N, Kim J.-H, Keshav S, Lempert W R , Mintusov E, Nishihara M, Samimy M, Uddi M 2009 *Plasma Sources Sc. and Tech.* **18**, p. 034018
[3] Starikovskiy A, Aleksandrov N 2013 *Prog. in En. and Comb. Sc.* **39**, I.1, pp. 61–110
[4] Vinogradov V A, Goldfeld M A, Starov A V 2013 *Comb., Explosion and Shock Waves* **49**, No. 4, pp. 383–391
[5] Zhang L, Yang V, Lin K.-C 2013 *AIAA* paper 2013-0118
[6] Li F, Yua X.-L, Tong Y.-G, Shen Y, Chen J, Chen L.-H, Chang X.-Y 2013 *Aerosp. Sc. and Tech.* **28**, pp. 72–78
[7] Shibkov V M, Aleksandrov A F, Chernikov V A, Ershov A P, Shibkova L V 2009 *J. prop and power.* **25**, No. 1

IOP Conf. Series: Materials Science and Engineering **249** (2017) 012022 doi:10.1088/1757-899X/249/1/012022

[8] Savelkin K V, Yarantsev D A, Adamovich I V, Leonov S B 2015 *Combust. Flame* **162**, I.3, pp. 825–835
[9] Firsov A A, Savelkin K V, Yarantsev D A, Leonov S B 2015 *Phil. Trans. A* **373**, I.2048
[10] Leonov S B, Kochetov I V, Napartovich A P, Sabel'nikov V A, Yarantsev D A 2011 *IEEE Trans. on Plasma Sc.* **39**, No. 2, pp. 781 - 787
[11] Kim W, Mungal M G, Cappelli M A 2010 *Combust. Flame* **157**, No. 2, pp. 374–383
[12] Leonov S B, Savelkin K V, Firsov A A, Yarantsev D A 2010 *High Temp.* **48**, I.6, pp.941-947
[13] Firsov A A, Yarantsev D A, Leonov S B 2015 *AIAA* paper 2015-2659
[14] Houpt A, Hedlund B, Leonov S, Ombrello T, Carter C 2017 *Exp Fluids* **58**, I.4, pp. 25-35

Acknowledgments
This work is supported by Russian President Grant for young scientists MK-1734.2017.8.